Persian-English
English-Persian
Biotechnology Glossary

Ali Akbar Pejman Aryan

Ramin Sarraf

Persian-English

English-Persian

Biotechnology Glossary

Ali Akbar Pejman Aryan

Ramin Sarraf

2006
Dunwoody Press

Persian-English English-Persian Biotechnology Glossary

All inquiries should be directed to:
Dunwoody Press
6525 Belcrest Rd, Suite 460
Hyattsville, MD 20782, U.S.A.

ISBN-13: 978-1-931546-24-9
ISBN-10: 1-931546-24-X
Library of Congress Control Number: 2006934992
Printed and bound in the United States of America

Table of Contents

Preface

This Persian-English English-Persian Biotechnology Glossary is one of a series of such glossaries in Arabic, Chinese, Hindi, Korean, Russian, and Urdu prepared by McNeil Technologies and Dunwoody Press to assist researchers in those languages to read and translate biotechnology writing.

Starting with a preliminary list of English Biotechnology terms developed by Dunwoody Press and making use of upwards of a dozen relevant tertiary sources, Persepolis Institute Persian research consultants Ramin Sarraf and Ali Akbar Pejman Aryan compiled this *Persian-English English-Persian Biotechnology Glossary*. The glossary presents 7,000+ English-Persian entries and 11,000+ Persian-English entries.

Dunwoody Press and Persepolis Institute welcome corrections and suggestions for the improvement, expansion, and updating of Persian-English English-Persian Biotechnology Glossary.

Thomas Creamer, Editorial Director, Dunwoody Press
Michael Craig Hillmann, President, Persepolis Institute

Bibliography

An Agricultural and Environmental Biotechnology Annotated Dictionary. By Susan Allender-Hagedorn and Charles Hagedorn. Salem (VA): Virginia Tech University. 20 June 2000. Online at filebox.vt.edu/cals/cses/chagedor /glossary/html.

Approved Terms, Chemistry (3). Prepared and certified by the Academy of Sciences, Chemistry Group. Tehran (Iran): The Academy of the Persian Language and Literature, Terminology Department. 2003.

Approved Terms, Veterinary Science (1). Prepared and certified by the Academy of Sciences, Veterinary Science Group. Tehran (Iran): The Academy of the Persian Language and Literature, Terminology Department. 2002.

Approved Terms, Biology, School Texts (1). Prepared and certified by the Academy of Sciences, Biology Committee. Tehran (Iran): The Academy of the Persian Language and Literature, Terminology Department. 2003.

Approved Terms, Medicine (1). Prepared and certified by the Academy of Sciences, Medical Terminology Committee. Tehran (Iran): The Academy of the Persian Language and Literature, Terminology Department. 2000.

Approved Terms, Energy and Environmental Engineering (1). Prepared and certified by the Academy of Sciences, Energy and Environmental Engineering Group. Tehran (Iran): The Academy of the Persian Language and Literature, Terminology Department. 2002.

Approved Terms, Agriculture, Crop Production and Breeding Technology (1). Prepared and certified by the Academy of Sciences, Agriculture, Crop Production and Breeding Technology Group. Tehran (Iran): The Academy of the Persian Language and Literature, Terminology Department. 2003.

Chemistry and Chemical Engineering Vocabulary, English-Persian and Persian-English. Second edition. By Ali Pourjavadi. Tehran: Iran University Press, 2000.

Chinese-English English-Chinese Biotechnology Glossary. Thomas Creamer and Wu Ying. Springfield (VA): Dunwoody Press. 2005.

A Dictionary of Biology, English-Persian. Second edition. Rahim Honarnejad. Gilan: Gilan University Press, 2003.

Dictionary of Chemistry, English-Persian and Persian-English. Reza Aghapour Moghaddam. Tehran: Karang Publishers, 2002.

The Dictionary of Electrical Engineering, Electronics, Computers, and Communication, English-Persian. Peter Klaus Budig. Translated by Nader Golestani. Tehran: Amir Kabir Publishing Corporation, 2002.

EU BioSociety Glossary. European Union BioSociety. 23 June 2005. Online at europa.eu.int/comm/research/biosociety/library/glossaryfind_en.cfm.

Farhang Moaser English-Persian Dictionary. Second edition. Mohammad Reza Bateni. Tehran: Farhang Moaser Publishers, 2003.

FarsiDictionary. http://www.farsidic.com/.

Glossary of Biotechnology for Food and Agriculture. A. Zaid, H.G. Hughes, E. Porceddu, and F. Nicholas. Rome (Italy): Food and Agriculture Organization of the United Nations (FAOUN), 2001.

Glossary of Biotechnology Terms. Manfred H. Fleschar and Kimbal R. Nill. Lancaster (PA): Technomic, Inc., 2001.

Online at biotechterms.org.

Glossary of Genetics and Biotechnology Terms, English-Persian. A. Movafagh. Tehran: Aeej Publishers, 2004.

The Language of Biotechnology: A Dictionary of Terms. John W. Walker and Michael Cox. Washington, D.C.: American Chemical Society, 1998.

Medical Dictionary, English-Persian. Mohammad Hooshmand Vijeh. Tehran: Bastan Publishers, 2000.

Principles of Physical Biochemistry. Kensal E. van Holde and W. Johnson. Translated by Mohammad Reza Housaindokht. Mashhad: Ferdowsi University of Mashhad, 2004.

Taber's Cyclopedic Medical Dictionary (Edition 20). Edited Donald Venes. Philadelphia: F.A. Davis Company, 2001.

Vocabulary of Biology, English-Persian and Persian-English. Reza Farzanpay. Tehran: Iran University Press, 1999.

Persian-English

آ

English	Persian
A (= adenine), A (= adenovirus)	آ.
aerosol	آئروسول
aerosolization	آئروسولیزه کردن
aerolysin	آئرولیزین
aeromonas	آئروموناس
aorta	آئورت
aortic	آئورتی
AE (= acrodermatitis enteropathica)	آ.ئی.
A-EB (= epidermolysis bullosa acquisita)	آ.-ئی.ب.
AF (= amniotic fluid)	آ.اف.
liquid	آب
ascites	آب آوردن شکم
restoration	آبادسازی
habitat restoration	آبادسازی زیستگاه
sterile water	آب استریل
flooding	آب انداختن
dehydration	آب برداری
hydrophilic	آب پذیر
hydration	آبپوشی
hydrated	آبپوشیده
hydrophilic	آب خواه
aqueous, hydrated	آبدار
hydration	آبدار کردن
vesicle	آبدانک
hydrophilic	آب دوست
hydration	آبدهی
hardening	آبدیده کردن
de-ionized water	آب دیونیزه شده
dehydration	آب زدایی
abzyme	آبزیم
gestation	آبستنی
hydration	آبش
mitogen-activated protein kinase cascade	آبشار پروتئین کیناز فعال شده توسط میتوژن
cascade	آبشار ماند
cascade	آبشاره

English	Persian
ascites	آب شکم
halophile	آب شور گرای
cascade	آبشیب
mitogen-activated protein kinase cascade	آبشیب پروتئین کیناز فعال شده توسط میتوژن
gene expression cascade	آبشیب تظاهر ژن (ی)
jasmonate cascade	آبشیب ژاسمونات
complement cascade	آبشیب مکمل
uncontaminated water	آب غیر آلوده
plating	آبکاری
replica plating	آبکاری همانند
hydrolysis	آب کافت
hydrolyzable	آبکافت شدنی
alkaline hydrolysis	آبکافت قلیایی
hydrolyze	آب کافت کردن
hydrolysis	آب کافتی
hydrolytic	آبکافتی
hydroponics	آب کشت
aquaculture	آب کشتی
aqueous	آبکی
hydrophilic	آبگرا
hydrophobic	آب گریز
hydrolysis	آب گسستگی
broth	آبگوشت (کشت)
liquid	آب گونه
hydration	آبگیری
vaccinia	آبله گاوی
monkeypox	آبله میمون
hydrophobic	آب ناپذیر
aqueous	آب نهاد
hydration	آب وندی
hydrophobic	آب هراس
aqueous	آبی
cyanobacteria	آبی ترکیزگان
methylene blue	آبی متیلنی
AP (= apurinic)	آ.پ.
apamin	آپامین
aptamer	آپتامر
aplastic anemia	آپلاستیک آنمی
apoenzyme	آپو انزیم
aporepressor	آپوریپرسور

English	Persian
apurinic	آپورینیک
apolipoprotein	آپولیپوپروتئین
apomorphine	آپومورفین
apomixis	آپو میکسی
ataxia telangiectasia	آتاکسیا تلانجی اکتازیا
Freidreich's ataxia	آتاکسی نوع فردریک
ATP (= adenosine triphosphate)	آ.ت.پ.
ATPase (= adenosine triphosphatase)	آ.ت.پ.آز
mitochondrial ATPase	آ.ت.پآز میتوکندریائی
atropine	آتروپین
overatropinization	آتروپین سازی بیش از حد
dentato rubral-pallidolusyian atrophy	آتروفی دنتاتو روبرال-پالائیدولوزیان
atrolysin	آترولایزین
atrioactivase	آتریوآکتیواز
atriopeptins	آتریوپپتین ها
pyrogen	آتشزا
atopic	آتوپیک
AC (= adenyl cyclase)	آ.ث.
milling	آجیده کاری
adaptor	آداپتور
adaptin	آداپتین
adalimumab	آدالیموماب
adamalysin	آدامالیزین
adamantane	آدامانتان
adjuvant	آدجووانت
adducin	آددوسین
address	آدرس
hilar adenopathy	آدنوپاتی هیلار
adenosine	آدنوزین
adenosine triphosphate (ATP)	آدنوزین تری فسفات
adenosine triphosphatase	آدنوزین تریفسفاتاز
adenosine deaminase	آدنوزین دآمیناز
double-stranded RNA adenosine deaminase	آدنوزین دآمیناز آر.ان.ای. دو رشته ای
adenosine diphosphate (ADP)	آدنوزین دی فسفات

English	Persian
adenosine monophosphate	آدنوزین مونو فسفات
cyclic AMP (= cAMP)	آدنوزین مونوفسفات حلقوی
adenoleucodystrophy	آدنولوکودیستروفی
adenomatous polyposis coli (gene)	آدنوماتوس پولیپوزیس کولای (ژن)
adenovirus	آدنو ویروس
adenoviridae	آدنوویریدا
adenylate	آدنیلات
adenilate cyclase	آدنیلات سیکلاز
adenylate cyclase	آدنیلات سیکلاز
adenyl cyclase	آدنیل سیکلاز
adenine	آدنین
adenin arabinucleoside triphosphate	آدنین آرابینوکلئوزید تریفسفات
nicotinamide adenine dinucleotide phosphate	آدنین دی نوکلئوتید فسفات نیکوتین امید
nicotinamide adenine dinucleotide	آدنین دی نوکلئوتید نیکوتین امید
adenine phosphoribosyl transferase	آدنین فسفوریبوسیل ترانسفراز
A II (= second meiotic anaphase)	آ.دو
adolase	آدولاز
ADH (=antidiuretic hormone), ADH (= alcohol dehydrogenase)	آ.د.هاش
adipsin	آدیپزین
adipose	آدیپوز
adipocyte	آدیپوسیت
R5P (= ribose-5-phosphate)	آر.5 پی.
rRNA (= ribosomal RNA)	آر.آر.ان.ای.
RE (= response element), RE (= RNA extraction)	آر.ئی.
RER (= rough endoplasmic reticulum)	آر.ئی.آر.
arabidopsis thaliana	آرابیدوپزیس تالیانا
rh	آر.اچ.
RSV (= respiratory syncytial virus)	آر.اس.وی.
arachin	آراشین
RF (= release factor)	آر.اف.

English	Persian
protein array	آرایه پروتئین
expression array	آرایه تظاهر
cDNA array	آرایه زنجیره واحد دی.ان.ای
live cell array	آرایه سلول زنده
taxonomy	آرایه شناسی
r-banding	آر.-بندینگ
RB (= retino blastoma), Rb (= retinoblastoma)	آر.بی.
RBS (= RNA binding site)	آر.بی.اس.
RBC (= red blood cell)	آر.بی.سی.
RP (= retinitis pigmentosa)	آر.پی.
rpm (= revolutions per minute)	آر.پی.ام.
RPA (= replication protein A)	آر.پی.ای.
arteriosclerosis	آرترواسکلرز
osteoarthritis	آرتروز استخوان
arthrogryposis multiplex congenita	آرتروگریپوز مالتیپلکس کنجنیتا
arthropathy	آرتریت
patent ductus arteriosus	آرتریوسوس مجرای باز
RT (= reverse transcription), rT (= room temprature)	آر.تی.
RT-PCR (= reverse transcriptase-PCR)	آر.تی.-.پی.سی.آر.
RTVL (= retroviral-like element)	آر.تی.وی.ال.
RDB (= reverse dot blot)	آر.دی.بی.
arginase	آرژیناز
arginine	آرژینین
arrestin	آرستین
rcp (= rearrangement)	آر.سی.پی.
RCV (= replication-competent virus)	آر.سی.وی.
archea	آرشیا
armyworm	آرمی وورم
fall armyworm	آرمی وورم پائیزی
bacillus amilolique facien RNase	آرناز باسیلوس آمیلولیک فاسین RNase
arenavirus	آرناویروس

English	Persian
RFLP (= restriction fragment length polymorphism)	آر.اف.ال.پی.
RFC (= replication factor C)	آر.اف.سی.
RLF (= replication licensing factor)	آر.ال.اف.
RMB (= Roche Molecular Biochemicals)	آر.ام.بی.
RNAse (= ribonuclease)	آر.ان.آز
RNA (= Ribo-Nucleic Acid)	آر.ان.ای.
antisense RNA	آر.ان.ای. آنتی سنس
information RNA	آر.ان.ای. اطلاعاتی
messenger RNA (mRNA)	آر.ان.ای. اطلاعاتی
transfer RNA (tRNA)	آر.ان.ای. انتقالی
stable RNA	آر.ان.ای. پایدار
transfer-messenger RNA	آر.ان.ای.پیامبر انتقالی
pre-messenger RNA	آر.ان.ای.پیش پیغمبر
messenger RNA (mRNA)	آر.ان.ای. پیک
transfer RNA (tRNA)	آر.ان.ای. حامل
nascent RNA	آر.ان.ای. در حال شکل گیری
ribosomal RNA (rRNA)	آر.ان.ای. ریبوزومی
catalytic RNA	آر.ان.ای. کاتالیزوری
short interfering RNA	آر.ان.ای. کوتاه تداخلی
small interfering RNA	آر.ان.ای. کوچک تداخلی
small cytoplasmic RNA	آر.ان.ای. کوچک سیتوپلاسم
small nucleolar RNA	آر.ان.ای. کوچک نوکلئولار
risk management RNA	آر.ان.ای. گرداننده ریسک
satellite RNA	آر.ان.ای. ماهواره ای
complementary RNA	آر.ان.ای. مکمل
heterogenous nuclear RNA	آر.ان.ای. هسته (ای) ناهمگن
RNP (= ribo-nucleo-protein)	آر.ان.پی.
RACE (= rapid amplification of cDNA ends)	آر.ای.سی.ئی.
configuration	آرایش
absolute configuration	آرایش مطلق
array, taxon	آرایه
tissue array	آرایه بافت
antibody array	آرایه پادتن

English	Persian	English	Persian
aromatic	آروماتیک	aspergillus fumigatus	آسپرژیلوس فومیگاتوس
arrhythmia	آریتمی	aspergillus niger	آسپرژیلوس نایجر
relaxed	آزاد	asepsis	آسپسی
accidental release	آزادسازی اتفاقی	astaxanthin	آستازانتین
covert release	آزادسازی پنهان	ascopore	آسکوپور
accidental release	آزادسازی تصادفی	axon	آسه
overt release	آزاد سازی علنی	milled	آسیا شده
deliberate release	آزاد سازی عمدی	milling	آسیا کاری
general release	آزادسازی کلی	milling	آسیا کردن
intended release	آزادسازی مورد نظر	allelopathic	آسیب رسان
controlled release	آزادسازی مهار شده	cytopathic	آسیب سلولی
unconfined release	آزادسازی نامحدود/مهار نشده	pathologic	آسیب شناختی
uncoupling	آزاد کردن	histopathologic	آسیب شناسی بافتی
azadirachtin	آزادیراکتین	transboundary harm	آسیب فرا مرزی/سرحدی
azaserine	آزاسرین	acyclovir	آسیکلوویر
ascus	آزکوس	acylcarnitine transferase	آسیل کارنیتین ترانسفراز
assay	آزمایش	acycloquanosine	آسیلوکوانوزین
aerosol vulnerability testing	آزمایش آسیب پذیری ذره پاش	pica	آشغال خوری
		express, overt, patent	آشکار
clinical trial	آزمایش بالینی	familiarity	آشنایی
pulse-chase experiment	آزمایش پالس-چیز	niche	آشیانه
single burst experiment	آزمایش تک انفجاری	initiation	آغاز
DNA diagnosis	آزمایش دی.ان.ای.	onset	آغاز بیماری
genetic testing	آزمایش ژنتیکی	activator, initiator	آغاز گر
technical testing	آزمایش فنی	initiation	آغازگری
reference laboratory	آزمایشگاه مرجع	pest, plague	آفت
confined field testing	آزمایش میدانی محدود	mid-oleic sunflowers	آفتاب گردان های میان-اولئیکی
biological indicator of	آزمایش نشانگر بیولوژیکی	enzootic	آفت در حیوانات
exposure study	سطح تماس	quarantine pest	آفت قرنطینه
avidity	آزمندی	biocide, pesticide	آفت کش
Ames test	آزمون آمز	genetically engineered	آفت کش میکروبی طراحی
cis/trans test	آزمون جذب-دفع	microbial pesticide	شده ژنتیکی
biomedical testing	آزمون زیست پزشکی	aflatoxin	آفلاتوکسین
blank test	آزمون شاهد	AK (= adenylate kinase	آ.کا.
cis/trans test	آزمون همسو-دگرسو	deficiency)	
azoospermia	آ زوئواسپرمی	actobindin	آکتوبیندین
asparagine	آسپاراژین	actomyosin	آکتومایوزین
aspartate transaminase	آسپارتات ترانس آمیناز	actidione	آکتیدیون
aspartokinase	آسپارتوکیناز	actin	آکتین
aseptic	آسپتیک	actinospectacin	آکتینواسپکتاسین
aspergillus flavus	آسپرژیلوس فلاووس	actinomycetes	آکتینومایست ها

6

English	Persian	English	Persian
actinomycin	آکتینومایسین	pollutant	آلاینده
actinomyosin	آکتینومیوزین	environmental pollutant	آلاینده محیط زیست
actinin	آکتینین	albumin	آلبومین
activin	آکتیوین	ovalbumin	آلبومین تخم
acrasin	آکرازین	serum albumin	آلبومین خون
acrodermatitis enteropathica	آکرودرماتیت آنتروپاتیکا	organ	آلت
acrosin	آکروزین	algicide	آلجیساید
acro-cyanosis	آکروسیانوز	aldehyde	آلدئید
acrolein	آکرولئین	aldose	آلدوز
acridine	آکریدین	aldosterone	آلدوسترون
acriflavine	آکریفلاوین	aldehyde	آلدهید
acrylamide	آکریلامید	allergy	آلرژی
accelerin	آکسلرین	algin	آلژین
aclacinomycin	آکلاسینومایسین	alginate	آلژینات
aquaporin	آکوا پورین	alpha amylase	آلفا آمیلاز
aconta	آکونتا	alpha amylase inhibitor-1	آلفا آمیلاز اینهیبیتور -1
Achondroplasia	آکوندروپلازی	alpha-antitrypsin	آلفا-آنتی تریپسین
aconitase	آکونیتاز	α-TIF (= α- trans - inducing factor)	آلفا-تی.آی.اف.
aconitin	آکونیتین		
agar	آگار	alpha-synuclein	آلفا-ساینوکلئین
agarase	آگاراز	alpha-solanine	آلفا سولانین
agarose gel electrophoresis	آگارز ژل الکتروفورز	alpha-chaconine	آلفا شاکونین
agarose	آگاروز	α- fetoprotein	آلفا-فیتوپروتئین
familiarity	آگاهی	maternal serum alfa fetoprotein (MSAFP)	آلفافیتوپروتئین سروم مادری
agraceutical (= nutraceuticals)	آگراسوتیکال		
		alphavirus	آلفا ویروس
aggressin	آگرسین	alfAFP (= alfalfa antifungal peptide)	آلف ای. اف. پی.
agrobacterium	آگروباکتریوم		
agrobacterium tumefaciens	آگروباکتریوم تومافاسین	alkaptonuria	آلکاپتونوریا
agropine	آگروپین	alcalase	آلکالاز
agglutinogen	آگلوتینوژن	alkaline	آلکالاین
agglutinin	آگلوتینین	alkaline hydrolysis	آلکالاین هیدرولایزیس
aglycon, aglycone	آگلیکون	alkaloid	آلکالوئید
agonists	آگونیست ها	tremorgenic indole alkaloid	آلکالوئید ترمورژنیک ایندولی
Ala (= alanine)	آلا	alkali	آلکالی
alarmone	آلارمون	alkaline phosphatase	آلکالین فسفاتاز
alamethicin	آلامتیسین	alkannin	آلکانین
allantoin	آلانتونین	allele	آلل
alanine	آلانین	multiple allele	آلل چند گانه
alanine aminotransferase	آلانین آمینوترانسفراز	dominant allele	آلل غالب
contamination	آلایش	recessive allele	آلل مغلوب

allelozyme	آللوزیم	organotroph	آلی پرور
allelic exclusion	آللیک اگزکلوژن	alicin, allicin (= alicin)	آلیسین
alloantigen	آلوآنتی ژن	amatoxin	آماتوکسین
alloisoleucine	آلوایزولئوسین	primer	آماده ساز
allopatric	آلوپاتریک	cocking	آماده شلیک کردن
allopurinol	آلوپورینول	statistics	آمار
allotype	آلوتیپ	demography	آمار نگاری
biological contaminant	آلودکننده زیستی	stochastic	آماری-تصادفی
contamination, infection,	آلودگی	tumor	آماس
infectiousness,		malignant edema	آماس بدخیم
infestation, pollution		arthritis	آماس مفصل
infestation	آلودگی به انگل	amantadine	آمانتادین
infestation	آلودگی به حشرات	amanitin	آمانیتین
biofouling, biological	آلودگی بیولوژیکی	synthesis	آمایش
contamination		amplicon	آمپلیکون
decontamination	آلودگی زدایی	amplimer	آمپلیمر
food contamination	آلودگی غذایی	ampicillin	آمپی سیلین
inapparent infection	آلودگی غیر واضح	illegal traffic	آمد وشد غیر قانونی
opportunistic infection	آلودگی فرصت طلب	amphetamine	آمفتامین
lytic infection	آلودگی کافنده	beta-aminoprionitrile	آمنیوپریونیتریل بتا
food contamination	آلودگی مواد غذایی	amniocentesis	آمنیوسنتز
contaminate, fouling	آلودن	amnion	آمنیون
smut	آلوده به سیاهک	amorph	آمورف
contaminate	آلوده کردن	educator	آموزگار
polluter	آلوده گر	amitosis	آمیتوز
aleurone	آلورون	conjugation, hybridization	آمیختگی
allozyme (= allosteric	آلوزیم	sexual conjugation	آمیختگی جنسی
anzyme)		cross-hybridization	آمیختگی متقاطع
allogeneic	آلوژنئیک	mixing, synthesizing	آمیختن
allostery	آلوستری	hybrid	آمیخته
allosterism	آلوستریزم	first filial hybrid	آمیخته نسل اول
aleukia	آلوسمی	amidase	آمیداز
alloxan	آلوکسان	deamidation	آمید زدایی
allograft	آلوگرافت	wide cross	آمیزش دور
allolactose	آلولاکتوز	dihybrid cross	آمیزش دی هیبرید
allometry	آلومتری	hybrid	آمیزه
allomone	آلومون	first filial hybrid	آمیزه نسل اول
alumina	آلومین	amygdalin	آمیگدالین
aluminum	آلومینیم	amylase	آمیلاز
aluminum	آلومینیوم	alpha amylase	آمیلاز آلفا
aliesterase	آلی استراز	amylopectin	آمیلو پکتین

8

amyloglucosidase	آمیلوگلوکزیداز	antibody	آنتی بادی
deamination	آمین زدایی	anti-idiotype	آنتی ایدیوتیپ
aminoacyl	آمینوآسیل	anti-idiotype antibody	آنتی ایدیوتیپ آنتی بادی
aminoacylase	آمینوآسیلاز	antibody	آنتی بادی
aminoacylation	آمینوآسیلاسیون	anti-o-polysaccharide antibody	آنتی بادی آنتی او.پلی ساکارید
aminoacyl-tRNA	آمینوآسیل-تی.آر.ان.ای.	anti-idiotype antibody	آنتی بادی آنتی ایدیوتیپ
aminoethoxyvinylglycine	آمینو اتوکسی وینیل گلیسین	antibody affinity chromatography	آنتی بادی افینیتی کروماتوگرافی
essential amino acid	آمینو اسید اساسی/ضروری		
amino peptidase	آمینو پپتیداز	anti-idiotype antibody	آنتی بادی ضد ایدیوتیپ
aminopterin	آمینوپترین	anti-o-polysaccharide antibody	آنتی بادی ضد پلی ساکاریدهای او.
amino purine	آمینو پورین		
aminopyridines	آمینوپیریدین ها	anti-o-polysaccharide antibody	آنتی بادی ضد چند قندی های او.
ornithine amino-transferase	آمینوترانسفراز اورنیتین		
aminotransferases	آمینو ترانسفراز ها	engineered antibody	آنتی بادی مهندسی شده
aminocyclopropane carboxylic acid	آمینو سیکلو پروپان کربوکسیلیک اسید	chimeric antibody	آنتی بادی نوترکیب
		cross-reacting antibody	آنتی بادی واکنشگر متقاطع
aminocyclopropane carboxylic acid synthase	آمینو سیکلو پروپان کربوکسیلیک اسید سینتاز	allotypic monoclonal antibodies	آنتی بادی های مونوکلونال آلوتیپیک
anabolite	آنابولیت	antibiotic	آنتی بیوتیک
AnabolisTM	آنابولیس	aminoglycoside antibiotic	آنتی بیوتیک آمینوگلیکوزید
anabolic	آنابولیک	beta-lactam antibiotic	آنتی بیوتیک بتا-لاکتام
anaphylatoxin	آنافیلاتوکسین	intravenous antibiotic	آنتی بیوتیک داخل وریدی
anaphylaxis	آنافیلاکسیس	antibiotic therapy	آنتی بیوتیک درمانی
base analogue	آنالوگ بازی	antibiotics	آنتی بیوتیک ها
nucleoside analog	آنالوگ نوکلئوزید	antiparallel	آنتی پارالل
analyte	آنالیت	antiproliferative	آنتی پرولیفراتیو
RFLP linkage analysis	آنالیز اتصال آر.اف.ال.پی.	antiporter	آنتی پورتر
anandamide	آناندامید	antithrombin III	آنتی ترومبین نوع سوم
anencephaly	آن انسفالی	antitoxin	آنتی توکسین
anaplasia	آن اپلازی	antigen, antigene	آنتی ژن
antagonism	آنتاگونیزم	streptococal antigen	آنتی ژن استرپتوکوکی
enterocyte	آنتروسیت	surface protein antigen A	آنتیژن ای. پروتئین سطحی
salmonella enteritidis	آنتریت سالمونلا	very late antigen	آنتی ژن خیلی دیر
antrin	آنترین	histocompatibility antigen	آنتی ژن سازگاری بافتی
anthocyanoside	آنتوسیانوزید	surface antigen	آنتی ژن سطحی
anthocyanidin	آنتو سیانیدین	antigenic switching	آنتی ژن سوئیچینگ
anthocyanin	آنتو سیانین	cacinoembryonic antigen	آنتی ژن کاسینو جنینی
antiangiogenesis	آنتی آنژیوژنز	capsular antigen	آنتی ژن کپسولی
anti-oncogene	آنتی آنکوژن	v antigen	آنتی ژن ویروسی
anti-o-polysaccharide	آنتی او. پلی ساکارید آنتی بادی	human leucocyte antigens	آنتی ژن های یاخته های سفید

English	Persian	English	Persian
	انسان	inducible enzyme	آنزیم سازش پذیر
antiserum	آنتی سرم	adaptive enzyme	آنزیم سازشی
antisense	آنتی سنس	cyclooxygenase	آنزیم سایکلو اکسیژن
antifolate	آنتی فولات	psychrophilic enzyme	آنزیم سرما خواه
anticoagulant	آنتی کوآگولانت	psychrophilic enzyme	آنزیم سرما دوست
anticooperativity	آنتی کوآپرتیویتی	membrane anchor cleaving enzyme	آنزیم شکافنده لنگر غشائی
anticodon	آنتی کودون		
antibody	آنتی کور	membrane anchor cleaving enzyme	آنزیم شکندی (شکنده)
engineered antibody	آنتی کور مهندسی شده		
anticholinesterase	آنتی کولین استراز	uncoating enzyme	آنزیم ضد روکش
antimetabolite	آنتی متابولیت	ferrochelatase	آنزیم فرو کلات
antimorph	آنتی مورف	metalloenzyme	آنزیم فلزی
antihistamine	آنتی هیستامین	inducible enzyme	آنزیم قابل القاء
androgen	آندروژن	repressible enzyme	آنزیم قابل سرکوب
andosterone	آندوسترون	agarase	آنزیم کافنده آگار
endometrium	آندومتر	holoenzyme	آنزیم کامل
enzyme	آنزیم	lipolytic enzyme	آنزیم کاهنده چربی
allosteric enzyme	آنزیم آلوستریک	xanthine oxidase	آنزیم گزانتین اکسیداز
exoglycosidase	آنزیم اکسوگلیکوسید	glucose oxidase	آنزیم گلوکوز اکسید
inducible enzyme	آنزیم القاء پذیر	glucocerebrosidase	آنزیم گلوکوسربروسید
dissociating enzyme	آنزیم انفکاکی	DNA glycosylase	آنزیم گلیکوسیل دی.ان.ای.
unwinding enzyme	آنزیم بازکننده پیچ وخم	DNA methylase	آنزیم متیله کننده دی.ان.ای.
incision enzyme	آنزیم برش	restrictive enzyme	آنزیم محدودکننده
BARnase (= bacillus amilolique facien RNase)	آنزیم بی.ای.آر.	modification enzyme	آنزیم مدیفیکاسیون
		marker enzyme	آنزیم نشانگر
proteolytic enzyme	آنزیم پروتئولیتیک	constitutive enzyme	آنزیم نهادی
proteolytic enzyme	آنزیم پروتئین کافت	harvesting enzyme	آنزیم هاروستینگ
cyclin dependent protein kinase	آنزیم پروتئین متکی به سیکلین	restriction enzymes	آنزیم های محدود کننده
		core enzyme	آنزیم هسته/اصلی
angiotensin-converting enzyme	آنزیم تبدیل آنژیوتنزین	homotropic enzyme	آنزیم همگرا
		homotropic enzyme	آنزیم هوموتروپیک
lipolytic enzyme	آنزیم تجزیه کننده چربی	enzymatic	آنزیمی
adaptive enzyme	آنزیم تطبیقی	angiostatin	آنژیواستاتین
regulatory enzyme	آنزیم تنظیمی	angiotensin	آنژیوتنزین
dissociating enzyme	آنزیم جدا ساز	angiotensinase	آنژیوتنزیناز
dicer enzyme	آنزیم خرد کننده	angiotensinogen	آنژیوتنزینوژن
dicer enzyme	آنزیم دایسر	angiotensinogenase	آنژیوتنزینوژناز
harvesting enzyme	آنزیم درو	angiogenin	آنژیوژنینآنژیوژنین
allosteric enzyme	آنزیم دگر ریختار	ensiling	آنسیل کردن
constitutive enzyme	آنزیم ساختمانی	ancrod	آنکرود
zymogen	آنزیم ساز	annexin	آنکسین

10

English	Persian
angstrom (A)	آنگستروم
anemia	آنمی
pernicious anemia	آنمی پرنیشیاس
sickle cell anemia	آنمی سلول داسی شکل
Fanconi anaemia	آنمی فانکونی
iron deficiency anemia	آنمی کمبود آهن
aneuploid	آنوپلوئید
aneuploidy	آنوپلوئیدی
anomer	آنومر
annealing	آنیلینگ
anion	آنیون
anionic	آنیونی
fame	آوازه
avidin	آویدین
starch	آهار
galvanized iron	آهن سفید
I (= isochromosome)	آی.
IR (= infrared)	آی.آر.
IRES (= internal ribosome entry site)	آی.آر.ئی.اس.
IRP (= iron response protein)	آی.آر.پی.
ISS (= inhibitor of sister chromatid separation)	آی.اس.اس.
ISCOM (= immune-stimulating complexes)	آی.اس.سی.او.ام.
IF (= initiation factor)	آی.اف.
IF2 (= initiation factor 2)	آی.اف.2
IMP (= inosine mono-phosphate)	آی.ام.پی.
INS (= insertion)	آی.ان.اس.
INC (= incomplete karyotype)	آی.ان.سی.
INV (= inversion), INV (= inverted)	آی.ان.وی.
IB (= immuno blotting)	آی.بی.
IBC	آی.بی.سی.
ITR (= inverted terminal repeat)	آی.تی.آر.
ITF (= integration factor)	آی.تی.اف.

English	Persian
ITP (= inosine tri-phosphate)	آی.تی.پی.
secretory IgA	آی.جی.آی. مترشح
IGHD (= isolated human growth hormone deficiency)	آی.جی.اچ.دی.
IGF (= insulin growth factor)	آی.جی.اف.
IgM (= Immunoglobin M)	آی جی. ام.
IgG (= immunoglobin G)	آی جی.جی.
Ider (= isoderivative chromosome)	آیدر
Idem (denotes the stemline karyotype in subclones)	آیدم
IDL (= intermediate-density lipoprotein)	آی.دی.ال.
IddM (= insulin-dependent diabetes mellitus), IDDM (= insulin-dependent diabetes mellitus)	آی.دی.دی.ام.
aerodynamic	آیروداینامیک
aerodynamic	آیرودینامیک
fallow	آیش
A I (= first meiotic anaphase)	آ.یک
IVSs (= intervening sequences)	آی.وی.اس. ها
IVF (= in vitro fertilization)	آی.وی.اف.

ئ

English	Persian
E1B (= E1B protein of adenovirus)	ئی.1.بی.
E2F (= transcription factor)	ئی.2. اف.
ER (= endoplasmic reticulum)	ئی.آر.
ERV (= endogenous retro-virus)	ئی.آر.وی.
EEO (= electro-end-osmosis value)	ئی.ئی.او.
EEC (=	ئی.ئی.سی.

English	Persian	English	Persian
ectrodactyly-ectodermal dysplasia-clefting)		supergene	ابر ژن
ES (= embryonic stem cell)	ئی.اس.	supercoil	ابر مارپیچ
EST (= expressed sequence tag)	ئی.اس.تی.	supercoiling	ابر مارپیچ شدن
		silk	ابریشم
EF (= elongation factor)	ئی.اف.	abrin	ابرین
EF1 (= elongation factor 1)	ئی.اف.1	device, organ	ابزار
ELISA (= enzyme-linked-immunos orbent assay)	ئی.ال.آی اس.ای.	abzyme	ابزایم
		abscisic acid	ابسیزیک اسید
		operator	اپراتور
		lac operon	اپرون لاک/لاکتوز
EM (= electron microscopy)	ئی.ام.	hut operon	اپرون هات
EMS (= ethyl methane sulfonate)	ئی.ام.اس.	applicator	اپلیکاتور
		sample applicator	اپلیکاتور نمونه گیری
EBV (= Epstein Barr virus)	ئی.بی.وی.	epithelial projection	اپیتلیال پروجکشن
EPR (= electron paramagnetic resonance)	ئی.پی.آر.	epithelium	اپی تلیوم
		epitope	اپی توپ
ETEC (= entro-toxigenic E.coli)	ئی.تی.ئی.سی.	epidermolysis bullosa acquisita	اپیدرمولیز بولوسا آکویزیتا
EGF (= epidermal growth factor)	ئی.جی.اف.	outbreak	اپیدمی
		behavioral epidemic	اپیدمی رفتاری
dehydrogenation	ئیدروژن گیری	epizootic	اپیزوئوتیک
EDS (= Ehlers-Danlos syndrome)	ئی.دی.اس.	episome	اپیزوم
		epigenetic	اپی ژنتیک
ECD (= endocardial cushion defect)	ئی.سی دی.	nutritional epigenetics	اپی ژنتیک تغذیه ای
		cancer epigenetics	اپی ژنتیک سرطان
e. coli (= Escherichia coli)	ئی.کولای	apigenin	اپیژنین
entro-toxigenic E.coli	ئی. کولای آنتروتوکسیژنیک	epistasis	اپیستازی
e. coli k12	ئی. کولای کی.12	capital femoral epiphysis	اپیفیز ران/فمورال اصلی
		epiphysitis	اپی فیزیت
I		myoclonic epilepsy with ragged red fibers	اپلپسی میوکلونیکی با فیبرهای قرمز آشفته
aa (amino acid)	آآ	epimer	اپیمر
eosinophils	ائوزین دوست ها	epimerase	اپی مراز
eosinophils	ائوزینوفیل ها	autacoid	اتاکوئید
infection, morbidity	ابتلاء	ethanol	اتانول
infectibility	ابتلاء پذیری	dilatation	اتساع
superantigen	ابر پادگن	ligate, linkage, linking	اتصال
negative supercoiling	ابرپیچش منفی	blunt-end ligation	اتصال انتهای صاف/کور
superovulation	ابرتخمکگذاری	splice (ing) junctions	اتصال بهم تابیده
carbon nanotube	ابر ریز لوله کربن(ی)	gap junction	اتصال بینابینی/گپ
supergene	ابر زاد	peptide linkage	اتصال پپتیدی

English	Persian
plasma protein binding	اتصال پروتئین پلاسما
protein splicing	اتصال پروتئینی
alternative splicing	اتصال جایگزین
alternative mRNA splicing	اتصال جایگزین پیک ار.ان.ای
G-banding (= Giemsa chromosome banding)	اتصال جی.
donor junction (acceptor junction site), linker	اتصال دهنده
polylinker	اتصال دهنده چند تائی
genetic linkage	اتصال ژنتیکی
Giemsa chromosome banding	اتصال کروموزومی جی امسا
recombination	اتصال مجدد
homopolymer tailing	اتصال هموپلیمری/هم پلیمری
hydrogen bond	اتصال هیدروژنی
accident, occurrence	اتفاق
stochastic	اتفاقی
atom	اتم
AtomizingTM	اتمایزینگ
meso carbon atom	اتم کربن مزو/میانی
ionized atom	اتم یونیده
autoantibody	اتوآنتی بادی
autoantigen	اتو آنتی ژن
autopolyploid	اتوپلی پلوئید
autoradiography	اتو رادیوگرافی
autosome	اتوزوم
autophagy	اتوفاژی
autophosphorylation	اتوفسفریلیشن
auto-claving	اتو کلاو (کردن)
auto-correlation	اتو کو ریلیشن
autograft	اتوگرافت
bioautography	اتوگرافی زیستی
autologous	اتولوگ
autolysis	اتولیز
autolysate	اتولیزات
autolysin	اتولیزین
ethidium bromide	اتیدیوم بروماید
label, tag	اتیکت
ethyl acetate	اتیل استات
ethylene	اتیلن
etioplast	اتیوپلاست

English	Persian
etiology	اتیولوژی
additive effect	اثر افزایشی
additive effect	اثر افزاینده
fingerprinting	اثر انگشت
founder effect	اثر بانی
efficacy	اثر بخشی
founder effect	اثر بنیان گذار
footprinting	اثر پا گرفتن
side effect	اثرجانبی
additive effect	اثر جمعی
edge effect	اثر حاشیه ای
eschar	اثر زخم
biological effect	اثر زیست شناختی
genetic effect	اثر ژنتیکی
side effect	اثرفرعی
effector	اثر کننده
synergistic effect	اثر گذاری دوگانه
synergistic effect	اثرگذاری مضاعف
greenhouse effect	اثر گلخانه ای
epistasis	اثر متقابل ژن ها
chronic effect	اثر مزمن/حاد/جدی
adverse effect	اثر معکوس
position effect	اثر موقعیت
adverse effect	اثر وارونه
position effect	اثر وضعیت
synergistic effect	اثر هم افزا
synergistic effect	اثر هم نیرو زادی
obligate	اجباری کردن
colony, community	اجتماع
climax community	اجتماع اوج
avoidance	اجتناب
lineage	اجداد
operator	اجرا کننده
coding parts of a gene	اجزاء رمز ساز ژن
non-coding parts of a gene	اجزاء غیر رمز ساز ژن
HRE (= hormone response element)	اچ.آر.ئی.
HIV (= human immunodeficiency virus)	اچ.آی.وی.
HE (= hereditary elliptocyptosis)	اچ.ئی.

HPP (= high profilerative potential)	اچ.اچ.پی.
HS (= hereditary spherocytosis)	اچ.اس.
HSR's (= homogeneously staining regions)	اچ.اس.آر. (ها)
HSC (= hemopoietic stem cells)	اچ.اس.سی.
HSV (= herpes simplex virus)	اچ.اس.وی.
Hfr	اچ.اف.آر.
HFI (= hereditary fructose intolerance)	اچ.اف.آی.
Hx (= hypoxanthine)	اچ. اکس.
HLA (= human leucocyte antigens)	اچ.ال.ای.
HMP (= hexose mono-phosphate)	اچ.ام.پی.
HMWK (= high molecular weight kininogen)	اچ.ام.دبلیو.کی.
HNPCC (= hereditary non-polyposis colorectal cancer)	اچ.ان.پی.سی.سی.
HOX1 (= homeo box region 1)	اچ.او.ایکس.1
HOX2 (= homeo box region 2)	اچ.او.ایکس.2
HOX3 (= homeo box region 3)	اچ.او.ایکس.3
HAEC (= human artificial episomal chromosome)	اچ.ای.ئی.سی.
HAT (= hypoxanthine–aminopterin-thymidine)	اچ.ای.تی.
Hb (= hemoglobin)	اچ.بی.
HBS (= heteroduplex binding site)	اچ.بی.اس.
Hb F (= fetal hemoglobin)	اچ.بی.اف.
HBA (= adult hemoglobin), HbA (= adult hemoglobin)	اچ.بی.ای.
HBD (= hypophospha-taemic bone disease)	اچ.بی.دی.
HBV (= hepatitis B virus)	اچ.بی.وی.
HPRH (= hypo plastic right heart)	اچ.پی.آر.اچ.
HPRT (= hypoxanthine phospho ribosyl transferase)	اچ.پی.آر.تی.
HPFH (= hereditary persistence of fetal hemoglobin)	اچ.پی.اف.اچ.
HPLC (= high-pressure/ high performance liquid chromatography)	اچ.پی.ال.سی.
hGH (= human growth factor)	اچ.جی.اچ.
HGPRT (= hypoxantine-guanine phospho-ribosyl transferase)	اچ.جی.پی.آر.تی.
HD (= Huntington's Disease)	اچ.دی.
HDL (= high-density lipoprotein)	اچ.دی.ال.
H.zea (= corn earworm)	اچ. زیا
HC (= hypertrophic cardiomyopathy)	اچ.سی.
hcG (= human chronic gonadotropin)	اچ.سی.جی.
HVS (= herpes virus saimiri)	اچ.وی.اس.
avoidance	احتراز
risk	احتمال خطر
sense	احساس
reduction, restoration, resuscitation	احیاء
denitrification, nitrate reduction	احیاء نیترات
proprietary	اختصاصی
specificity	اختصاصی بودن
diffusion, mixing	اختلاط
diversity, variation	اختلاف

English	Persian	English	Persian
breakdown	اختلال	use value	ارزش کاربرد
monogenic disorder/trait	اختلال تک ژنی	passive use value	ارزش کاربرد غیر فعال
inborn error	اختلال توارثی	total economic value	ارزش کامل اقتصادی
hemostatic derangement	اختلال درانعقاد خون	total environmental value	ارزش کامل زیست محیطی
autoimmune disorder	اختلال در سیستم خود ایمنی	valuation	ارزش گذاری
notification	اخطاریه	bequest value	ارزش میراث/واگذاری
pyuria	ادرار چرک دار	ceiling value	ارزش نهائی
aedes albopictus	ادس آلبوپیکتوس	existence value	ارزش وجودی
assimilation	ادغام	ethical values	ارزش های اخلاقی
cell fusion	ادغام پاخته ای	evaluation	ارزشیابی
cyclic adenosine monophosphate	ادنوزین مونو فسفات چرخه ای/متناوب/سیکلیک/گردشی	assessment, evaluation, valuation	ارزیابی
cyclic	ادواری	risk assessment	ارزیابی احتمال خطر
advanced informed agreement	ادوانس اینفورم اگریمنت	risk assessment	ارزیابی ریسک
small RNA	ار.ان.ای. کوچک	organelle	ارگانل
small nuclear RNA (snRNA)	ار.ان.ای. کوچک هسته ای	organotroph	ارگانوتروف
liaison, linkage	ارتباط	pathogenic organism	ارگانیزم بیماری زا
synapse	ارتباط دو نرون	coliform organism	ارگانیزم کولی باسیل شکل
ortholog	ارتولوگ	parental organism	ارگانیزم والد
heredity, inheritable	ارثی	target organism	ارگانیزم هدف
inheritance	ارثیه	organism	ارگانیسم
valuation	ارزش	ergotamine	ارگوتامین
private value	ارزش اختصاصی	oropharyngeal	اروفارینژیال
direct use value	ارزش استفاده مستقیم	erythropoietin	اریتروپوئیتین
electro-end-osmosis value	ارزش الکترو-اند-اسمز(ی)	erythropoiesis	اریتروپوئیزیس
option value	ارزش انتخاب	erythrocyte	اریتروسیت
primary value	ارزش اولیه	sheep erythrocytes	اریتروسیت گوسفندی
insurance value	ارزش بیمه گذاری	erythritol	اریتریول
development value	ارزش پیشبرد/توسعه	ase	از-
secondary value	ارزش ثانویه	lyse	از بین بردن
conservation value	ارزش حفاظت	knockout	از پا در آوردن
critical value of dissolved oxygen concentration	ارزش حیاتی غلظت اکسیژن محلول	incapacitate	از توان انداختن
net present value	ارزش خالص کنونی	incapacitation	از توان اندازی
c value	ارزش سی .	azotobacter	ازتوباکتر
quasi-option value	ارزش شبه گزینه	swarming	ازدحام کردن
non-use value	ارزش غیر کاربردی	dehydration	از دست رفتن آب بدن
non-consumptive value	ارزش غیر مصرفی	lymphocytosis	ازدیاد لنفوسیت ها در خون
		attenuate	از شدت چیزی کاستن
		infiltrate	از صافی رد کردن
		repression	از کارافتادگی تظاهر ژنی
		knockout	از کاراندازی

English	Persian	English	Persian
incapacitation	از کار اندازی	SAR (= scaffold attachment region)	اس.ای.آر.
ozone-depleting potential (ODP)	ازن کاهی	SAHH (= S-adeno homocysteine hydrolase)	اس.ای.اچ.اچ.
ozonolysis	ازنولایزیس	SASD (= sialic acid storage disease)	اس.ای.اس.دی.
breakdown	از هم پاشیدن		
S (= S phase), S (= sedimentation coefficient), s (second)	اس.	SINE (= short interspersed nuclear element)	اس.ای.ان.ئی.
SRF (= serum response factor)	اس.آر.اف.	SAP (= stress activated protein)	اس.ای.پی.
SRBC (= sheep erythrocytes)	اس.آر.بی.سی.	SIV (= simian immunodeficiency virus)	اس.ای.وی.
SRP (= signal recognition particle)	اس.آر.پی.	spreader	اسپردر
		sperm	اسپرم
substantially equivalent	اساسا برابر	spermatocyte	اسپرماتوسیت
SSLP (= simple sequence length polymorphism)	اس.اس.ال.پی.	spermatogonium cell	اسپرماتوگونی
		spermatid	اسپرماتید
SSB (= single-strand binding protein)	اس.اس.بی.	spermatium	اسپرماتیوم
		spermiogenesis	اسپرم زائی
ssDNA	اس.اس. دی.ان.ای.	spermophilus	اسپرموفیلوس
SSCP (= single-strand conformational polymorphism)	اس.اس.سی.پی.	spermiogenesis	اسپرمیوژنز
		aerosol, atomizer, spray, sprayer	اسپری
SFC (= Spot-Formi cell)	اس.اف.سی.	portable spray	اسپری دستی
SLE (= systemic lupus erythematosus)	اس.ال.ئی.	aerosolize(d)	اسپری شده
		aerosolize(d)	اسپری کردن
SLD (= short limbed dwarfism)	اس.ال.دی.	Mercury Knapsack mistblower	اسپری مرکوری ناپساک
SMG (= sub-mandibular glands)	اس.ام.جی.	seed-specific promoter	اسپسیفیک پروموتر دانه ای
		uv-absorbance spectroscopy	اسپکترواسکوپی جذب یو.وی.
SMD (= somatostatin)	اس.ام.دی.	ultraviolet spectroscopy	اسپکترواسکوپی فرا بنفش
snRNA (= small nucleolar RNA)	اس.ان.آر.ان.ای.	combined gas chromatography-mass spectrometre	اسپکترومتر انبوه رنگ نگاری گازی ترکیب شده
snRNP (= small nuclear ribonucleoprotein)	اس.ان.آر.ان.پی.		
		spectrin	اسپکترین
SNP (= single nucleotide polymorphism)	اس.ان.پی.	spectinomycin	اسپکتینومایسین
		spliceosome	اسپلایسوزوم
SOD (= super-oxide dismutase)	اس.او.دی.	spore	اسپور
		sporangiospore	اسپورانژیوسپور
SA (= streptococal antigen), SA (= surface antigen)	اس.ای.	sporangium	اسپورانژیوم
		endospore	اسپور داخلی

English	Persian
sporulation	اسپور زایی
sporozoite	اسپوروزوئیت
SP1 (= transcription factor)	اس.پی.1
SP3 (= spermatogenesis factor 3)	اس.پی.3
SPM (= suspended particulate matter)	اس.پی.ام.
SP-B (= surfactant-associated protein B)	اس.پی.-بی.
spiramycin	اسپیرامایسین
spirochete	اسپیروکت
spin	اسپین
spina bifida	اسپینا بیفیدا
spin trapping	اسپین تراپینگ
Spinosad TM	اسپینوزاد
spinosyns	اسپینوسین ها
stearate	استارات
stearoyl-acp desaturase	استاراویل ای.سی.پی. دیساتیوراز
stearidonate	استاریدونات
osteogenesis imperfecta	استئوژنز ایمپرفکتا
osteocalcin	استئوکالسین
statin	استاتین
staurosporin(e)	استارواسپورین
staphylococin	استافیلوکوسین
staphylococcus	استافیلو کوک
staphylococcus aureus	استافیلو کوک اورئوس
staphylococcal	استافیلو کوکی
staphylokinase	استافیلوکیناز
stacchyose (= stachyose)	استاکیوز
standard	استاندارد
stanol ester	استانول استر
genomic exclusion	استثناء ژنومیکی
RNA extraction	استخراج آر.ان.ای.
site-directed mutagenesis	استخلاف
skeleton	استخوان بندی
ester	استر
stanol fatty acid ester	استر اسید چرب استانل
streptavidin	استرپتا ویدین
streptothricin	استرپتوتریسین
streptogenin	استرپتوژنین
streptococcus	استرپتوکوک
streptococcus mutans	استرپتوکوک میوتان (ها)
streptococcal	استرپتوکوکی
streptokinase	استرپتوکیناز
Streptolydigins	استرپتولایدیزین ها
streptolysin	استرپتولایزین
streptomyces	استرپتومایس ها
streptonigrin	استرپتونیگرین
phorbol ester	استر فوربول
strychnine	استرکنین
steroid	استروئید
steroidogenesis	استروئیدوژنز
estrogen	استروژن
phytoestrogen	استروژن گیاهی
sterol	استرول
plant sterol	استرول گیاهی
phytosterols	استرول های گیاهی
stroma	استروما
stromelysin	استروملیسین
esterification	استری شدن
strychnine	استریکنین
streaking	استریکینگ
sterile	استریل
stereoisomer	استریو ایزومر
ascites	استسقای شکم
aptitude, capacity, competency	استعداد
establishment potential	استعداد استقرار
high profilerative potential	استعداد انتشار بالا
marrow repopulation ability	استعداد بازسازی مغز استخوان
genetic susceptibility	استعداد ژنتیکی
phosphorylation potential	استعداد فسفات افزایی
use	استعمال
application, use	استفاده
user	استفاده کننده
persistence	استقامت
induction	استقرا
gene stacking	استک کردن ژن (ی)
persistence	استمرار
inhalation	استنشاق

17

English	Persian	English	Persian
aortic stenosis	استنوز آئورتی	SCT (= secondary constriction)	اس.سی.تی.
pulmonary stenosis	استنوز ریوی	spheroplast	اسفروپلاست
solid	استوار	hereditary spherocytosis	اسفروسیتوز ارثی
persistence	استواری	sphingolipid	اسفنگولیپید
graduated cylinder	استوانه مدرج	sphingomyelin	اسفنگومیلین
acetone	استون	eschar	اسکار
stevioside	استوویوزید	scrubbing	اسکراب کردن
STRs (= short tandem repeats)	اس.تی.آر.اس. ها	scrapie	اسکراپی
STIC (= serum-trypsin-inhibitory capacity)	اس.تی.آی.سی.	scaffold	اسکلت
		cytoskeleton	اسکلت سلولی
		sclerobasidium	اسکلروبازیدیوم
STS (= sequence tagged site)	اس.تی.اس.	scleroprotein	اسکلروپروتئین
		sclerotium	اسکلروتیوم
STDs (= sexually transmitted disease)	اس.تی.دی.	amyotrophic lateral sclerosis	اسکلروز آمیوتروفیک جنبی
stigmasterol	استیگماسترول	sclerin	اسکلرین
prevalence	استیلا	base excision sequence scanning	اسکن توالی خارج سازی/برداشت بازی
acetyl spiramycin	استیل اسپیرامایسین		
acetyl carnitine	استیل کارنیتین	squalamine	اسکوالامین
acetyl-CoA	استیل کو آ.	squalene	اسکوالن
acetyl-CoA carboxylase	استیل کو آ. کربوکسیلاز	scotophobin	اسکوتوفوبین
acetyl-coenzyme A, acetyl co-enzyme A	استیل کوآنزیم آ	upflow sludge blanket	اسلاج بلانکت آپفلو
		gun, weapon	اسلحه
acetylcholine	استیل کولین	osmosis	اسمز
acetylcholinesterase	استیل کولین استراز	reverse osmosis	اسمز معکوس
SDS (= sodium dodecylsulfate)	اس.دی.اس.	osmotins	اسموتین ها
		snurposome	اسنورپوزوم
SC (= secretory component)	اس.سی.	SV (= specific volume)	اس.وی.
scRNA (= small cytoplasmic RNA)	اس.سی.آر.ان.ای.	SV40 (= simian virus 40)	اس.وی.40
SCID (= severe combined immuno-deficiency disease)	اس.سی.آی.دی.	SVI (= sludge volume index)	اس.وی.آی.
		acid	اسید
		apurinic acid	اسید آپورینیک
SCE (= sister chromatid exchange)	اس.سی.ئی.	apyrimidinic acid	اسید آپیریمیدینیک
		adenylic acid	اسید آدنیلیک
SCA (= sickle cell anemia)	اس.سی.ای.	arachidonic acid	اسید آراشیدونیک
SCARMD (= severe, childhood, autosomal, recessive muscular dystrophy)	اس.سی.ای.آر.ام.دی.	aspartic acid	اسید آسپارتیک
		acrylic acid	اسید آکریلیک
		alginic acid	اسید آلژینیک
		alpha-rumenic acid (=	اسید آلفا رومنیک

conjugated linoleic acid)	اسید دوکوساهگزانوئیک
acid alpha-glucosidase	اسید آلفا گلوکوزیداز
a-linolenic acid	اسید آ.-لینولنیک
gamma-aminobutyric acid	اسید آمینوبوتریک گاما
amino acid	اسید آمینه
essential amino acid	اسید آمینه اساسی/ضروری
nonessential amino acid	اسید آمینه غیر ضروری
glucogenic amino acid	اسید آمینه قندزا
stearidonic acid	اسید استئاریدونیک
stearic acid	اسید استئاریک
acetylcholinesterase acid	اسید استیل کولین استراز
ascorbic acid	اسید اسکوربیک
oxalic acid	اسید اگسالیک
itaconic acid	اسیدایتاکونیک
eicosapentaenoic acid	اسید ایکوزاپنتا انوئیک
eicosapentanoic acid	اسید ایکوزاپنتانوئیک
eicosatetraenoic acid	اسید ایکوزاتترا انوئیک
boletic acid	اسید بولتیک
palmitic acid	اسید پالمیتیک
pantothemic acid	اسید پانتوتمیک
propionic acid	اسید پروپیونیک
acidophilic	اسید پسندی
pyruvic acid	اسید پیروویک
periodic acid	اسید تناوبی
fatty acid	اسیدچرب
free fatty acid	اسید چرب آزاد
essential fatty acid	اسید چرب اساسی/ضروری
saturated fatty acid	اسید چرب اشباع شده
unsaturated fatty acid	اسیدچرب اشباع نشده
omega-3 fatty acid	اسید چرب امگا-3
essential polyunsaturated fatty acid	اسید چرب پر اشباع نشده اساسی
trans fatty acid	اسید چرب تبدیلی
monounsaturated fatty acid	اسید چرب تک اشباع نشده
highly unsaturated fatty acid	اسید چرب شدیدا اشباع نشده
acidophilic	اسید خواهی
deoxyribo nucleic acid	اسید داکسی ریبونوکلئیک
dioctadectylamidoglycylspe rmine tetra fluoroacetic acid	اسید دایوکتادکتیلآمیدوگلیسیلسپ رمین تترا فلوئورواستیک

docosahexanoic acid (DHA)	اسید دوکوساهگزانوئیک
retinoic acid	اسید رتینوئیک
rosemarinic acid	اسید رزمارینیک
rumenic acid	اسید رومنیک
ribonucleic acid, Ribo-Nucleic Acid	اسید ریبونوکلئیک
messenger RNA (mRNA)	اسید ریبونوکلئیک پیامبر
ribosomal ribonucleic acid	اسید ریبونوکلئیک ریبوزومی
transfer RNA (tRNA)	اسید ریبونوکلئیک ناقل
xanthylic acid	اسید زانتیلیک
jasmonic acid	اسید ژاسمونیک
salicylic acid	اسید سالیسیلیک
sorbic acid	اسید سوربیک
sialic acid	اسید سیالیک
citric acid	اسید سیتریک
bile acid	اسید صفراوی
fusaric acid	اسید فوساریک
fumaric acid	اسید فوماریک
phytic acid	اسید فیتیک
kainic acid	اسید کاینیک
kojic acid	اسید کوجیک
keto acid	اسید کیتو
acidophilic	اسیدگرائی
glutamic acid	اسید گلوتامیک
lauric acid	اسید لوریک
linoleic acid	اسید لینولئیک
conjugated linoleic acid (CLA)	اسید لینولئیک هم یوغ/مزدوج
linolenic acid	اسید لینولنیک
malic acid	اسید مالیک
mercapturic acid	اسید مرکاپتوریک
mevalonic acid	اسید موالونیک
nucleic acid	اسید نوکلئیک
end-labelled nucleic acid	اسید نوکلئیک برچسب خورده در انتها
acidotropic	اسیدوتروپیک
acidosis	اسیدوز
acidophilic	اسیدوفیلیک
excitatory amino acids	اسیدهای آمینه تحریک شونده
polyunsaturated fatty acids	اسید های چرب پر اشباع نشده

English	Persian
hyaluronic acid	اسید هیالورونیک
urocanic acid	اسید یوروکانیک
dissemination	اشاعه
nosocomial spread	اشاعه بیمارستانی
disseminating	اشاعه دادن
disseminator	اشاعه دهنده
quenching	اشباع کردن
activator	اشتارکن
Escherichia coli (E. coli)	اشریشیاکولای
radiation	اشعه
x-ray	اشعه ایکس
radiotherapy	اشعه درمانی
ultraviolet rays	اشعه های فرا بنفش
uv rays	اشعه یو.وی.
saponin	اشنان
breed, origin	اصل
restoration	اصلاح
habitat restoration	اصلاح بوم
precautionary principle	اصل احتیاطی
gene repair	اصلاح ژن (ی)
mismatch repair	اصلاح نامناسب
breeding	اصلاح نژاد
marker-assisted breeding	اصلاح نژاد به کمک نشانه
breed	اصلاح نژاد کردن نسل
convergent improvement	اصلاح همگرا
strong sustainable development principle	اصل توسعه نیرومند پایدار
lineage	اصل و نسب
emergency	اضطرار
extinction	اضمحلال
paravertebral	اطراف مهره ای
notification	اطلاع
data	اطلاعات
genetic information	اطلاعات ژنتیکی
baseline data	اطلاعات مقدماتی
notification	اطلاعیه
validation	اعتبار بخشی
homeostasis	اعتدال مزاج
lineage	اعقاب
prior informed consent	اعلام موافقت قبلی
F (= Faraday constant), F (= phenylalanine)	اف.
F1P (= fructose-1-phosphate)	اف. 1 پی.
F6P (= fructose-6-phosphate)	اف.6 پی.
F8C (= clotting factor VIII)	اف.8 سی.
F8VWF (= von Willebrand factor disease)	اف.8 وی.دبلیو.اف.
F9 (= clotting factor IX)	اف.9
FRAX (= fragile X syndrome)	اف.آر.ای.اکس.
FRDA (= Freidreich's ataxia)	اف.آر.دی.ای.
FISH (= fluorescence in situ hybridization)	اف.آی.اس.اچ.
FIGE (= field inversion gel electrophoresis)	اف.آی.جی.ئی.
FEN (= flap endo nuclease)	اف.ئی.ان.
FH (= familial hypercholesterolaemia)	اف.اچ.
FMR1 (= fragile X mental retardation)	اف.ام.آر.1
FMN (= flavin mono-nucleotide)	اف.ام.ان.
FOS (= oncogene FOS)	اف.او.اس.
FA (= Fanconi anaemia)	اف.ای.
FAP (= familial adenomatous polyposis)	اف.ای.پی.
FAD (= flavin adenine dinucleotide)	اف.ای.دی.
FADH2 (= reduced flavin adenine dinucleotide)	اف.ای.دی.اچ.2
FBP (= fructose-1,6-bis-phosphate)	اف.بی.پی.
FBPase (= fructose bis-phosphatase)	اف.بی.پی.آز
depression	افت
FGF (= fibroblast growth factor)	اف.جی.اف.
FGA (= fibrinogen, alpha	اف.جی.ای.

chain)		water activity	اکتیویته آب
F-duction	اف. داکشن	ecdysone	اکدیزون
Fd (= ferrodoxin)	اف. دی.	ochratoxin	اکراتوکسین
F-duction	اف. رسانی	XR (= X- linked recessive)	اکس.آر.
technology	افزار سازی	XR-MR (= X-linked recessive mental retardation)	اکس.آر.-ام.آر.
amplification, enhancement, growth	افزایش		
immuno-enhancing	افزایش ایمنی	XXY (= 47XXY)	اکس.اکس.وای.
gene amplification	افزایش ژن (ی)	oxalate	اکسالات
scale-up	افزایش مقیاس	oxalate oxidase	اکسالات اکسیداز
additive	افزایشی	XLH (= X-linked hypophosphatemic rickets)	اکس.ال.اچ.
additive	افزودنی		
redundancy	افزونگی		
redundancy	افزونی	XMP (= xanthosine mono phosphate)	اکس.ام.پی.
depression	افسردگی		
FCS (= fetal calf stem)	اف.سی.اس.	XO (= 45XO)	اکس.او.
sprayer	افشان-	oxidation	اکسایش
aerosolizing, spray	افشاندن	beta oxidation	اکسایش بتا
nozzle	افشانک	biological oxidation	اکسایش زیست شناختی
aerosol, atomizer, spray, sprayer	افشانه	redox	اکسایش و کاهش
		oxidant	اکساینده
infectious aerosol	افشانه آلوده کننده	express	اکسپرس
inhaler	افشانه تنفسی	expressed sequence tag (EST)	اکسپرس سیکوئنس تگ
TOPAS aerosol	افشانه تی.او.پی.ای.اس.		
aerosolization	افشانه سازی کردن	bacterial expressed sequence tag (= expressed sequence tag)	اکسپرس سیکوئنس تگ باکتریائی
aerosolize(d)	افشانه شده		
portable spray	افشانه قابل حمل		
aerosol spray	افشانه گردپاش	acute exposure	اکسپوژر حاد
spray	افشک	XP (= xeroderma pigmentation)	اکس.پی.
F Met	اف.مت		
aphidicolin	افیدیکولین	extremozyme	اکسترموزایم
aphicidin	افیسیدین	external-beam radiation	اکسترنال بیم رادیشن
contact precaution	اقدام احتیاطی تماسی	reciprocal externality	اکسترنالیتی متقابل
airborne precaution	اقدام احتیاطی (ذرات) هوابرد	XD (= X-linked dominant)	اکس.دی.
control measure	اقدام مهار کننده	x-gal	اکس.-گال
acclimatization	اقلیم پذیری	oxidant	اکسنده
biome	اقلیم زیست	XYY (= 47XYY)	اکس.وای.وای.
meteorology	اقلیم شناسی	auxotroph	اکسوتروف
acclimatization	اقلیمی شدن	exoglycosidase	اکسوگلیکوسیداز
ectromelia	اکتروملی	exon	اکسون
ectoplasm	اکتوپلاسم	ex vivo	اکس ویوو

English	Persian
excipient	اکسیپینت
oxytetracycline	اکسی تترا سایکلین
oxytocin	اکسی توسین
oxidase	اکسیداز
alcohol oxidase	اکسیداز الکل
urate oxidase	اکسیداز اورات
monoamine oxidase	اکسیداز مونوآمین
acid oxide	اکسید اسیدی
oxidation	اکسیداسیون
fatty acid oxydation	اکسیداسیون اسید چرب
redox	اکسیداسیون و احیاء
oxidant	اکسیدان
oxidant	اکسید کننده
nitric oxide	اکسید نیتروژن
oxidation, reduction	اکسیده شدن
oxidation	اکسیده کردن
oxygenase	اکسیژناز
mixed-function oxygenase	اکسیژناز کارکرد آمیخته
singlet oxygen	اکسیژن تک دانه
microaerophile	اکسیژن خواه
oxymyoglobin	اکسی میوگلوبین
auxins	اکسین
oxyhemoglobin	اکسی هموگلوبین
aequorin	اکوئورین
ecotourism	اکوتوریزم
ecosystem	اکوسیستم
ecology	اکولوژی
agroecology	اکولوژی زراعی
exobiology	اگزوبیولوژی
exogenous	اگزوژنی
exocytosis	اگزوسیتوز
exon	اگزون
exonuclease	اگزونوکلئاز
agglomeration	اگلومراسیون
LIF (= leukemia inhibitory factor)	ال.آی.اف.
LINE (= long interspersed nuclear element)	ال.آی.ان.ئی.
LE (= lupus erythematosus)	ال.ئی.
LEP (= leptotene)	ال.ئی.پی.
LH (= luteinizing hormone)	ال.اچ.

English	Persian
ellagic acid	الاژیک اسید
ellagic tannin	الاژیک تانن
elastase	الاستاز
elastin	الاستین
LA (= left atrium), LA (= linkage analysis)	ال.ای.
LAT (= latency-associated transcript)	ال.ای.تی.
LPS (= lipo-poly saccharide)	ال.پی.اس.
LPL (= lipo-protein lipase)	ال.پی.ال.
epididymo-orchitis	التهاب بیضه و بربخ
salmonella enteritidis	التهاب روده باریک ناشی از سالمونلا
pachymeningitis	التهاب سخت شامه
myelitis	التهاب طناب نخاعی
prostatitis	التهاب غده پروستات
lymphadenitis	التهاب گره های لنفی
necrotizing lymphadenitis	التهاب گره های لنفی کشنده نسج
cervical lymphadenitis	التهاب گره های لنفی گردنی
necrotizing lymphadenitis	التهاب گره های لنفی نکروز آور
cystitis	التهاب مثانه
viral encephalitis	التهاب مغز ویروسی
arthritis, arthropathy	التهاب مفصل
vaginosis	التهاب مهبل
myelitis	التهاب نخاع
hemorrhagic mediastinitis	التهاب و خونریزی میان سینه
LT (= labile toxin)	ال.تی.
LTR (= long terminal repeat)	ال.تی.آر.
LTC-IC (= long-term culture-initiating cells)	ال.تی.سی.-آی.سی.
LGMD2A (= limb-girdle muscular dystrophy type 2A)	ال.جی.ام.دی.2.آ.
conjugation	الحاق
gene insertion	الحاق ژن (ی)
cell fusion	الحاق یاخته ای
gene introgression	الحاق یک ژن درون مخزن ژنی یک جمعیت

LDH (= lactate de-hydrogenase)	ال.دی.اچ.	zone electrophoresis	الکتروفوروز ناحیه ای
LDL (= low-density lipoprotein)	ال.دی.ال.	capillary zone electrophoresis	الکتروفوروز ناحیه مویرگی
LCR (= locus control region)	ال.سی.آر.	electrofusion	الکتروفیوژن
LCAT (= lecithin-chole-sterol-acyl-transferase)	ال.سی.ای.تی.	electrolyte	الکترولیت
inducer	القا کننده	electronegative	الکترومنفی
interferon inducer	القاگر اینترفرون	electronegative	الکترون کشان
induction	القاء	electronegativity	الکترون کشانی
sexduction	القاء جنسی	biomolecular electronics	الکترونیک زیست ملکولی
inducer	القاء گر	bioelectronics	الکترونیک زیستی
induction	القایش	methanol	الکل تقلیبی
electrostatic	الکترواستاتیک	molecular sieve	الک ملکولی
electroporesis	الکتروپورزیس	mold, standard, template	الگو
electroporation	الکتروپوریشن	transcription	الگو برداری
glass electrode	الکترود شیشه ای	algorithm	الگوریتم
electrodialysis	الکترودیالیز	modelling	الگو سازی
electrochemistry	الکتروشیمی	vicariant pattern	الگوی جانشین
electrochemical	الکتروشیمیائی	x-ray diffraction pattern	الگوی شکست اشعه ایکس.
electrophoresis	الکتروفورز	diamond	الماس
gel electrophoresis	الکترو فورز ژلی	elution	الوشن
pulse field gel electrophoresis	الکتروفورز ژلی پالس فیلد (میدان ضربه ای)	hereditary elliptocyptosis	الیپتوسیپتوز وراثتی
poly-aclamide gel electrophoresis	الکتروفورز ژلی پلی آکلامیدی	oligo (dT) cellulose	الیگو دی.تی. سلولز
pulse-field gradient gel electrophoresis	الکتروفورز ژلی شیب میدان ضربه ای	LUC (= luciferase)	ال.یو.سی.
agarose gel electrophoresis	الکتروفوروز ژل آگاروز	M (= male), M (= mitosis)	ام.
denaturant gradient gel electrophoresis	الکتروفوروز ژل با شیب تقلیب کننده	M I (= first meiotic metaphase)	ام.1
denaturing polyacrylamide gel electrophoresis	الکتروفوروز ژل تغییر ماهیت دهنده پلی آکریلامید	M13	ام. 13
denaturing gradient gel electrophoresis	الکتروفوروز ژل تغییر ماهیت دهنده گرادیان	M II (= second meiotic metaphase)	ام.2
polyacrylamide gel electrophoresis (PAGE)	الکتروفوروز ژلی پلی آکریل آمید	MR (= mental retardation)	ام.آر.
field inversion gel electrophoresis	الکتروفوروز ژلی فیلد اینورژن	mRNA (= messenger RNA)	ام.آر.ان.ای.
capillary electrophoresis	الکتروفوروز مویرگی	polycistronic mRNA	ام.آر.ان.ای. چند سیسترونی
		MRA (= marrow repopulation ability)	ام.آر.ای.
		MRV (= mucosa rota virus)	ام.آر.وی.
		Mi (= microphtalmic)	ام.آی.
		MIT (= monoiodo-tyrosine)	ام.آی.تی.
		MERRF (= myoclonic epilepsy with ragged red fibers)	ام.ئی.آر.آر.اف.
		MELAS (= mitochondrial	ام.ئی.ال.اس.

English	Persian
myopathy, encephalopathy, lactic acidosis and stroke-like)	
MEN (= multiple endocrine neoplasia)	ام.ئی.ان.
MHC (= major histocompatibilty complex)	ام.اچ.سی.
MS (= multiple sclerosis)	ام.اس.
MSAFP (= maternal serum AFP)	ام.اس.ای.اف.پی.
MSD (= multiple sulfatase deficiency)	ام.اس.دی.
MSUD (= maple syrup urine disease)	ام.اس.یو.دی.
MLD (= meta chromatic leano dystrophy)	ام.ال.دی.
MLV (= murine leukemia virus)	ام.ال.وی.
MMR (= methyl-directed mismatch repair)	ام.ام.آر.
MMTV (= mouse mammary tumor virus)	ام.ام.تی.وی.
MOS (= mannanoli-gosacchariddes)	ام.او.اس.
MO-MLY (= Moloney murine leukemia virus)	ام.او-ام.ال.وای.
MAR (= matrix-associated region marker chromosome)	ام.ای.آر.
MALT (= mucosa-associated lymphoid tissue)	ام.ای.ال.تی.
MAP (= mitogen activated protein)	ام.ای.پی.
MAC (= mammalian artificial chromosome)	ام.ای.سی
MACE (= membrane anchor cleaving enzyme)	ام.ای.سی.ئی.
Mb (= megabase pairs), Mb (= myoglobin)	ام.بی.

English	Persian
MBP (= myelin basic protein)	ام.بی.پی
MPS (= muco-poly-saccharidosis)	ام.پی.اس.
MPOD (= myelo-peroxidose deficiency)	ام.پی.او.دی.
conjugation, fusion, mixing	امتزاج
protoplast fusion	امتزاج پروتوپلاست
gene fusion	امتزاج ژن (ی)
cell fusion	امتزاج سلولی
asset	امتیاز
MTX (= Mrthotrexate)	ام.تی.اکس.
MTG (= mono-thio-glycerol)	ام.تی.جی.
MTV (= mammary tumor virus)	ام.تی.وی.
MGMT (= methyl-guanine-DNA methyl-transferase)	ام.جی.ام.تی.
MDS (= myelo-dysplastic syndrome)	ام.دی.اس.
MDP (= muramyl di-peptide)	ام.دی.پی.
biological diseases	امراض زیستی
MZ (= mono zygote)	ام.زد.
MCH (= mean corpuscular hemaglobin)	ام.سی.اچ.
MCHC (= mean corpuscular hemaglobin concentration)	ام.سی.اچ.سی.
MCG (= multiple cloning group)	ام.سی.جی.
MCK (= muscle creatine kinase)	ام.سی.کی.
MCV (= mean corpuscular volume)	ام.سی.وی.
metabonomic signature resource	امضای متابونومیکی
establishment potential	امکان(ات)
	امکان استقرار
MYC (= MYC gene/oncogene)	ام.وای.سی.
emulsion	امولسیون

n (= nano =10-9)	ان.	natural selection	انتخاب طبیعی
NER (= nucleotide excision repair)	ان.ئی.آر.	disruptive selection	انتخاب گسلنده
		species selection	انتخاب گونه
NH2 (= amino group)	ان.اچ.2	normalizing selection	انتخاب متعادل کننده
NHEJ (= non-homologus end joining)	ان.اچ.ئی.جی	artificial selection	انتخاب مصنوعی
		interferon	انترفرون
NS (= nephrotis syndrome)	ان.اس.	alpha interferon	انترفرون آلفا
NF (= neuro-fibromatosis)	ان.اف.	immune interferon	انترفرون ایمنی
NGF (= nerve growth factor)	ان.اف.جی.	beta interferon	انترفرون بتا
		genetically engineered interferon	انترفرون مهندسی (ژنتیک) شده
nm (= nanometer)	ان.ام.		
NMR (= nuclear magnetic resonance)	ان.ام.آر.	enterotoxin	انتروتوکسین
		enterocyte	انتروسیت
enantiopure	انانتیوپیور	intron	انترون
enantiomer	انانتیومر	entrainment	انترینمنت
NADH (= NAD, reduced form)	ان.ای.دی.اچ.	diffusion, dispersal, dispersion, dissemination, transmission	انتشار
NADPH (= NADP, reduced form)	ان.ای.دی.پی.اچ.		
		immunodiffusion	انتشار ایمنی
NADP+ (= nicotinamide adenine dinucleotide phosphate)	ان.ای.دی.پی.مثبت	facilitated diffusion	انتشار تسهیل شده
		intended release	انتشاردلخواه
		spreader	انتشار دهنده
NAD+ (= nicotinamide adenine dinucleotide)	ان.ای.دی. مثبت	gel diffusion	انتشار ژلی
		gene flow	انتشار ژن (ی)
agglomeration	انبار	transmission of infection	انتشار عفونت
seed bank	انبار بذر	transboundary release	انتشارفرا مرزی/سرحدی
accumulation, agglomeration	انباشت	covert release	انتشار مخفی
		intended release	انتشار مطلوب
accumulation, agglomeration	انباشتگی	Mendelian transmission	انتشار مندلی
		rational expectation	انتظار منطقی
stacking	انباشتن	locomotion, transfer, transition, translocation, transmission	انتقال
dilatation	انبساط		
accumulation	انبوهی		
selection	انتخاب	transamination	انتقال آمین
mutation breeding	انتخاب از طریق جهش	horizontal disease transmission	انتقال افقی بیماری
mass selection	انتخاب انبوه		
improved mass selection	انتخاب انبوه بهینه	oxygen transfer	انتقال اکسیژن
marker-assisted selection	انتخاب به کمک نشانه	fluorescence resonance energy transfer	انتقال انرژی تشدیدی فلوئورسانس
postnatal selection	انتخاب پس از تولد		
directional selection	انتخاب جهت دار		
randomized	انتخاب شده تصادفی	safe transfer	انتقال ایمن

English	Persian	English	Persian
transamination	انتقال بنیان آمین	lysis	انحلال
disease transmision	انتقال بیماری	knockout	انداختن
abortive transduction	انتقال بی نتیجه	sizing	اندازه بندی
biological transport	انتقال بیولوژیکی	meter	اندازه گیر
lateral transfer	انتقال جانبی	measurement, metering, sizing	اندازه گیری
linear energy transfer	انتقال خطی انرژی		
intracellular transport	انتقال درون سلولی	ambient measurement	اندازه گیری امبیانت
periodic transfer	انتقال دوره ای	genetic distancing	اندازه گیری فاصله ژنتیکی
ABC transporter	انتقال دهنده ای.بی.سی.	ambient measurement	اندازه گیری محیط/محیطی
vesicular transport	انتقال ریز کیسه ای	cell size	اندازه یاخته
biologic transmission	انتقال زیست شناختی	organ	اندام
acoustic gene transfer	انتقال ژن آکوستیکی/صوتی	xenogenic organ	اندام بیگانه سرشت
vertical gene transfer	انتقال ژن عمودی	gland	اندام ترشحی
gene translocation	انتقال ژن (ی)	organogenesis	اندام ریخت گیری
vertical transmission	انتقال عمودی	organogenesis	اندام زایی
membrane transport	انتقال غشایی	xenogenetic organ	اندام زنوژنتیک
indirect transmission	انتقال غیر مستقیم	modelling	اندام سازی
transboundary transfer	انتقال فرا مرزی/سرحدی	organelle	اندام سلولی
active transport	انتقال فعال	morphology	اندام شناسی
transamination	انتقال گروه آمین	organelle	اندامک
near-infrared transmission	انتقال مادون قرمز	organelle	اندام کوچک
direct transfer	انتقال مستقیم	organism	اندامگان
Mendelian transmission	انتقال مندلی	xenogeneic organ	اندام نا مشابه
nuclear transfer	انتقال هسته ای	photophore	اندام نوری
integration	انتگراسیون	endoenzyme	اندوآنزیم
integrin	انتگرین	endopeptidase	اندوپپتیداز
3' end	انتهای '3	endothelin	اندوتلین
3'-OH terminus	انتهای '3-او.اچ.	endothelium	اندوتلیوم
5' end	انتهای '5	endotoxic	اندوتوکسیک
5'-P terminus	انتهای '5-پی.	endotoxin	اندوتوکسین
pter	انتهای بازوی کوتاه	delta endotoxin	اندوتوکسین دلتا
cohesive end	انتهای چسبنده	endorphin	اندورفین
sticky end (= cohesive end)	انتهای چسبنده/ناصاف	endosome	اندوزوم
distal	انتهایی	endosperm	اندوسپرم
NTDS (= neural tube defects)	ان.تی.دی.اس.	endospore	اندوسپور
		endostatin	اندوستاتین
proliferating cell nuclear antigen	انتی ژن هسته سلولی تکثیر شونده	endogamy	اندوگامی
		endoglycosidase	اندوگلیکوسیداز
variation	انحراف	endonuclease	اندونوکلئاز
standard deviation	انحراف معیار	AP endonuclease	اندونوکلئاز ای.پی.
attenuation	انحطاط	Pst1 endonuclease	اندونوکلئاز پی.اس.تی.1

restriction endonuclease	اندونوکلئاز محدود کننده
indicator	اندیکاتور
paternity index	اندیکس پدری
NDV (= newcastle disease virus)	ان.دی.وی.
free energy	انرژی آزاد
Gibbs free energy	انرژی آزاد گیبز
radiation	انرژی تابشی
biomass energy	انرژی زیست توده
bioenergy	انرژی زیستی
activation energy	انرژی فعال سازی
phosphate-group energy	انرژی گروه فسفات
isolation	انزوا
regulatory enzyme	انزیم تنظیم کننده
injector	انژکتور
ensiling	انسایل کردن
ensiling	انسایلینگ
coronary thrombosis	انسداد شریان های اکلیلی
bovine spongiform encephalopathy	انسفالوپاتی اسپونجیفورم گاوی
encephalopathic	انسفالوپاتیک
insulin	انسولین
branch	انشعاب
adaptation, compensation, conformation	انطباق
protein conformation	انطباق پروتئین(ی)
staggered conformation	انطباق پله ای
native conformation	انطباق طبیعی
ecological succession	انعطاف پذیری بوم شناختی
coagulation, gelation, ligation	انعقاد
hemostasis	انعقاد خون
disseminated intravascular coagulation	انعقاد درون آوندی گسترده
disseminated intravascular coagulation	انعقاد درون رگی گسترده
lipoprotein-associated coagulation	انعقاد ناشی از لیپوپروتئین
nuclear magnetic resonance	انعکاس مغناطیسی هسته ای
penetrance	انفاذ
infarction	انفارکتوس

explosive	انفجاری
medical informatics	انفورماتیک پزشکی
breakdown, extinction	انقراض
excision	انقطاع
telomere-associated chromosome fractionation	انکسار کروموزومی وابسته به تلومر
refraction	انکسار (نور)
enkephalin	انکفالین
oncostatin M	انکواستاتین ام.
oncogene FOS	انکوژن اف.او.اس.
v-oncogene	انکوژن وی.
oncology	انکولوژی
oncomodulin	انکومدولین
onconavirus	انکوناویروس
NK (= natural killer)	ان.کی.
fingerprinting	انگشت نگاره
fingerprinting	انگشت نگاری
genetic fingerprinting	انگشت نگاری ژنتیکی
molecular fingerprinting	انگشت نگاری ملکولی
parasite	انگل
parasitism	انگل بودن
host	انگل پذیر
host	انگل دار
parasitoid	انگل مانند
parasitic, parasitism	انگلی
induction	انگیختگی
homogenetic induction	انگیختگی همگن
EnviromistTM	انوایرومیست
enoyl-acyl protein reductase	انول-آسیل پروتئین رداکتاز
enolpyruvil shikimate	انولپایروویل شیکیمیت
enolpiruvil shikimate	انولپیرووبل شیکیمات
disposal, lysis	انهدام
carbon dioxide	انیدرید کربنیک
ORI (= origin of DNA replication)	او.آر.آی.
ORF (= open reading frame)	او.آر.اف.
ORC (= origin recognition complex)	او.آر.سی.
OI (= osteogenesis	او.آی.

27

imperfecta)		oligosaccharide	اولیگو ساکارید
OFAGE (= orthogonal field	او.اف.ای.جی.ئی.	oligofructan	اولیگوفروکتان
alternation gel)		oligofructose	اولیگو فروکتوز
OFD (= oral-facial-digital)	او.اف.دی.	oligomycin	اولیگومایسین
OAT (= ornithine	او.ای.تی.	oligomer, telomere	اولیگومر
amino-transferase)		oligonucleotide	اولیگونوکلئوتید
ubiquinol (=reduced	اوبیکوئینول	allele-specific	اولیگونوکلئوتید مختص به آلل
coenzyme Q)		oligonucleotide	
ubiquitin	اوبیکیتین	first meiotic anaphase	اولین آنافاز میوتیک
ubiquitinated	اوبیکیتین شده	first meiotic prophase	اولین پروفاز میوتی
ubiquinone	اوبیکینون	first meiotic metaphase	اولین متافاز میوتی
opines	اوپاین ها	olivomycin	اولیوومایسین
operon	اوپرون	ovalbumin	اووالبومین
opsonin	اوپسونین	OVA (= Ov-albumin)	او.وی.ای.
opsonization	اوپسونین سازی	donor	اهدا کننده
eutrophication	اوتروفیکیشن	domestication	اهلی سازی
otocephaly	اوتوسفالی	domesticate, domestication	اهلی کردن
OTC (= ornithine	او.تی.سی.	import	اهمیت
transcarbamylase		AR (= autosomal recessive)	ای.آر.
deficiency)		ARS (= autonomous	ای.آر.اس.
ODP (= ozone-depleting	او.دی.پی.	replicating segment),	
potential)		ARS (= autonomous	
ODD (= oculo-dental-	او.دی.دی.	replicating sequence)	
digital syndrome)		ARMS-PCR	ای.آر.ام.اس-پی.سی.آر.
aureofacin	اورئوفاسین	ARCD (= acquine renal	ای.آر.سی.دی.
uracil	اوراسیل	cystic disease)	
aura virus	اورا ویروس	AID (= artificial	ای.آی.دی.
orbital	اوربیتال	insemination by donor)	
hormone	اورمون	AE (= acrodermatitis	ای.ئی
ornithine	اورنیتین	enteropathica)	
orosomucoid	اوروسموکوئید	A-EB (= epidermolysis	ای.-ئی.بی.
urease	اوره آز	bullosa acquisita)	
uridine	اوریدین	AS (= Angelman	ای.اس.
ossein	اوسئین	syndrome), AS (=	
oscillin	اوسیلین	antigene), AS (= aortic	
oleate	اولئات	stenosis)	
oleosome	اولئوزوم	ASO (= allele-specific	ای.اس.او.
oleic acid	اولئیک اسید	oligonucleotide)	
oligopeptide	اولیگو پپتید	ASOH (= allele- specific	ای.اس.او.اچ.
oligotrophic	اولیگوتروفیک	oligonucleotide	
oligos	اولیگوز	hybridization)	

English	Persian
ASC (= antibody-secreting cell)	ای.اس.سی.
AF (= amniotic fluid)	ای.اف.
AFP (= α- Fetoprotein)	ای.اف.پی.
ALS (= amyotrophic lateral sclerosis)	ای.ال.اس.
ALL (= acute lymphocytic leukemia)	.ال.ال.
Ala (= alanine)	.ال.ای.
ALB (= albumin)	ای.ال.بی.
ALD (= adenoleucodystrophy)	ای.ال.دی.
ALV (= avian leukemia virus)	ای.ال.وی.
AMH (= anti-mullerian hormone)	ای.ام.اچ.
AML (= acute myeloblastic leukemia)	ای.ام.ال.
AMMOL (= acute myelomonacytic leukemia)	ای.ام.ام.او.ال.
AMP (= adenosine monophosphate)	ای.ام.پی.
AMD (= acid maltase deficiency)	ای.ام.دی.
AMC (= arthrogryposis multiplex congenita)	ای.ام.سی.
ANLL (= acute nonlymphocytic leukemia)	ای.ان.ال.ال.
AAT (= alpha-antitrypsin)	ای.ای.تی.
AAG (= acid alpha-glucosidase)	ای.ای.جی.
AAV (= adeno-associated virus)	ای.ای.وی.
ab (= antibody)	ای.بی.
abl (=Abelson strain of murine(,ABL (=Abelson strain of murine)	ای.بی.ال
ABO (= ABO blood group)	ای.بی.او.
AP (= apurinic)	ای.پی.
APRT (= adenine phosphoribosyl transferase)	ای.پی.آر.تی.
APL (= acute promyelocytic leukemia)	ای.پی.ال.
APTT (= activated partial thromboplastin time)	ای.پی.تی.تی.
APC (= adenomatous polyposis coli (gene)), APC (= antigen presenting cell)	ای.پی.سی.
A II (= second meiotic anaphase)	ای.تو.
AT (= ataxia telangiectasia)	ای.تی.
AT3 (= antithrombin III)	ای.تی.3.
AT-AC intron	ای.تی.-ای.سی. اینترون
AT3 (= antithrombin III)	ای.تی. تری
ATD (= asphyxiating thoracic dysplasia)	ای.تی.دی.
ATY (= atopic diathesis)	ای.تی.وای.
volume rendering	ایجاد ظرفیت
ideogram	ایدئو گرام
AIDS = (acquired immune deficiency syndrome)	ایدز
AD (= autosomal dominant)	ای.دی.
ADH (= alcohol dehydrogenase)	ای.دی.اچ.
a-DNA	ای.-دی.ان.ای.
ADA (= adenosine deaminase)	ای.دی.ای.
ADP (= adenosine diphosphate)	ای.دی.پی.
ADPKD (= autosomal dominant polycystic kidney)	ای.دی.پی.کی.دی.
ADCC (= antibody dependant cellular cytotoxicity), ADCC (= apurinic cellular citotoxicity)	ای.دی.سی.سی.
idioblast	ایدیو بلاست

English	Persian
idiotroph	ایدیو تروف
idiotope	ایدیوتوپ
idiophase	ایدیوفاز
idiochromatin	ایدیو کروماتین
idiochromosome	ایدیو کروموزوم
idiogram	ایدیوگرام
idiovariation	ایدیو واریاسیون
airbrush	ایر براش
aerobic	ایروبیک
AZO (= azoospermia)	ای.زد.او.
isozyme	ایزو آنزیم
isoprene	ایزوپرن
isograft	ایزو-پیوند
isotachophoresis	ایزو تاکو فورز
capillary isotachophoresis	ایزوتاکوفورسیس کاپیلاری
capillary isotachophoresis	ایزوتاکوفورسیس موئین/مویسان
capillary isotechophoresis	ایزوتکوفورسیس کاپیلاری
capillary isotechophoresis	ایزوتکوفورسیس موئین/مویسان
isotope	ایزوتوپ
radioactive isotope, radioisotope	ایزوتوپ پرتوزا
radioactive isotope	ایزوتوپ رادیواکتیو
isothiocyanates	ایزوتیوسیانات ها
isodisomy	ایزودیزومی
isozyme	ایزوزیم
isoflavone	ایزوفلاون
isoflavonoid	ایزوفلاوونوئید
isoflavin	ایزوفلاوین
isochromosome	ایزوکروموزوم
isograft	ایزوگرافت
isoleucine, iso-leucine	ایزولوسین
isomer	ایزومر
isomerase	ایزومراز
stereoisomer	ایزومر فضایی
enantiomer	ایزومر نوری
cis/trans isomerism	ایزومری سیس-ترانس
persistence	ایستادگی
quick-stop	ایست سریع
AC (= adenyl cyclase)	ای.سی.
ACH (=	ای.سی.اچ.

English	Persian
acetylcholinesterase)	
ACS (= autonomous consensus sequence)	ای.سی.اس.
ACP (= acyl carrier protein)	ای.سی.پی.
eicosanoid	ایکوزانوئید
AK (= adenylate kinase deficiency)	ای.کی.
immunotherapy	ایمن درمانی
adoptive immunization	ایمن سازی انتخابی
immunization	ایمن کردن
immunity	ایمنی
cell-mediated immunity	ایمنی با واسطه سلول
cell mediated immunity	ایمنی با واسطه سلولی
adaptive immunity	ایمنی تطبیقی
humoral immunity	ایمنی خونی
natural immunity	ایمنی ذاتی
immunogen	ایمنی زا
immunogenic	ایمنی زایی
passive immunity	ایمنی زودگذر
adaptive immunity	ایمنی سازشی
active immunization	ایمنی سازی فعال
immunoassay	ایمنی سنجی
radioimmunoassay	ایمنی سنجی پرتویی
chemiluminescent immunoassay (CLIA)	ایمنی سنجی نورافشانی شیمیایی
immunologic	ایمنی شناختی
natural immunity	ایمنی طبیعی
passive immunity	ایمنی غیر فعال
active immunity	ایمنی فعال
immunocompetent	ایمنی قابل قبول
immunosuppressive	ایمنی کاه
active immunity	ایمنی کنش ور
immunosuppressive	ایمنی کوب
humoral-mediated immunity	ایمنی متعادل خونی
safe safety	ایمنی مطمئن
immunomagnetic	ایمنی-مغناطیسی
passive immunity	ایمنی ناکنش ور
immunoadsorbent	ایمونو ادسوربانت
immunoadhesin	ایمونو ادهسین
immunoaffinity	ایمونو افینیتی

rocket immunoelectrophoresis	ایمونو الکتروفورز موشکی	inhaler	اینهیلر
immuno blotting	ایمونو بلاتینگ	initiator codon	اینیشیتور کدون
immunoblotting	ایمونوبلاتینگ	A I (= first meiotic anaphase)	ای.وان.
immunobiology	ایمونوبیولوژی	unit	ایوه
immunopotentiator	ایمونوپتانسیاتور	Ava I	ای.وی.ای. 1
immunocyte	ایمونوسیت	AU4000	ای. یو. 4000
immunoconjugate	ایمونو کانجوگیت	AUG (initiator codon)	ای.یو.جی.
immunoglobolin , immunoglobulin	ایمونوگلوبولین		
immunoglobin M	ایمونوگلوبین ام.		**ب**
immunoglobin G	ایمونوگلوبین جی.	portal	باب
immunology	ایمونولوژی	slough	باتلاق
imidazole	ایمیدازول	entrainment	با خود بردن
intein	اینتئین	entrainment	با خودبری
intasome	اینتازوم	self-incompatibility	باخود ناسازگاری
interband	اینترباند	wind	باد
interphase	اینتر فاز	puffer	باد بزن
interferon	اینترفرون	solanaceae	بادنجانیان
interferon-beta	اینترفرون بتا	fructification	بارآوری
gamma interferon	اینترفرون گاما	piezoelectric	بارابرقی
interkinesis	اینترکینزیس	acid rain	باران اسیدی
interleukin	اینترلوکین	gestation	بارداری
internalin	اینترنالین	vernalization	باردهی
zygotic induction	اینداکشن زیگوتی	dominant	بارز
homogenetic induction	اینداکشن همگن	BAR gene	بار ژن
sludge volume index (SVI)	ایندکس حجم گل و لای	genetic load	بار ژنتیکی
induration	ایندیوریشن	bio-bar code	بار کد زیستی
in-silico	این سیلیکو	BARnase (= bacillus amilolique facien RNase)	بار ناز
in-silico screening	این سیلیکو اسکرینینگ		
in-silico biology	این سیلیکو بیولوژی	fertilization	بارور سازی
informosome	اینفورموزوم	zygote	بارورشده
informofer	اینفورموفر	fecundity, fertility	باروری
incontinentia pigmenti	اینکونتیننتیا پیگمانتی	base	باز
invasin	اینوازین	rehydration	باز آبدهی
inositol	اینوزیتول	reannealing	باز آبکاری
inositol lipid	اینوزیتول لیپید	rearrangement	باز آرائی
inosine	اینوزین	surrogate market	بازار جایگزین
inosine tri-phosphate	اینوزین تری فسفات	nitrogenous base	باز ازت ساز
inosine mono-phosphate	اینوزین مونو-فسفات	rhizoremediation	باز اصلاحی ریشه ای
inulin	اینولین	SOTE (= standard oxygen	بازاک استاندارد

English	Persian	English	Persian
transfer efficiency)		inhibitor	سیکلین
AOTE (= actual oxygen	بازاک واقعی	fusion inhibitor	بازدارنده همجوشی
transfer efficiency)		inhibition, prevention	بازداری
ecosystem rehabilitation	باز پروری بوم سازگان	down processing	بازداری واکنش عادی
ecosystem rehabilitation	باز پروری زیستگاه	expiration	بازدم
genetic recombination	بازپیوستگی ژنتیکی	basophils	باز دوست ها
total internal reflection	بازتابش شب نمایی درونی	standard oxygen transfer	بازده انتقال اکسیژن استاندارد
fluorescence	کامل	efficiency (SOTE)	
recombinant	بازترکیب	actual oxygen transfer	بازده انتقال اکسیژن واقعی
high-frequency recombinant	باز ترکیب پر بسآمد	efficiency (AOTE)	(بازاک واقعی)
recombination	بازترکیبی	annealing	باز ریخت
additive recombination	باز ترکیبی افزایشی	reperfusion	باز ریزش
genetic recombination	باز ترکیبی ژنتیکی	regeneration	باززایی
homologous recombination	باز ترکیبی مشابه	regeneration	بازسازی
recirculation	باز چرخش/چرخانی	ecosystem restoration	باز سازی بوم سازگان
inhibition, repression	بازدارندگی	extension	بازشدگی
catabolite repression	بازدارندگی کاتابولیت	minor base	باز فرعی
inhibitor, repressor,	بازدارنده	reactivate	بازفعال سازی
suppressor		host-cell reactivation	باز فعالی سلول میزبان
alpha amylase inhibitor-1	بازدارنده آلفا آمیلاز نوع اول	uncoupling	باز کردن
enzyme inhibitor	بازدارنده آنزیم	conversion, relapse	بازگشت
angiotensin-converting	بازدارنده آنزیم تبدیل آنژیوتنزین	renaturation	بازگشت طبع
enzyme inhibitor		sequela	بازمانده
enzyme inhibitor	بازدارنده آنزیمی	reuse	بازمصرف
fusion inhibitor	بازدارنده امتزاج	mapping	بازنمایی
immunosuppressive	بازدارنده ایمنی	peptide mapping	بازنمایی پپتید
proteasome inhibitor	بازدارنده پروتئازوم	transduction mapping	بازنمایی ترا رسانی
trypsin inhibitor	بازدارنده تریپسین	receptor mapping	بازنمایی گیرنده
cowpea trypsin inhibitor	بازدارنده تریپسین از نوع لوبیا	nitrogen base	باز نیتروژن دار
	چشم بلبلی	nitrogenous base	باز نیتروژن ساز
Bowman-Birk trypsin	بازدارنده تریپسین بومن-برک	nitrogen base	باز نیتروژنه
inhibitor		reperfusion	باز وامیختگی
Kunitz trypsin inhibitor	بازدارنده تریپسین کونیتز	basophilic	بازوفیل
tyrosine kinase inhibitor	بازدارنده تیروزین کیناز	basophils	بازوفیل (ها)
protein tyrosine kinase	بازدارنده تیروزین کیناز پروتئینی	basophilic	بازوفیلی
inhibitor		rare bases	بازهای کمیاب
renin inhibitor	بازدارنده رنین	nucleic base	باز هسته ای
transcriptional repressor	بازدارنده رونویسی	ecosystem rehabilitation	بازیابی اکوسیستم
CI repressor	بازدارنده سی.آی.	recycling	باز یافت/یابی
Cro repressor	بازدارنده کرو	bacillus	باسیل
cyclin-dependent kinase	بازدارنده کیناز وابسته به	bacillus anthracis	باسیل سیاهزخم

32

tuberculosis	باسیل مرض سل	bacteremia	باکترمی
bacillus thuringiensis	باسیلوس تورینژنسیس	bacteremic	باکترمیائی
bacillus subtilis	باسیلوس سوبتیلیس	bacteremic	باکترمیک
bacillus licheniformis	باسیلوس لایکنیفورمیس	bacterium	باکتری
waste	باطله	iron bacteria	باکتری آهن
garden	باغ	erwinia uredovora	باکتری اروینیا اوره دوورا
garden	باغبانی کردن	erwinia caratovora	باکتری اروینیا کاراتوورا
garden	باغچه	Escherichia coliform	باکتری اشرشیا کولیفورم
garden	باغستان	bacterium tularense	باکتری تب خرگوشی
garden	باغی	bacterium tularense	باکتری تب مگس آهو
tissue	بافت	bacterium tularense	باکتری تولارمی
histopathologic	بافت آسیب شناسی	bacterium tularense	باکتری تولارنس
granulation tissue	بافت التیامی	acetic acid bacteria	باکتری جوهر سرکه
biopsy	بافت برداری	bacteriophage, phage	باکتری خوار
epithelium	بافت پوششی	transducing phage	باکتری خوار تبدیل کننده
connective tissue	بافت پیوندی	transducing phage	باکتری خوار ترارسان
histoblast	بافت تنده	lambda bacteriophage,	باکتری خوار لاندا
endothelium	بافت توپوشی	lambda phage	
germ plasm	بافت تولید مثلی	temperate phage	باکتری خوار ملایم
RT-PCR out of	بافت جایگزین	marine bacteria	باکتری دریائی
paraffin-embedded tissue	آر.تی.-پی.سی.آر خارج از	bacillus anthracis	باکتری سیاهزخم
	پارافین	bacteriology	باکتری شناسی
paraffin-embedded tissue	بافت جایگزین در پارافین	anaerobic bacteria	باکتری غیر هوازی
histocompatibility	بافت سازگاری	bacterium tularense	باکتری فرانسیسلا تولارنسیس
bronchial-associated	بافت لنفوئیدی نایژه ای	photolithotrophic bacteria	باکتری فوتولیتوتروف
lymphoid tissue		sulfate reducing bacterium	باکتری کاهنده سولفات
mucosa-associated	بافت لنفوئیدی وابسته به	coccus	باکتری کروی شکل
lymphoid tissue	موکوس	bacteride, bacteriocide	باکتری کش
necrosis	بافت مردگی	bactericidal	باکتری کشانه
caseous necrosis	بافت مردگی پنیری	coliform bacteria	باکتری کولی باسیل شکل
cellular necrosis	بافت مردگی سلولی	thermophilic bacteria	باکتری گرما دوست
caseous necrosis	بافت مردگی نکروز	gram-positive bacteria	باکتری گرم مثبت
chimera	بافت ناهمسان	gram-negative bacteria	باکتری گرم منفی
DNA chimera	بافت نا همسان دی.ان.ای.	slime bacteria	باکتری گل ولای
neoplasm	بافت نوساخته	vibrio cholerae	باکتری مژکدار وبا
periodontium	بافتهای پوشاننده دندان	nitrate bacteria	باکتری نیترات
histiocyte	بافت یاخته	bacteriostat	باکتریواستات
TE buffer	بافر تی.ئی	bacteriocin	باکتریوسین
lupine	باقلا مصری	bacteria	باکتری ها
conserved	باقی	ferrobacteria	باکتری های آهن دار
bakanae	باکانائه	extremophilic bacteria	باکتریهای افراط دوست/گرا

English	Persian	English	Persian
extremophilic bacteria	باکتریهای اکسترموفیلیک	malnutrition	بد غذائی
bifidobacteria	باکتری های دو شاخه ای	scaffold	بدنه پروتئینی
denitrifying bacteria	باکتری های شوره زدا	refractile body	بدنه شکننده/منکسر کننده نور
nitrifying bacteria	باکتری های شوره ساز	Golgi body	بدنه گلژی
photolithotrophic bacteria	باکتری های نور کانی پرور	somatic	بدنی
aerobic bacteria	باکتری های هوازی	aseptic	بدون آلودگی
metered	با کنتور اندازه گیری شده	sessile	بدون پایک
baculovirus	باکولوویروس	relaxed	بدون پیچ
hypernatremia	بالا بودن سدیم خون	afebrile	بدون تب
hypercholesterolemia	بالا بودن کلسترول خون	sessile	بدون دمگل
upstream	بالادست	oil-free	بدون روغن
volumetric flask	بالن حجم سنجی	amorph	بدون ریخت
clinical	بالینی	acentric	بدون سانترومر
banjo	بانجو	aseptic	بدون میکروب
high resolution chromosome banding	باندینگ کروموزومی با دقت بالا	malnutrition	بدی تغذیه
gene-bank	بانک ژن (ی)	bioseeds	بذر مهندسی شده
receptor-mediated	با واسطه گیرنده	carrier	بر-
codon bias	بایاس کدون	adsorption	برآشامی
conserved	بایسته	node, tumor	برآمدگی
bifidus	بایفیدوس	nodule	برآمدگی مخچه
B-thalassemia, β-thalassemia	بتا تالاسمی	assessment, evaluation	برآورد
beta-DNA	بتا دی.ان.ای.	diversity estimation	برآورد تنوع
beta-secretase	بتا سکرتاز	needs assessment	برآوردنیاز
beta carotene	بتا کاروتن	homology	برابری
β-galactosidase	بتا گالاکتوسیداز	bradyrhizobium japonicum	برادیریزوبیوم جاپونیکوم
beta-Lactamase	بتا لاکتاماز	bradykinin	برادی کینین
rash	بثورات جلدی	brazzein	برازئین
mist	بخار	brassica campestre	براسیکا کمپستر
smoke	بخار بیرون دادن	brassica campestris	براسیکا کمپستریس
mist	بخار گرفتن	brassica napus	براسیکا ناپوس
randomized	بختی سازی شده	irritability	برانگیختگی
randomized	بختین	excitation, resuscitation	برانگیزش
organ	بخش	solid	بربسته
cistron	بخش فعال ژن	incubation	برتخم نشینی
donor	بخشنده	dominant	برتر
inhalation	بخور	sex influenced dominance	برتری متاثر از جنسیت
malignant	بدخیم	sessile	برجا
malformation	بد ریختی	epithelial projection	برجستگی پوششی/روپوشه ای
malformation	بد شکلی	epiglottis	برچاکنای
		label, labling, tag	برچسب

34

affinity tag	برچسب افینیتی	powered	برقی
bacterial expressed	برچسب باکتریائی توالی	adsorption	برکشی
sequence tag (=	اکسپرس	translation	برگردان
expressed sequence tag)		chlorophyll	برگ سبزینه
expressed sequence tag	برچسب توالی اکسپرس	relapse, reversion	برگشت
(EST)		renaturation	برگشت به حالت طبیعی
affinity tag	برچسب خویشاوندی	ecological resilience	برگشت پذیری بوم شناختی
affinity tag	برچسب دار	retrograde	برگشت دهنده/کننده
radiolabeled	برچسب رادیواکتیو	total anomalous pulmonary	برگشت کاملا غیر عادی
labling, tag, tagging	برچسب زدن	venous return	سیاهرگیری ریوی
tandem affinity purification	برچسب زدن پاکسازی کشش	total anomalous pulmonary	برگشت کامل وریدی ریوی غیر
tagging (TAP tagging)	متوالی	venous return	عادی
quantum tag	برچسب کوانتومی	inversion, transversion	برگشتگی
differential labeling	برچسب گذاری دیفرانسیلی	uptake	برگیری
DNA shuffling	بر خوردن/زدن دی.ان.ای.	heredity	برماند
randomized	بر خورده	heredity	برماندی
vector	بردار	disaster planning	برنامه ریزی بلایا/حوادث
bifunctional vector	بردار دو کاربردی/بایفانکشنال	daffodil rice	برنج دافودیل
uptake	برداشت	rare cutter	برنده کمیاب
harvesting	برداشت کردن	accession, expressivity,	بروز
colony lift	برداشت کولونی	incidence	
excision	برداشتن	gene expression	بروز ژن (ی)
tolerance	بردباری	bromodeoxyuridine	بروموداکسی اوریدین
burn-through range	برد برن ترو	bromoxynil	بروموکسی نیل
host range	برد میزبانی/انگل پذیری	ethidium bromide	برومید اتیدیوم
prevalence survey	بررسی شیوع	cyanogen bromide	برومید سیانوژن
prospective study	بررسی قریب الوقوع	ectoderm	برون پوست
observational study	بررسی مشاهده ای	ectodermal, ectodermic	برون پوستی
case study	بررسی موردی	epithelium	برون پوش
cohort study	بررسی همزادگان	extraocular	برون چشمی
berseem (=berseem clover)	برسیم	outcrossing	برون چلیپائی
staggered cut	برش پله ای	exogenous	برون خاست
fulminant	برق آسا	exocytosis	برون رانی
electrophoresis	برق بری	exogenous, supergene	برون زا
capillary electrophoresis	برق بری مویرگی	exogenous	برون زاینده
capillary zone	برق بری ناحیه مویرگی	exotoxin	برون زهر
electrophoresis		protein exotoxin	برون زهر پروتئین
electrochemistry	برق شیمی	exobiology	برون زیست شناسی
electrochemical	برق شیمیائی	extracellular	برون سلولی
electrolyte	برق کافته	pericardium	برون شامه قلب
shock	برق گرفتگی	outbreeding	برون نژادگیری

English	Persian	English	Persian
exocytosis	برون یاختگی	mucoid	بلغمی
extracellular	برون یاخته ای	volume	بلندی صدا
brevetoxin	بروه توکسین	blower	بلوئر
segregation distorter (SD)	برهم زننده تفکیک	crystallization	بلور شدگی
protein-protein interaction	بر هم کنش پروتئین-پروتئین	crystalloid	بلور مانند
excision	بریدن	x-ray crystallography	بلورنگاری اشعه ایکس
aorta	بزرگ سرخرگ	crystalloid	بلور نما
aortic	بزرگ سرخرگی	accelerated maturation	بلوغ شتابدار/تسریع شده
macromolecule	بزرگ مادیزه	bomb	بمب
biomagnification	بزرگ نمایی زیستی	microwave bombardment	بمباران ریز موجی
lymphogranuloma	بزرگی غدد لنفاوی	bomb	بمباران کردن
transformation frequency	بسامد تغییرشکل	microwave bombardment	بمباران میکرو ویو
gene frequency	بسامد ژن (ی)	dirty bomb	بمب کثیف
polymer	بسپار	construct	بنا
addition polymer	بسپار افزایشی	ligation, linkage, node	بند
linkage	بست	hemostasis	بند آمدگی خون
bed	بستر	joining (J) segment	بند اتصال
biological filter bed	بستر صافی زیستی	coagulation	بندایش
stroma, substrate	بستره	Delaney clause	بند دیلانی
coagulation, gel	بستن	port	بندر
ligation	بستن رگ	synovial	بندی
totipotency	بس توانی	legume	بنشن
packaging	بسته بندی	methyl violet	بنفش متیلی
gelation	بسته شدن	substrate	بنیاد
amplification	بسط	chromogenic substrate	بنیاد رنگ زا
amplify	بسط دادن	systeomics	بنیادی
beaker	بشر	free radical	بنیان آزاد
epithelium	بشره غشاء مخاطی	oxygen free radical	بنیان بی اکسیژن
aerosolize(d)	بصورت اسپری در آمده	heterosis	بنیه دورگه
bottle	بطری	wind	بو
bottle	بطری کردن	botulism	بوتولیزم
conservation	بقاء	botulinum	بوتولینوم
recruitment	بکار گماری	mold, yeast	بوزک
DNA backbone	بکبون دی.ان.ای.	garden	بوستان
parthenocarpic	بکبارده	habitat	بوم
parthenocarpy	بکباردهی	acclimatization	بوم پذیری
parthenogenetic	بکرزا	ecotourism	بوم جهانگردی
parthenogenesis	بکر زایی	ecology	بوم سازگاری
backcross	بک کروس	ecosystem	بوم سازگان
blot	بلات	ecosystem	بوم سامانه
ingestion	بلع	ecotoxicology	بوم سم شناسی

English	Persian	English	Persian
			سازی
ecology	بوم شناخت	BRCA (= breast cancer gene)	بی.آر.سی.ای.
ecology	بوم شناسی	innocuousness	بی آسیبی
genecology	بوم شناسی ژن (ی)	deamidation	بی آمید شدن
agroecology	بوم شناسی کشاورزی	BSE (= bovine spongiform encephalopathy)	بی.اس.ئی.
ecotourism	بوم گردشگری	BSP (= sulphobromo phtahalein)	بی.اس.پی.
ecosystem	بوم نظام	BFGF (= basic fibroblast growth factor)	بی.اف.جی.اف.
indigenous	بومی	BXN gene	بی.اکس.ان. ژن
introduction	بومی سازی	BLV (= bovine leukemia virus)	بی.ال.وی.
bunyavirus	بونیا ویروس	BLUP (=best linear unbiased prediction)	بی.ال.یو.پی.
bunyaviridae	بونیا ویروس ها	BMT (= bone marrow transplantation)	بی.ام.تی.
inheritance of acquired characteristic	به ارث بردن مشخصه اکتسابی	BMD (= Becker muscular dystrophy)	بی.ام.دی.
inherited	به ارث برده	expression	بیان
vernalization	بهارگی	BALT (= bronchial-associated lymphoid tissue)	بی.ای.ال.تی.
vernalization	بهاره کردن		
sustainable	به اندازه (مصرف)		
valuation	بهاء	BAPN (= beta-aminoprionitrile)	بی.ای.پی.ان
enhancement	بهبود		
immuno-enhancing	بهبود ایمنی	BAC (= bacterial artificial chromosome)	بی.ای.سی.
best linear unbiased prediction (BLUP)	بهترین پیش بینی بی طرفانه خطی		
powered	به توان رسیده	sessile	بی پایه
idiotype	به جور	aseptic	بی پلشت
renature	به حالت طبیعی باز گرداندن	deprotection	بی پناه کردن
immunocompromised	به خطر افتاده از لحاظ ایمنی	BP (= bP)	بی.پی.
environmental health	بهداشت محیط	BPG (= blood pressure gauge)	بی.پی.جی.
code	به رمز در آوردن		
code	به رمز نوشتن	afebrile	بی تب
use	به کاربردن	BTWC (=Biological and Toxin Weapons Convention)	بی.تی.دبلیو.سی.
euploid	بهگان		
euploid	بهلاد		
differential splicing	بهم تابیدگی دیفرانسیلی	fluctuant, labile, liquid	بی ثبات
splicing	بهم تابیدن	immobilization	بی جنبش سازی
RNA splicing	بهم تابیدن آر.ان.ای.	immobilization	بی حرکت سازی
in-situ	به موضع		
eugenics	به نژاد شناسی		
breeding, eugenics	به نژادی		
mutation breeding	به نژادی جهشی		
agglutination	به هم چسبندگی		
metabolic disturbance	به هم خوردگی متابولیکی		
metabolic disturbance	به هم ریختگی سوخت و		

English	Persian
anesthesia	بی حسی
deprotection	بی حفاظ کردن
benign	بی خطر
BWS (= Beckwith-Weideman syndrome)	بی.دبلیو.اس.
abrachius	بی دست مادر زاد
abrachiatism	بی دستی مادرزاد
armyworm	بید سرباز
fall armyworm	بید سرباز پائیزی
sessile	بی دمبرگ
innocuousness	بی زیانی
sessile	بی ساقه
abrachiocephalus	بی سر و دست مادر زاد
abrachiocephalia	بی سرو دستی مادر زاد
fructose bis-phosphatase	بیس-فسفاتاز فروکتوز
BCR (= breakpoint cluster region)	بی.سی.آر.
bcr-abl gene	بی.سی.آر.-ای.بی.ال. ژن
bce4	بی.سی.ئی. 4
BCP (= blue cone pigment)	بی.سی.پی.
b sitostanol	بی. سیتوستانول
hypersensitivity	بیش حساسیتی
hyperpolarization	بیش قطبیت
amorph, amorphous	بی شکل
hyperthermophilic	بیش گرما خواه
hyperthermophilic	بیش گرما دوست
innocuousness	بی ضرری
intergenerational equity	بی طرفی بین نسلی
inactivation	بی فعالیتی
acardiac	بی قلب مادر زاد
acardia	بی قلبی مادر زاد
b-conglycinin	بی. کونگلیسینین
unidirectional externality	بیگانگی یک جهتی
xenograft	بیگانه پیوند
macrophage, phagocyte	بیگانه خوار
phagocytosis	بیگانه خواری
phagocytize	بیگانه خواری کردن
phagocytize	بیگانه خوردن
xenogenesis	بیگانه زایی
antixenosis	بیگانه گریزی
bilirubin	بیلی روبین

English	Persian
risk patient	بیمار خطر پذیر
associate hospital	بیمارستان وابسته
morbidity	بیماری
Alzheimer's disease	بیماری آلزایمر
case-finding	بیمار یابی
maple syrup urine disease (MSUD)	بیماری ادرار میپل سیروپی/شیره افرائی
hypophosphataemic bone disease	بیماری استخوانی هیپوفسفاتائمیک
chronic-obstructive-pulmonary disease	بیماری انسداد/مسدود کننده ریه مزمن
parasitic disease	بیماری انگلی
endemic	بیماری بومی
Tay-Sachs disease	بیماری تای-ساکز
sialic acid storage disease	بیماری تجمع اسید سیالیک
lysosomal storage disease	بیماری تجمع لیزوزومی
poly-cystic ovarian disease	بیماری تخمدان پلی-سیستیک
Mad Cow Disease	بیماری جنون گاوی
pandemic	بیماری جهانگیر
multifactorial disease	بیماری چند عاملی
acute illness	بیماری حاد
enzootic	بیماری حیوانی
autoimmune disease	بیماری خود ایمن
Kennedy disease	بیماری دیستروفی میوتونیک
pathogenic, virulent	بیماری زا
pathogenesis, pathologic, virulence	بیماری زایی
genetic disease	بیماری ژنتیکی
SARS	بیماری سارز
tuberculosis	بیماری سل
acquine renal cystic disease	بیماری سیست کلیوی اسبی
von Willebrand factor disease	بیماری عامل فون ویل براند
infection, infectious disease	بیماری عفونی
coronary heart disease	بیماری قلبی کرونر
congenital heart disease	بیماری قلبی مادر زادی
chronic heart disease	بیماری قلبی مزمن
diabetes	بیماری قند
liver	بیماری کبد
white mold disease	بیماری کفک/کپک سفید
Achondroplasia	بیماری کوتاهی دست و پا

English	Persian	English	Persian
chronic granulomatous disease	بیماری گرانولوماتوز مزمن	biotherm	بیوترم
X-linked disease	بیماری مرتبط با اکس	bioterrorism	بیو تروریزم
chronic disease	بیماری مزمن	bioterrorist	بیو تروریست
communicable disease, contagious disease	بیماری مسری	biotron	بیوترون
		biotechnology	بیو تکنولوژی
arthropathy	بیماری مفصلی	environmental biotechnology	بیو تکنولوژی محیط زیست
sexually transmitted disease (STD)	بیماری مقاربتی/آمیزشی	biotelemetry	بیو تله متری
severe combined immuno-deficiency disease	بیماری نقص ایمنی مرکب حاد	biotope	بیوتوپ
		biotype	بیوتیپ
		biotin	بیوتین
Norum's disease	بیماری نوروم	biotinylation	بیوتینیلاسیون
contagious disease	بیماری واگیر	biocheck 60	بیوچک 60
contagion	بیماری واگیر دار	biodisc	بیو دیسک
hereditary disease	بیماری وراثتی	biodyne	بیودین
diabetes mellitus optic atrophy deafness	بیماری ولفرام	membrane filter bioreactor	بیورآکتور فیلتر غشائی
		biorythm	بیوریتم
Wolman's disease	بیماری وولمن	bios	بیوس
virus disease	بیماری ویروسی	bioscience	بیوساینس
wilson's disease	بیماری ویلسون	biosynthesis	بیو سنتز
Huntington's disease	بیماری هانتینگتون	biocenology	بیو سنولوژی
lymphogranuloma	بیماری هوچکین	biofuel	بیو سوخت
drift	بی مقصد حرکت کردن	biosorbent	بیوسوربنت
spectrum	بیناب	biochemistry	بیو شیمی
arrhythmia	بی نظمی	biochemical	بیوشیمیائی
monogenic disorder/trait	بی نظمی تک زایی	biomembrane	بیو غشاء
hemostatic derangement	بی نظمی در بند آمدن خون	biophotolysis of water	بیو فتو لیز آب
arrhythmia	بی نواختی	biophysics	بیو فیزیک
biosilk	بیو ابریشم	biofilter	بیوفیلتر
bioscrubber	بیواسکرابر	activated biofilter	بیو فیلتر فعال شده
bioscrubbing	بیواسکرابینگ	biofilm	بیوفیلم
bioelectronics	بیو الکترونیک	bioconversion	بیوکانورژن
bioMEMS (= bio microelectromechanichal systems)	بیو ام.ئی.ام.اس.	biocrat	بیوکرات
		biogas	بیوگاز
		biology	بیولوژی
bioseeds	بیوبذر	structural biology	بیولوژی ساختاری
biopreparat	بیوپرپارات	biologics	بیولوژیک
biomedicine	بیو پزشکی	biological	بیولوژیکی
biopsy	بیوپسی	biolistics	بیولیستیک
biopolymer	بیو پولیمر	active biomass	بیوماس آکتیو
		biomaterial	بیومواد

bionics	بیونیک
anaerobic	بی هوازی
facultative anaerobe	بی هوازی اختیاری
anaerobic	بی هوازیستی
chloroform	بیهوش کردن با کلروفورم
anesthesia, narcosis	بی هوشی
acellular	بی یاخته

پ

downstream	پائین دست
papain	پاپائین
papaverine	پاپاورین
patent ductus arteriosus	پاتنت دوکتوس آرتریوسوس
pathogenesis	پاتوژنز
pathogenic	پاتوژنیک
pathologic	پاتولوژیک
patulin	پاتولین
antibody	پاد تن
anti-idiotype antibody	پادتن آنتی ایدیوتیپ
humanized antibody	پادتن انسانی شده
direct fluorescent antibody	پادتن پرتو افشان مستقیم
polyclonal antibody	پادتن چند کلونی
anti-idiotype antibody	پادتن ضد ایدیوتیپ
catalytic antibody	پادتن فروکافتی
catalytic antibody	پادتن کاتالیزوری
semisynthetic catalytic antibody	پادتن کاتالیزوری نیمه مصنوعی
magnetic antibody	پادتن مغناطیسی
engineered antibody	پادتن مهندسی شده
allotypic monoclonal antibodies	پادتن های تک بنیانی دگر مونه ای
monoclonal antibodies	پادتن های تک دودمانی
antitoxin	پاد زهر
antibody, antidote	پادزهر
antitoxin	پادزهرابه
diphtheria antitoxin	پادزهر دیفتری
engineered antibody	پادزهر مهندسی شده
antibiotic	پادزی
antibiotic	پاد زیست
biocide	پادزیست

intravenous antibiotic	پادزیست درون سیاهرگی
antibiotic	پادزیو
antisera	پاد سرم ها
antagonist	پادکردار
antagonism	پادکرداری
antigen	پادگن
neoantigen	پادگن جدید
very late antigen	پادگن خیلی دیر
cacinoembryonic antigen	پادگن کاسینو جنینی
tumor-associated antigen	پادگن مربوط به تومور
chimeric antibody	پادگن نوترکیب
neoantigen	پادگن نوساخته
neoantigen	پادگن نوظهور
v antigen	پادگن ویروسی
human leukocyte antigens	پادگن های گلبول سفید انسانی
parabiosis	پارابیوز
parapatric speciation	پاراپاتریک اسپشیشن
paraprotein	پاراپروتئین
parataxonomist	پاراتاکسونومیست
parathormone	پاراتورمون
parathormone	پاراتیرین
pararetrovirus	پارارتروویروس
parafusin	پارافوزین
paracrine	پاراکرین
paraquat	پاراکوات
paramyxovirus	پارامایوکسوویروس
paramyxoviridae	پارامایوکسوویروس(ها)
paramyosin	پارامیوزین
tyrosine	پاراهیدروکسی فنیل آلانین
parthenocarpy	پارتنوکارپی
paresis	پارزی
parvovirus	پارو ویروس
lysis	پاره گی
particle	پاریزه
response	پاسخ
immune response	پاسخ ایمنی
primary immune response	پاسخ ایمنی اولیه
secondary immune response	پاسخ ایمنی ثانویه
humoral immune response	پاسخ ایمنی خونی
response	پاسخ به محرک

English	Persian		English	Persian
humoral response	پاسخ خونی		persistence, stability	پایداری
cellular response	پاسخ سلولی		pyrethrins	پایرترین
antibody-mediated immune response	پاسخگویی ایمنی توسط آنتی بادی		pyrexia	پایرکسیا
			filopodia	پای کاذب
sprayer	پاش-		conservation	پایندگی
nozzle	پاشنده		bed	پایه
spray	پاشیدن		peptone	پپتون
aseptic	پاک		peptid, peptide	پپتید
protein purification	پاکسازی پروتئین		peptidase	پپتیداز
paclitaxel	پاکلیتاکسل		transit peptide	پپتید انتقالی
pachytene	پاکیتن		leader peptide	پپتید پیشرو
decontamination	پاکیزه سازی		T-cell modulating peptide	پپتید تعدیل کننده سلول تی.
pachynema (= pachytene)	پاکینما		peptide T	پپتید تی.
palladium	پالادیوم		peptidergic	پپتیدرژیک
filtration	پالایش		cecropin a peptide	پپتید سکروپین آ
sewage treatment	پالایش فاضلاب		cecropin a peptide	پپتید سکروپین آلفا
palmitate	پالمیتات		cecropin a peptide	پپتید سکروپین ای.
palytoxin	پالی توکسین		alfalfa antifungal peptide	پپتید ضد قارچ آلف آلفا
palindrome	پالیندروم		signal peptide (= leader peptide)	پپتید علامتی
pantetheine	پانتتئین			
pancreastatin	پانکرااستاتین		transit peptide	پپتید گذرا
pancreatin	پانکراتین		chloroplast transit peptide	پپتید ناقل کلروپلاست
pancreas	پانکراس		peptidoglycan	پپتیدوگلیکان
panmixis	پانمیکسی		peptido-mimetic	پپتیدو-میمتیک
panmixia	پانمیکسیا		atrial peptides	پپتیدهای دهلیزی
panose	پانوز		peptidyl transferase	پپتیدیل ترانسفراز
chain terminator	پایان بخش زنجیره		pepsin	پپسین
helix terminator	پایان بخش مارپیچ		potassium	پتاسیم
terminator	پایان دهنده		establishment potential	پتانسیل استقرار
rho-dependent terminator	پایان دهنده متکی به رو		pterostilbenes	پتروستیل بن (ها)
acral	پایانکی		petrochemistry	پتروشیمی
N-terminal	پایانه ان.		diffusion, dispersal, dissemination, distribution, transmission	پخش
NOS terminator	پایانه ان.او.اس.			
telomere	پایانه پار			
cohesive end	پایانه چسبنده		light-scattering	پخش سبک
telophase	پایانه چهر		disseminating	پخش کردن
terminalization	پایانه سازی		spreader	پخش کن
flush end	پایانه صاف/کور/پخ		spreader	پخش کننده
cohesive termini	پایانه های چسبنده		biological defense	پدافند بیولوژیکی
persistence	پایایی		parents	پدر و مادر
sustainable	پایدار		penetrance	پدیدار شوندگی

English	Persian	English	Persian
phenotype	پدیدگان	irradiation	پرتوتابی
acceptor	پذیرا	actinomycetes	پرتوجلد ها
acceptor	پذیرفتنی	irradiation, radiotherapy	پرتو درمانی
acceptor, receptor	پذیرنده	irradiation	پرتو دهی
epidermal growth factor receptor	پذیرنده عامل رشد روپوستی	external-beam radiation	پرتو دهی خارجی
		radioactive	پرتوزا
parataxonomist	پرا آرایه شناس	radioactivity	پرتوزایی
variance	پراش	radioactive	پرتو کنش ور
peroxidase	پر اکسیداز	exposure	پرتوگیری
horseradish peroxidase	پر اکسیداز ترب کوهی	acute exposure	پرتو گیری حاد
hydrogen peroxide	پراکسید هیدروژن	ultraviolet rays	پرتو های فرا بنفش
diffusion, dispersal, dispersion	پراکندگی	standard	پرچم
		hypersensitivity	پر حساسیتی
gel diffusion	پراکندگی ژلی	delayed hypersensitivity	پر حساسیتی عقب/به تاخیر افتاده
dispersion, variance	پراکنش		
primer	پرایمر	RNA processing	پردازش آر.ان.ای.
probe (= DNA probe)	پرب	bioprocessing	پردازش زیستی
nucleic acid probe	پرب اسید نوکلئیک	prednisolone	پردنیسولون
dissolved oxygen probe	پرب اکسیژن محلول	mask, membrane	پرده
oligonucleotide probe	پرب اولیگونوکلئوتید	polyribosome	پر رناتن
multi-locus probe	پرب چند نقطه ای	polymorphism	پر ریختی
DNA probe	پرب دی.ان.ای	polygenic	پر زاد
genetic probe	پرب ژنتیکی	polygenic	پر ژنی
gene probe	پرب ژن (ی)	fill	پر شدن
fluorogenic probe	پرب فلوئوروژنیک/فلوئورسانس زا/شب	perforin	پرفورین
	نمایی/پرتوافشانی/شارندگ ی زا	polysaccharide	پر قند
		fill	پر کردن
		percutaneous	پرکوتانوس
RNA probes	پرب های آر.ان.ای.	colony	پرگنه
hybridization probe	پرب هیبریداسیون	polyploid	پر لاد
polymer	پر پار	flying	پرنده
delivery	پرتاب	polynucleotide	پر نوکلئوتید
luminescence	پرتو	proanthocyanidin	پروآنتوسیانیدین
fluorescence, irradiation, radiation	پرتو افشانی	fat(s)	پرواری
		flying	پرواز کردن
bright greenish-yellow fluorescence	پرتو افشانی درخشان زرد مایل به سبز	monarch butterfly	پروانه مونارک
		proopiomelanocortin	پرواوپیوملانوکورتین
external-beam radiation	پرتوافکنی اشعه خارجی	probe (= DNA probe)	پروب
external-beam radiation	پرتوافکنی اکسترنال بیم	nucleic acid probe	پروب اسید نوکلئیک
totipotency	پر توانی	dissolved oxygen probe	پروب اکسیژن محلول
x-ray	پرتو ایکس	proband	پروباند

English	Persian
oligonucleotide probe	پروب اولیگونوکلئوتید
probiotics	پروبایوتیکز
multi-locus probe	پروب چند نقطه ای
DNA probe	پروب دی.ان.ای
bioprobe	پروب زیستی
genetic probe	پروب ژنتیکی
gene probe	پروب ژن (ی)
fluorogenic probe	پروب فلوئوروژنیک/فلوئورسانس زا/شب نمایی/پرتوافشانی/شارندگی زا
heterologous probing	پروب ناهمگن
RNA probes	پروب های آر.ان.ای.
hybridization probe	پروب هیبریداسیون
proplastid	پروپلاستید
propolypeptide	پروپلی پپتید
propylene glycol	پروپیلین گلیکول
propioyl-CoA carboxylase	پروپیول-کوآ کاربوکسیلاز
neutral protease	پروتآز خنثی
protease, proteolytic enzyme	پروتاز
acid protease	پروتاز اسیدی
proteasome	پروتازوم
proteolytic	پروتئولیتیک
proteolysis	پروتئولیز
proteome	پروتئوم
proteomics	پروتئومیکز
protein	پروتئین
Protein 53	پروتئین 53
nonheme-iron protein	پروتئین آهن غیر همی
E1B protein of adenovirus	پروتئین ئی. 1 بی. آدنوویروس
attachment protein	پروتئین اتصال
zona pellucida sperm-binding protein	پروتئین اتصال اسپرم زونا پلوسیدا
poly-adenylate binding protein	پروتئین اتصال پلی آدنیلات
single-strand binding protein	پروتئین اتصال تک رشته
TATA-binding protein (TBP)	پروتئین اتصال تی.ای.تی.ای.
DNA-binding-protein	پروتئین اتصال دی.ان.ای.
cartilage link protein	پروتئین اتصال غضروف
methyl-CpG-binding protein	پروتئین اتصال متیل-سی.پی.جی.
proteinase	پروتئیناز
protein S	پروتئین اس.
SOS protein	پروتئین اس.او.اس.
nonhiston protein	پروتئین اسیدی
whey acidic protein	پروتئین اسیدی کشک
intrinsic protein	پروتئین اصلی/ذاتی/درونی
fusion protein	پروتئین امتزاج
storage protein	پروتئین انبار
membrane transporter protein	پروتئین انتقالگر غشایی
storage protein	پروتئین اندوخته
finger protein	پروتئین انگشتی
zinc finger protein	پروتئین انگشتی روی
immunogenic protein	پروتئین ایمنی زا
unwinding protein	پروتئین باز کننده مارپیچ
myelin basic protein	پروتئین بازی مایلین
odorant binding protein	پروتئین بو
minimized protein	پروتئین به حداقل رسیده
helix-destabilizing proteins	پروتئین بی ثبات کننده مارپیچ
surfactant-associated protein B	پروتئین بی. وابسته به سورفکتانت
thermal hysteresis protein	پروتئین پسماند حرارتی
plasma protein	پروتئین پلاسما
plasma protein	پروتئین پلاسمایی
coat protein	پروتئین پوششی
signaling protein	پروتئین پیام دهنده
amyloid precursor protein	پروتئین پیش درآمد آمیلوئید
early protein	پروتئین پیشین
visible fluorescent protein	پروتئین تابناک مرئی
late protein	پروتئین تازه
invasin, transmembrane protein	پروتئین ترا شامه ای
trans-acting protein	پروتئین ترانس اکتینگ
single-cell protein	پروتئین تک یاخته
cell-differentiation protein	پروتئین تمایز سلولی
transmembrane regulator protein	پروتئین تنظیم کننده تراشامه ای
sonic hedgehog protein	پروتئین جوجه تیغی صوتی

English	Persian
homologous protein	پروتئین جور
mutator protein	پروتئین جهنده
adhesion protein	پروتئین چسبندگی
DNA-binding protein	پروتئین چسبنده دی.ان.ای.
acyl carrier protein	پروتئین حامل آسیل
minimized protein	پروتئین حداقل
motor protein	پروتئین حرکتی
Desert Hedgehog Protein (Dhh)	پروتئین خارپشت بیابانی/کویری
hedgehog protein	پروتئین خار پشت/جوجه تیغی
scaffolding protein	پروتئین داربستی
deoxyribonucleoprotein	پروتئین داکسی ریبونوکلئیک
heterologous protein	پروتئین دگر ساخت
storage protein	پروتئین ذخیره
ras protein	پروتئین راس
retinoblastoma protein	پروتئین رتینوبلاستوما
pancreatic thread protein	پروتئین رشته ای پانکراتیکی
recombinant protein	پروتئین رکمبینان
coat protein	پروتئین روکشی
guanine nucleotide releasing protein	پروتئین رها ساز نوکلئوتید گوانین
bone morphogenetic protein	پروتئین ریخت زای استخوانی
simple protein	پروتئین ساده
syk protein	پروتئین سایک
green fluorescent protein	پروتئین سبز فلوئورسنتی
scleroprotein	پروتئین سخت
tumor-suppressor protein	پروتئین سرکوبگر تومور
protein C	پروتئین سی.
C-reactive protein	پروتئین سی.-رآکتیو
cis-acting protein	پروتئین سیز- فعال
chaperone protein	پروتئین شاپرون
chaperone protein	پروتئین شاپرون ملکولی
CTD-associated SR-like protein	پروتئین شبه اس.آر. وابسنه به سی.تی.دی.
proteomics	پروتئین شناسی
structural proteomics	پروتئین شناسی ساختاری
heat-shock protein	پروتئین شوک حرارتی
antifreeze protein	پروتئین ضد انجماد
cold-shock protein	پروتئین (ضد) شوک سرمائی
integral membrane protein	پروتئین غشای اصلی

English	Persian
nonchromosomal protein	پروتئین غیر کروموزومی
nonhiston protein	پروتئین غیر هیستونی
acute-phase protein	پروتئین فاز حاد
transactivating protein	پروتئین فعال ساز ترانس
viral transactivating protein	پروتئین فعال ساز ترانس ویروسی
GTPase activating protein	پروتئین فعال ساز جی.تی.پی.آز
mitogen activated protein	پروتئین فعال شده توسط میتوژن
stress activated protein	پروتئین فعال شونده با تنش
activator protein	پروتئین فعال کننده
catabolite activator protein	پروتئین فعال کننده فرآورده تجزیه
metalloprotein	پروتئین فلزی
proteolytic	پروتئین کافت
krüppel protein	پروتئین کروپل
minimized protein	پروتئین کمینه
minimized protein	پروتئین کوچک شده
chimeric protein (chimera)	پروتئین کیمریک
protein kinase	پروتئین کیناز
protein kinase C	پروتئین کیناز سی.
mitogen-activated protein kinases (MAPK)	پروتئین کیناز فعال شده توسط میتوژن
high mobility group protein	پروتئین گروهی شدیدا متحرک
globular protein	پروتئین گویچه ای
cAMP receptor protein (= catabolite activator protein)	پروتئین گیرنده سی.ای.ام.پی.
lux protein	پروتئین لاکس
zinc leucine protein	پروتئین لوسین روی
nuclear matrix protein	پروتئین ماتریکس هسته ای
chaperone protein	پروتئین مراقب
conjugated protein	پروتئین مزدوج
homologous protein	پروتئین مشابه
minimized protein	پروتئین مینیم شده
heterologous protein	پروتئین نا همساخت
chimeric protein (chimera), recombinant protein	پروتئین نوترکیب
lux protein	پروتئین نور
iron response protein	پروتئین واکنش آهن
C-reactive protein	پروتئین واکنش پذیر سی.
transport proteins	پروتئین های انتقالی

44

stress proteins	پروتئینهای تنشی	anabolic process	پروسه آنابولیکی
wnt proteins	پروتئین های دبلیو.ان.تی.	acid saccharification process	پروسه ساکاریفیکاسیون اسیدی
switch proteins	پروتئین های سوئیچ	Symba process	پروسه سیمبا
Cry Proteins	پروتئین(های) کرای	microbial film process	پروسه فیلم میکروبی
pathogenesis related proteins	پروتئین های وابسته به عوامل بیماری زا	prophase	پروفاز
heterologous protein	پروتئین هترولولوگ	prophage	پروفاژ
nucleoprotein	پروتئین هسته	cryptic prophage	پروفاژ نهان ساز
replication protein A	پروتئین همانند سازی ای.	proflavin	پروفلاوین
fusion protein	پروتئین همجوشی	profilin	پروفیلین
homologous protein	پروتئین همگون	prokephalin	پروکفالین
thermal hysteresis protein	پروتئین هیسترزیس ترمال	procollagen	پروکولاژن
protamine	پروتامین	prolactin	پرولاکتین
prothrombin	پروترومبین	prolamine	پرولامین
protocol	پروتکل	proline	پرولین
Biosafety Protocol	پروتکل بیوایمنی	inducible promoter	پروموتور القاء پذیر
Biosafety Protocol	پروتکل زیست ایمنی	provirus	پرو ویروس
proto-oncogene	پروتوانکوژن	swinging-bucket rotor	پره سطل آونگین
protoplast	پروتوپلاست	swinging-bucket rotor	پره سوئنگینگ-باکت
protoplasm	پروتوپلاسم	avoidance	پرهیز
nucleoplasm	پروتوپلاسم هسته سلول	Pre-RC (= pre-replication complex)	پری-آر.سی.
prototroph	پروتوتروف		
protozoa	پروتوزوا	prebiotics	پریبایوتیکز
protoxin	پروتوکسین	preproinsulin	پریپرو انسولین
protomer	پروتومر	preprohormone	پری پروهورمون
protista	پروتیستا	periplasm	پریپلازم
breed	پروردن	peritoneal	پریتونیال
cultivar	پرورده	nick	پریدگی
culture	پرورش	perikaryon	پریکاریون
silviculture	پرورش جنگل	primase (= DNA primase)	پریماز
breeding	پرورش حیوانات	DNA primase	پریماز دی.ان.ای.
cultivar	پروره	primordium	پریم اوردیوم
trophic	پروره ای	primaverose	پریماوروز
progesterone	پروژسترون	primosome	پریموزوم
Human Genome Project	پروژه ژنوم انسان	prion	پریون
prostaglandin	پروستا گلاندین	clinician	پزشک بالینی
prostaglandin endoperoxide synthase	پروستا گلاندین اندو پر اکسید سنتاز	alternative medicine, complementary and alternative medicine	پزشکی غیر سنتی
processing	پروسس کردن	complementary and alternative medicine,	پزشکی مکمل و غیر متعارف
processivity	پروسسیتیویته		
vectorial processing	پروسسینگ ناقلی		

45

English	Persian	English	Persian
complementary medicine		plastid	پلاستید
wastewater	پساب	plastidome	پلاستیدوم
sewerage	پساب داری	blood plasma, plasma	پلاسما
sewer	پساب رو	plasmacyte	پلاسماسیت
dehydration	پسابش	plasmapheresis	پلاسما فرسیز
postexposure	پس از تماس	plasmalemma	پلاسمالما
viral retroelement	پسا عنصر ویروسی	plasmalogen	پلاسمالوژن
vulgaris	پست	blood plasma	پلاسمای خون
postaglandis A	پستاگلاندیس ای.	plasmoptysis	پلاسموپتیز
postaglandin	پستاگلاندین	plasmodesma	پلاسمودسما
nozzle	پستانک	plasmodium	پلاسمودیوم
Micronair spray nozzle	پستانک اسپری مایکرون ایر	plasmogamy	پلاسموگامی
metaphase	پس چهره	plasmolysis	پلاسمولیز
retrograde	پسرو	plasmon	پلاسمون
inbreeding depression	پس روی خویش آمیزی	surface plasmon	پلاسمون سطحی
epigenetic	پس زاد شناسی	plasmid	پلاسمید
nutritional epigenetics	پس زاد شناسی تغذیه ای	2-micron plasmid	پلاسمید ۲ میکرونی
epigenetic	پس زایشی	yeast episomal plasmid	پلاسمید اپیزومی مخمر
immunological rejection	پس زدن ایمونولوژیکی	yeast integrative plasmid	پلاسمید اینتگرتیو مخمری
waste	پسماند	low calcium response plasmid	پلاسمید پاسخگو به کلسیم کم
waste management	پسمانداری		
high level waste	پسماند پر پرتو	Ti plasmid	پلاسمید تی.آی.
solid waste	پسماند جامد	multi-copy plasmid	پلاسمید چند نسخه ای
low level waste	پسماند کم پرتو	relaxed circle plasmid	پلاسمید دایره ای/مدور سست
residue	پس مانده		
chlorine residual	پس مانده کلر	ri plasmid	پلاسمید ری
psoralen, psoralene	پسورالن	stringent plasmid	پلاسمید سخت
retropharyngeal	پشت حلقی	relaxed plasmid	پلاسمید سست
solid support	پشتیبان جامد	runaway plasmid	پلاسمید فراری
support	پشتیبانی	supercoiled plasmid	پلاسمید فوق مارپیچ
mosquito	پشه	tumor-inducing plasmid	پلاسمید مسبب تومور
pleocytosis	پلئوسیتوز	yeast replication plasmid (YPR)	پلاسمید هماند سازی مخمری
pleiotropy	پلئیتروپی		
pleiotropism	پلئیتروپیزم	conjugating plasmid	پلاسمید هم یوغ/مزدوج
pleistocene	پلئیستوسن	plasmin	پلاسمین
polarity	پلاریته	plasminogen	پلاسمینوژن
plastid	پلاست	severe, childhood, autosomal, recessive muscular dystrophy	پلاسیدگی عضلانی اتوزومی مغلوب حاد کودکان
plastogene	پلاستوژن		
plastocyanin	پلاستوسیانین		
plastoquinone, plasto-quinone	پلاستوکینون	plaque	پلاک
		amyloid plaque	پلاک آمیلوئید

platelet	پلاکت	vent DNA polymerase	پلیمرآز دی.ان.ای. منفذ
blood platelet	پلاکت خونی	antithrombogenous polymer	پلی مر آنتی ترومبوژنوس
plakoglobin	پلاکوگلوبین	polymerase	پلیمراز
pellagra	پلاگرا	polymerase I	پلیمراز 1
plug flow digester	پلاگ فلو دایجستر	RNA polymerase	پلیمراز آر.ان.ای.
PlantibodiesTM	پلانتیبادیز	DNA-dependent rna polymerase	پلیمراز آر.ان.ای. وابسته به دی.ان.ای.
plankton	پلانکتون	tag polymerase	پلیمراز برچسبی
phytoplankton	پلانکتون گیاهی	poly(A) polymerase	پلیمراز پلی آدنین
disulphydryl bridge	پل دی سولفیدریل	Thermus Aquaticus Polymerase	پلیمراز ترموس آکواتیکوس
sepsis	پلشتی	taq polymerase(= Thermus Aquaticus Polymerase)	پلیمراز .تی.ای.کیو.
plecksterin	پلک استرین		
molecular bridge	پل ملکولی	taq DNA polymerase (= Thermus Aquaticus Polymerase)	پلیمراز تی.ای.کیو. دی.ان.ای.
DNA bridges	پل های دی.ان.ای.		
polyadenylation	پلی آدنیله سازی		
polyadenylated	پلی آدنیله شده	T4 DNA polymerase	پلیمراز دی.ان.ای. تی.4.
polyacrylamide gel	پلی آکریل آمید ژل	addition polymer	پلیمر افزایشی
polyethylene glycol (PEG)	پلی اتیلن گلیکول	biopolymer	پلیمر زیستی
poly-L-lysine	پلی ال.لیزین	antithrombogenous polymer	پلی مر ضد ترومبوژنوس
polypeptide	پلی پپتید	dendritic polymer	پلی مر های دندریتی/دندرایتی
vasoactive intestinal polypeptide	پلی پپتید روده ای وازواکتیو	polymerization	پلیمریزاسیون
polyprotein	پلی پروتئین	hommopolymer	پلیمر یکنواخت
polyploid	پلیپلوئید	amplified fragment length polymorphism (AFLP)	پلی مورفیزم طولی قطعه تقویت شده
polyploidy	پلی پلوئیدی		
familial adenomatous polyposis	پلیپوز آدنوماتوسی خانوادگی	single-strand conformational polymorphism	پلی مورفیسم تک رشته ای تطبیقی
polydactyl	پلی داکتیل		
polyribosome	پلی ریبوزوم	simple sequence length polymorphism	پلی مورفیسم طولی توالی ساده
polysome (= polyribosome)	پلی زوم		
polygene	پلی ژن	polynucleotide	پلی نوکلئتید
polygenic	پلی ژنیک	polynucleotide phosphorylase	پلی نوکلئوتید فسفریلاز
polysaccharide	پلی ساکارید		
non-starch polysaccharide	پلی ساکارید بدون نشاسته/فاقد نشاسته	polyvalent	پلی والنت
		polyhydroxyalkanoate	پلی هیدروکسی آلکانوآت
microbial polysaccharide	پلی ساکارید میکروبی	polyhydroxyalkanoic acid	پلی هیدروکسی آلکانوئیک اسید
polyphenol	پلی فنل		
polycation conjugate	پلیکیشن کانجوگیت	polyhydroxylbutylate	پلی هیدروکسیل بوتیلات
polygalacturonase	پلی گالاکتوروناز	gun	پمپ دستی
polylinker	پلی لینکر	shelter-in-place	پناهگاه-در-محل
polymer	پلی مر	pentose	پنتوز
polymer	پلیمر		

epithelium	پوشش سنگفرشی	five prime	پنج پریم
nuclear envelope	پوشش هسته ای	covert	پنهان(ی)
capsular, epithelial	پوششی	renin	پنیر مایه
capsid	پوشه	extended spectrum penicillin	پنی سیلین دارای طیف گسترده
capsule	پوشینه	penicillium patulum	پنیسیلیوم پاتولوم
capsular	پوشینه ای	penecillium citrinum	پنیسیلیوم سیترینوم
Bowman's capsule	پوشینه بومن	powder	پودر
plaque, washer	پولک	Chapin plant and rose powder duster	پودر پاش چاپین
psi	پوند بر اینچ مربع	powdered	پودر زده
dynamics	پویایی	alicin	پودر سیر
dynamics	پویایی شناسی	powdered	پودر شده
pH	پ. هاش	atomizing	پودر کردن
optimum pH	پ.هاش بهینه	powder	پودر کردن/زدن
optimum pH	پ. هاش مطلوب	podophyllotoxin	پودوفیلوتوکسین
bed	پی	port-a-cath	پورتاکات
P (= phosphate), p (= pico =10-12), P (= short arm of chromosome)	پی.	porpho-bilino-gen	پورفوبیلینوژن
		porphyrins	پورفیرین
P1 (= first meiotic prophase), P1 (= phosphatidyl-inositol)	پی.1	puromycin	پورومایسین
		mash	پوره
p53 (= Protein 53)	پی.53	porin, purine	پورین
PRPP (= 5-phospho-ribosyl-1-pyro phosphate)	پی.آر.پی.پی.	skin	پوست
		acrodermatitis	پوست آماس پایانکی
		scab	پوست زخم
PRV (= pseudo-rabies virus)	پی.آر.وی.	skin	پوست کندن
		case, membrane, skin	پوسته
acroneurosis	پی آسیبی پایانکی	T-shell	پوسته تی.
PI (= paternity index)	پی.آی.	plasma membrane	پوسته خارجی سلول
PE (= phosphatidyl-ethanolamine)	پی.ئی.	scab	پوسته زخم
		percutaneous	پوستی
PEP (= phospho-enol-pyruvate)	پی.ئی.پی.	decay	پوسیدگی
		gibberella ear rot	پوسیدگی ذرت ژیبرلائی/جیبرلائی
PEPCK (= phospho enol-pyruvate carboxylase)	پی.ئی.پی.سی.کی.		
		brown stem rot	پوسیدگی قهوه ای ساقه
PET (= paraffin-embedded tissue)	پی.ئی.تی.	soft rot	پوسیدگی نرم
		protective clothing	پوشاک محافظ
PET-PCR (= RT-PCR out of paraffin-embedded tissue)	پی.ئی.تی.-پی.سی.آر	mask	پوشانه
		mask	پوشش
		capsid	پوشش پروتئینی
PEG (= poly-ethylene	پی.ئی.جی.	capping	پوشش دادن

glycol)

Ph (= Philadelphia chromosome) — پی.اچ.

PHA-LCM (= phyto-hem-agglutinin-stimulated lymphocyte conditioned medium) — پی.اچ.ای.-ال.سی.ام.

PHP (= pseudo-hypo-parathyroidism) — پی.اچ.پی.

PS (= phosphatidyl-serine), PS (= pulmonary stenosis) — پی.اس.

psi — پی.اس.ای.

Pst1 (= Pst1 endonuclease) — پی.اس.تی.1

PFGE (= pulse field gel electrophoresis) — پی.اف.جی.ئی.

PXE (= pseudo-xanthoma elasticum) — پی.اکس.ئی.

PLP (= pyridoxal-5-phosphate) — پی.ال.پی.

PMFS (= phenylmethylsulfonylfluoride) — پی.ام.اف.اس.

myoelectric signal — پیام الکتریکی ماهیچه

environment impact assessment — پیامد سنجی

PNP (= purine nucleoside phosphorylase) — پی.ان.پی.

Pi (= inorganic phosphate) — پی.اندیس آی.

PAS (= periodic acid) — پی.ای.اس.

PAGE (= poly-aclamide gel electrophoresis) — پی.ای.جی.ئی.

PADP (= poly-adenylate binding protein) — پی.ای.دی.پی.

PAC (= P1-derived artificial chromosome) — پی.ای.سی.

neuroglia — پی بان

pBR322 — پی.بی.آر.322

PBS (= phosphate-buffered saline) — پی.بی.اس.

PBL (= peripheral blood lymphocyte)

PBMNC (= peripheral blood mono-nuclear cells) — پی.بی.ام.ان.سی.

PBG (= porpho-bilino-gen) — پی.بی.جی.

pipette — پیپت

PP (= protein purification) — پی.پی.

PPHP (= pseudo-pseudo-hypo-parathyroidism) — پی.پی.اچ.پی.

PPi (= pyro-phosphate ion) — پی.پی.اندیس آی.

Pitt-3 — پیت-3

pytalin — پیتالین

PTRF (= polymerase I and transcript release factor) — پی.تی.آر.اف.

PTH (= parathormone), PTH (= phenyl-thio-hydration) — پی.تی.اچ.

PTC (= phenyl-thio-carbomide) — پی.تی.سی.

PG (= prostaglandin) — پی.جی.

streptococcus — پیچ گوییزه

PWS (= Prader-Willi syndrome) — پی.دابلیو.اس.

exogenous — پیدا زا

generation — پیدایش

initiation, occurrence — پیدایش

bacteremia — پیدایش باکتری در خون

spontaneous generation — پیدایش خودبخودی

PDHC (= pyruvate de-hydrogenase complex) — پی.دی.اچ.سی.

PDA (= patent ductus arteriosus) — پی.دی.ای.

PDGF (= platelet-derived growth factor) — پی.دی.جی.اف.

paraclinical — پیرابالینی

periderm — پیراپوست

periplasm — پیرادشته

periodontium — پیرا دندان

pyralis — پیرالیس

pyranose — پیرانوز

perikaryon — پیراهسته

English	Persian
senescence	پیرشـدگی
aging	پیر شدن
pyrenoid	پیرنوئید
thiamine pyrophosphate, thyamine pyrophosphate	پیروفسـفات تیامین
pyrrolizidine alkaloid	پیرولیزیدین آلکالوئید
pyronin Y	پیرونین وای.
pyruvate kinase	پیروویت کیناز
aging, senescence	پیری
pyridoxal	پیریدوکسـال
pyridoxamine	پیریدوکسـامین
pyridoxine	پیریدوکسـین
progeria	پیری زود رس
cellular aging	پیری سـلولی
piericidin	پیریسیدین
pyrimidine	پیریمیدین
piezoelectric	پیزوالکتریک
neurotoxin	پی زهرابه
neurobiologist	پی زیست شناس
neurobiology	پی زیست شناسی
PC (= phosphatidyl-choline)	پی.سی.
PCR (= polymerase chain reaction)	پی.سی.آر.
nested PCR	پی.سی.آر. جایگزین
multiplex PCR	پی.سی.آر. مرکب
PCNA (= proliferating cell nuclear antigen)	پی.سی.ان.ای.
PCOD (= poly-cystic ovarian disease)	پی.سی.او.دی.
PCd (= premature centromere division)	پی.سی.دی.
PCC (= premature chromosome condensation)	پی.سی.سی.
proenzyme, zymogen	پیش آنزیم
mitogen-activated protein kinase cascade	پیشار پروتئین کیناز فعال شـده توسط میتوژن
pre-mRNA	پیش ام.آر.ان.ای.
proinsulin	پیش انسـولین
cryptic prophage	پیش باکتری خوار نهان ساز
promoter	پیش برنده

English	Persian
Cis-acting sequence in RNA polymerase II promoters	پیش برنده توالی سی.آی.اس.-عملگر در پلیمراز 2 آر.ان.ای.
ab initio gene prediction	پیشبینی ژن اب اینیتیو
propeptide	پیش پپتید
prototroph	پیش پرور
protoderm	پیش پوست
protodermal	پیش پوستی
progeria	پیش پیری
precursor	پیش تاز
zymogen	پیشتاز آنزیم
pre-tRNA	پیش تی.آر.ان.ای.
prophase	پیش چهر
prodrome	پیش در آمد
preclinical	پیش درمانگاهی
protoplast	پیش دش
protoplasm	پیش دشته
proplast	پیش دیسه
leader	پیشرو
prodrome	پیش زمینه
protozoa	پیش زیان
proenzyme	پیش زیما
precursor	پیش ساز
precursor	پیش شرط
prepotency	پیش قوت
preclinical	پیش کلینیکی
procollagen	پیش کولاژن
proband	پیش گدازه
prevention	پیش گیری
chemoprophylaxis	پیشگیری (از بیماری یا عفونت) با دارو
chemoprophylaxis	پیشگیری شیمیائی
precursor	پیش ماده
plasma, protoplasm	پیش مایه
inducible promoter	پیش محرک القاء پذیر
strong promoter	پیش محرک قوی
precursor	پیش مرحله
precursor	پیش نیاز
provitamin	پیش ویتامین
provirus	پیش ویروس
pronucleus	پیش هسته

procaryotes, prokaryotes	پیش هسته ای ها
prophase	پیش هنگام
prohormone	پیش هورمون
protoplast	پیش یاخته
pleistocene	پیشین پدید
pedigree	پیشینه
second messenger	پیک ثانویه
configuration	پیکر بندی
absolute configuration	پیکر بندی مطلق
Barr body	پیکره بار
refractile body	پیکره شکننده/منکسر کننده نور
ketone body	پیکره کیتون
Golgi body	پیکره گلژی
somatic	پیکری
pico	پیکو-
picorna	پیکورنا
picornavirus	پیکورناویروس
picornaviridae	پیکورناویروس(ها)
picogram	پیکوگرم
PK (= pyruvate kinase)	پی.کی.
PQ (= plasto-quinone)	پی.کیو.
PKU (= phenylketonuria)	پی.کی.یو.
pigmentation	پیگمنتاسیون
xeroderma pigmentation	پیگمنتاسیون زرودرما
contact tracing	پیگیری اشاعه دهندگان/افراد
	ساری/ناقل
pilus	پیلوس
serum	پیماب
blood serum	پیماب خون
serotype	پیماب مونه
immune sera	پیماب های مصونی
convention	پیمان
measurement	پیمایش
biological measurement	پیمایش بیولوژیکی
emulsion	پیمایه
pinocytosis	پینوسیتوز
callus	پینه
affinity, linkage	پیوستگی
partial linkage	پیوستگی جزئی
gene linkage	پیوستگی ژن (ی)
partial linkage	پیوستگی ناقص

syndactyly	پیوسته انگشتی
syncytium	پیوسته یاخته
affinity, bond, graft, linkage, transplantation	پیوند
allograft	پیوند آلو
covalent bond	پیوند اشتراکی
covalent coordinate bond	پیوند اشتراکی یکسویه
blunt-end ligation	پیوند انتهای صاف/کور
non-homologus end joining	پیوند انتهای غیر همسان
xenograft, xenotransplant	پیوند بیگانه
peptide bond	پیوند پپتیدی
π- bond	پیوند پی
disulphide bond	پیوند دو گوگردی
disulfide bond	پیوند دی سولفید
xenograft	پیوند دیگر سرشت
graf	پیوند زدن
xenotransplant	پیوند زنو
xenogeneic transplantation	پیوند زنوژنئیکی
genetic linkage	پیوند ژنتیکی
gene splicing	پیوند ژن (ی)
sigma bond	پیوند سیگما
transplantation	پیوند عضو
phosphodiester bond	پیوند فسفو دی استر
graf	پیوندک
synapse	پیوندگاه نرونی
reassociation	پیوند مجدد
bone marrow transplantation	پیوند مغز استخوان
nuclear transplantation	پیوند هسته ای
synapsis	پیوند همور
hydrogen bond	پیوند هیدروژنی
graf	پیوندی
linkage	پیوند یافتگی
adipose	پیه دار
adipose	پیه مانند
tracer	پی یاب
neuron	پی یاخته
PUBS (= percutaneous umbilical blood sampling)	پی.یو.بی.اس.
PUVA (= plus ultra-violet	پی.یو.وی.ای.

English	Persian	English	Persian
light of the A wavelength)		DNA bending	تاشدگی دی.ان.ای.
		protein folding	تا شدن پروتئین
		cloning	تاک سازی
ت		directional cloning	تاک سازی جهت یافته
		taxol	تاکسول
occupation theory of agonist action	تئوری اشغال فعالیت آگونیست	taxon	تاکسون
		taxonomy	تاکسونومی
non-equilibrium theory	تئوری غیر تعادلی	gene taxi	تاکسی ژن (ی)
teosinte	تئوزینت	tachykinin	تاکی کینین
target validation	تائید هدف	thalassemia	تالاسمی
spinning	تاب دادن	solid	تام
radiation	تابش	tannin	تانن
internal radiation	تابش داخلی/درونی	vesicle	تاول
anneal, annealing	تابکاری	vesicle	تاولچه
phosphorus	تابنده	Tay-Sachs	تای-ساکز
tabun	تابون	validation	تایید کردن
selective estrogen effect	تاثیر استروژن گزینشی	approvable letter	تایید نامه نهایی
labile	تاثیر پذیر	temperature	تب
irritability	تاثیر پذیری	pyrogen	تب آور
late effect	تاثیر تازه	crossing-over	تبادل
deterministic effect	تاثیر جبری	information exchange	تبادل اطلاعات
mild effect	تاثیر خفیف	crossing-over	تبادل ژنتیکی
adverse effect	تاثیر زیانبار	sister chromatid exchange	تبادل کروماتید خواهر
deterministic effect	تاثیر گمارشی	lineage, pedigree	تبار
drug interaction	تاثیر متقابل دارو	phylogenetic	تبار زایشی دودمانی
chronic effect	تاثیر مزمن/حاد/جدی	antipyretic	تب بر
adverse effect	تاثیر مضر	conversion, transformation	تبدیل
mild effect	تاثیر ملایم	transamination	تبدیل آمین
corona	تاج	blast transformation	تبدیل بلاست
corona	تاج خورشید	aerosolization	تبدیل به افشانه کردن
strand	تار	gelatinization	تبدیل به ژلاتین شدن/کردن
spirochete	تار پیچان	weaponize	تبدیل به سلاح کردن
spinning	تارتنی	bioconversion	تبدیل زیستی
spinning	تار ریسی	atomizing	تبدیلسازی به ذرات ریز
30 nm chromatin fibre	تارکروماتینی 30 نانومتری	weaponization	تبدیل سازی به سلاح
fog, turbidity	تاری	transposition	تبدیل شدن
life history	تاریخچه زندگی (از لقاح تا مرگ)	chromatin modification	تبدیل کروماتینی
expiration	تاریخ مصرف	transformant	تبدیل کننده
sex pilus	تاژک جنسی	nitrification	تبدیل نیتروژنی
flagella	تاژکها	pyrogen	تب زا
facilitated folding	تاشدگی تسهیل شده	pulse shape discrimination	تبعیض شکل ضربه

discriminator	تبعیض کننده
tuberculosis	تب لازم
crystallization	تبلور
viral hemorrhagic fever	تب ناشی از خونریزی ویروسی
enzyme denaturation	تبه گونی آنزیم
tetraploid	تتراپلوئید
tetrad	تتراد
tetrasome	تترازوم
tetrasomic (= tetrasome)	تترازومی
tetracycline	تتراسایکلین
tetralogy of Fallot	تترالوژی فالوت
tetrahydrofolic acid	تترا هیدروفولیک اسید
tetro-hydro-folate	تترو-هیدرو-فولات
tektin	تتکین
immobilization	تثبیت
nitrogen fixation	تثبیت ازت
proprietary	تجارتی
homology	تجانس
regeneration, resuscitation	تجدید حیات
empricial	تجربی
dialysis, fragmentation, lysis	تجزیه
hydrolysis	تجزیه آبی
enzyme analysis	تجزیه آنزیم
catabolic, lytic	تجزیه ای
biodegradation	تجزیه بیولوژیکی
pedigree analysis	تجزیه شجره ای
breakdown	تجزیه شدن/کردن
assay	تجزیه شیمیایی
atomizing	تجزیه کردن
analysis of variance	تجزیه واریانس
catabolism	تجزیه و تخریب مواد
ionization	تجزیه یونی
gene expression	تجلی ژن (ی)
accumulation, agglomeration, aggregation, concentration	تجمع
critical micelle concentration	تجمع بحرانی ذرات کلوئیدی
critical micelle	تجمع بحرانی میسل ها

concentration	
platelet aggregation	تجمع پلاکت
spontaneous assembly	تجمع فوری/خودبخودی
swarming	تجمع کردن
human equivalent concentration	تجمع هم ارز انسانی
equipment	تجهیزات
Croplands equipment	تجهیزات کراپلند
pressure	تحت فشار قرار دادن
vagility	تحرک
excitation, initiation	تحریک
irritability	تحریک پذیری
excitatory	تحریک شونده
excitatory	تحریکی
probe (= DNA probe)	تحقیق
linkage analysis	تحلیل اتصال
RFLP linkage analysis	تحلیل اتصال آر.اف.ال.پی.
pest risk analysis	تحلیل احتمال خطر آفت
comparative analysis	تحلیل تطبیقی
expression analysis	تحلیل تظاهر
gene expression analysis	تحلیل تظاهر ژن (ی)
nearest neighbor sequence analysis	تحلیل توالی نزدیکترین همسایه
deterministic analysis	تحلیل جبر باورانه
metabolic flux analysis	تحلیل جریان سوخت و ساز
DNA analysis	تحلیل دی.ان.ای .
population viability analysis	تحلیل زیستایی جمعیت
southern blot analysis	تحلیل ساترن بلات
metabolic flux analysis	تحلیل سیلان متابولیکی
population viability analysis	تحلیل قابلیت زیست جمعیت
gene function analysis	تحلیل کار ژن (ی)
gene function analysis	تحلیل کارکرد ژن (ی)
protein interaction analysis	تحلیل کنش متقابل پروتئین
qualitative spectrometric analysis	تحلیل کیفی اسپکترومتری
qualitative spectrometric analysis	تحلیل کیفی به کمک طیف سنج
deterministic analysis	تحلیل گمارشی
serial analysis of gene expression	تحلیل متوالی تظاهر ژن
restriction analysis	تحلیل محدودیت

53

transcript analysis	تحلیل نسخه	fermentation	تخمیر
analysis of variance	تحلیل واریانس	stormy fermentation	تخمیر طوفانی
lytic	تحلیلی	microbial fermentation	تخمیر میکروبی
tolerance	تحمل	assessment	تخمین
aluminum tolerance	تحمل آلومینیم	needs assessment	تخمین مایحتاج
acquired tolerance	تحمل اکتسابی	RNA interference	تداخل آر.ان.ای.
drought tolerance	تحمل خشکسالی	interference	تداخل عمل
drug tolerance	تحمل دارو	indirect interaction	تداخل غیر مستقیم
cold tolerance	تحمل سرما	consistency	تداوم
salinity tolerance	تحمل شوری	device	تدبیر
cross tolerance	تحمل متقاطع	chronic, insidious	تدریجی
salt tolerance	تحمل نمک	formulation	تدوین
evolution, transition	تحول	threonine	ترئونین
cancer epigenetics	تحولات پیدایشی سرطان	transfection	ترا آلودگی
delivery	تحویل	acute transfection	ترا آلودگی حاد
gene delivery	تحویل ژن (ی)	percutaneous, transdermal	تراپوستی
actinobacillosis	تخته زبانی	teratogen (= teratology)	تراتوژن
narcosis	تخدیر	teratogenesis	تراتوژنز
denaturation	تخریب	teratology	تراتولوژی
mold deterioration	تخریب کپک	teratoma	تراتوما
specialization	تخصص	translocation	تراجایی
quantom speciation	تخصیص کمی	gene translocation	تراجایی ژن (ی)
mitigation	تخفیف	sequence	ترادف
evacuation	تخلیه	high-frequency transduction	ترارسانی پر بسآمد
zygote	تخم	signal transduction	ترارسانی علامتی
zygote	تخم بارور	generalized transduction	ترا رسانی کلی
capsule	تخمدان	abortive transduction	ترارسانی ناموفق
spore	تخم قارچ	specialized transduction	ترارسانی ویژگی یافته
ovum, zygote	تخمک	transformation	تراریخت سازی
oogenesis	تخمک زایی	transformation	ترا ریختی
breeding	تخم کشی	level	ترا
inbreeding	تخم کشی بسته	trophic level	تراز پروره ای
marker-assisted breeding	تخم کشی به کمک نشانه	balance	ترازو
breed	تخم کشی کردن	microbalance	ترازوی حساس
oocyte	تخمک نابالغ	bench-scale	ترازوی میزی
oocyte	تخمک نارس	microbalance	ترازوی میکرو
oocyte	تخمک یاخته	transgene	تراژن
zygote	تخم گشنیده	transgenic	تراژنی
spore	تخم میکروب	proteome chip	تراشه پروتئوم/محتوای پروتئین سلولی
oocyte	تخم نرسیده		
oocyte	تخم یاخته	protein chip	تراشه پروتئین

DNA chip	تراشه دی.ان.ای.	nsferase	لیستئین-کلسترول-آسیل
microfluidic chip	تراشه ریز سیال	hypoxanthine phospho	ترانسفراز هیپوزانتین فسفو
biochip	تراشه زیستی	ribosyl transferase	ریبوزیل
gene chip	تراشه ژن (ی)	hypoxantine-guanine	ترانسفراز هیپوزانتین-گوانین
transfer	ترافرست	phospho-ribosyl	فسفو-ریبوسیل
illegal traffic	ترافیک غیر قانونی	transferase	
transplantation	تراکاشت	transferin, transferrin	ترانسفرین
dialysis	تراکافت	bacterial transformation	ترانسفورمیشن باکتریائی
dialysis	تراکافتی	transcapsidation	ترانس کپسیداسیون
accumulation,	تراکم	transketolase	ترانس کتولاز
agglomeration,		transcriptase	ترانسکریپتاز
aggregation,		RNA transcriptase	ترانسکریپتاز آر.ان.ای.
concentration, density		reverse transcriptase-PCR	ترانسکریپتاز-پی.سی.آر.
threshold concentration	تراکم آستانه ای		معکوس
platelet aggregation	تراکم پلاکت	reverse transcriptase	ترانسکریپتاز معکوس
biomass	تراکم حیوانات زنده	transcriptome	ترانسکریپتوم
deagglomeration	تراکم زدایی	transcutaneous	ترانسکوتانوس
biomass	تراکم زیست	(percutaneous)	
densitometer	تراکم سنج	transgalacto-oligosac-	ترانس گالاکتو-اولیگوساکارید
reference concentration	تراکم مرجع	charides	
benchmark concentration	تراکم معیار	transposition	ترانهش
human equivalent	تراکم هم ارز انسانی	infiltration, osmosis	تراوش
concentration		osmosis	تراوندگی
diapedesis	تراگذری	secretor	تراونده
transmission	تراگسیل	cholinergic	تراونده استیل کولین
transmission of infection	تراگسیل آلودگی	infiltrate	تراویدن
neurotransmitter	تراگسیلنده عصبی	turbidimetry	تربیدیمتری
transversion	تراگشت	trypsinogen	تریپسینوژن
terramycin	ترامایسین	terpene	ترین
transaldolase	ترانس آلدولاز	terpenoid	ترین نما
transaminase	ترانس آمیناز	sequence	ترتب
transport mechanism	ترانسپورت مکانیسم	sequence	ترتیب
transposase	ترانسپوزاز	amino acid sequence	ترتیب اسید آمینه
transposon	ترانسپوزون	translation	ترجمه
specialized transduction	ترانسداکشن تخصیص یافته	cell-free translation	ترجمه بدون سلولی
transgene	ترانس ژن	reciprocal translation	ترجمه دوطرفه
transgenesis	ترانس ژنز	transposition	ترجمه شدن
transgenosis	ترانسژنوز	nick translation	ترجمه شکاف
transferase	ترانسفراز	reversion	ترجمه مجدد
terminal transferase	ترانسفراز پایانه	vectorial translation	ترجمه ناقلی
lecithin-cholesterol-acyl-tra	ترانسفراز	acidotropic	ترشاپرور

English	Persian	English	Persian
acidophilic	ترشا خواهی	regeneration	ترمیم اندام
salicylic acid	ترشای بید	base excision repair	ترمیم برداشت بازی
oleic acid	ترشای روغن	nucleotide excision repair	ترمیم برش نوکلئوتیدی
suppuration	ترشح چرک	DNA repair	ترمیم دی.ان.ای.
succus entericus	ترشح روده ای	methyl-directed mismatch repair	ترمیم ناهمآهنگ توسط متیل
suppuration	ترشح ریم	mismatch repair	ترمیم ناهماهنگ
secretor	ترشح کننده	cell turnover	ترن اوور سلولی
conformation	ترکیب	exudative	ترواش کننده
amphoteric compound	ترکیب آمفوتریک	exudative	ترواشی
organic compounds	ترکیبات آلی	tropomyosin	تروپومیوزین
homologous compounds	ترکیبات هم رده	troponin	تروپونین
beta conformation	ترکیب بتا	tropism	تروپیسم
xenobiotic compound	ترکیب بیگانه زی	chemical terrorism	تروریزم شیمیائی
mutagenic compound	ترکیب جهش زا	bioterrorism	تروریزم میکروبی
amphoteric compound	ترکیب خنثی	bioterrorist	تروریست میکروبی
xenobiotic compound	ترکیب دگرزی	thrombosis, thrombus, trombose	ترومبوز
binary compound	ترکیب دوتایی		
toxin	ترکیب زهردار	coronary thrombosis	ترومبوز کرونر
synthesizing	ترکیب سازی	thrombomodulin	ترومبومودولین
morphology	ترکیب شناسی	thrombin	ترومبین
physiologically active compound	ترکیب فعال فیزیولوژیکی	dissemination	ترویج
		disseminator	ترویج کننده
chiral compound	ترکیب کایرال	trehalase	تر هالاز
synthesizer	ترکیب کننده	trehalose	ترهالوز
gauche conformation	ترکیب گاش	medium chain triacyglyceride	تریاسیگلیسرید های زنجیره محیط
recombination	ترکیب مجدد		
meso compound	ترکیب مزو (میانی)	triacylglycerol	تری اسیل گلیسرول
host vector system	ترکیب ناقل-میزبان	triacyglycerides	تریاسیل گلیسرید ها
teratogenic compound	ترکیب ناهنجاری زایی	tryptophan	تریپتوفان
coordination compound	ترکیب هم آرا	trypsin	تریپسین
teratogenic compound	ترکیب هیولا سازی	triploid	تریپلوئید
combinatorial	ترکیبی	trisomy	تریزومی
bacteremia	ترکیز خونی	x-trisomy (= 47XXX)	تریزومی اکس.
bacteria	ترکیزگان	partial trisomy	تریزومی ناقص
aerobic bacteria	ترکیزگان هوازی	inositol 1,4,5-tri-phosphate	تری فسفات-اینوزیتول،4،1،5
bacterium	ترکیزه	cytidine triphosphate	تریفسفات سایتیدین
phage	ترکیزه خوار	uridine tri-phosphate	تری فسفات یوریدین
bacteremia	ترکیزه خونی	trichothecene	تریکوتسن
bactericide	ترکیزه کش	trichothecene mycotoxin	تریکوتسن مایکوتوکسین
thermus aquaticus	ترموس آکواتیکوس	trichothene mycotoxin	تریکوتن مایکوتوکسین
compensation, restoration	ترمیم		

tricothene mycotoxin	تریکوتن مایکوتوکسین
trichoderma harzianum	تریکو درما هارزیانوم
tricho-rhino-phalangeal	تریکو-رینو-فالانژیال
trichosanthin	تریکو سانتین
triglyceride	تری گلیسرید
medium chain triglyceride	تری گلیسرید دارای 8-10 اتم کربن .
hepatic triglyceride-lipase	تریگلیسرید لیپاز هپاتیکی
glyphosate-trimesium	تریمزیوم گلیفوسات
trihybrid	تری هایبرید
injector	تزریق کننده
testosterone	تستوسترون
mitigation	تسکین
palliative	تسکین دهنده
sequence	تسلسل
homology	تشابه
DNA fingerprinting (DNA profiling)	تشخیص اثر انگشت دی.ان.ای.ی
retrospective diagnosis	تشخیص بازنگرانه
retrospective diagnosis	تشخیص پس نگر
prenatal diagnosis	تشخیص پیش از تولد
prenatal diagnosis	تشخیص پیش زایشی
DNA typing	تشخیص تیپ دی.ان.ای.
sex determination	تشخیص جنسیت
DNA diagnosis	تشخیص دی.ان.ای.
nanopore detection	تشخیص ریز منفذ
rapid microbial detection	تشخیص سریع میکروب
pulse shape discrimination	تشخیص شکل ضربه
DNA typing	تشخیص هویت از راه آنالیز دی.ان.ای.
surface plasmon resonance	تشدید پلاسمون سطحی
dissection	تشریح
protocol	تشریفات
irradiation, radiation	تشعشع
implant radiation	تشعشع درون کاشت
adaptive radiation	تشعشع سازشی
microbial mat	تشکچه میکروبی
occurrence	تشکیل
spermiogenesis	تشکیل اسپرم
oogenesis	تشکیل تخمک یا اووم
ulceration	تشکیل زخم

organogenesis	تشکیل عضو
gametogenesis	تشکیل گامت
speciation	تشکیل گونه
thrombosis	تشکیل لخته
coronary thrombosis	تشکیل لخته درون سرخرگهای قلب
proprietary	تصاحب گرانه
accident	تصادف
stochastic	تصادفی
seizure	تصرف
filtration	تصفیه
wastewater treatment	تصفیه پساب/فاضلاب
membrane filtration	تصفیه غشایی
millipore filteration	تصفیه میلیپور
arteriosclerosis, atherosclerosis	تصلب شرائین
cystic fibrosis	تصلب کیستی بافتها
antagonism	تضاد
dilution	تضعیف
attenuated	تضعیف شده
attenuation	تضعیف کردن
process validation	تضمین فرآیند
beta conformation	تطابق بتا
conformation	تطبیق
adaptor	تطبیق دهنده
gauche conformation	تطبیق گاش
expression	تظاهر
protein expression	تظاهر پروتئین
gene expression	تظاهر ژن (ی)
homeostasis	تعادل
hysteresis	تعادل رطوبتی دوگانه
genetic equilibrium	تعادل ژنتیکی
mutualism	تعاون
population	تعداد
variable number of tandem repeats	تعدادمتغیر تکرار های توام
inhibition, mitigation	تعدیل
biomodulator	تعدیل کننده زیستی
small ubiquitin-related modifier (SUMO)	تعدیل کننده کوچک وابسته به یوبیکویتین
putrefaction	تعفن

aerosolize(d)	تعلیق مایع/جامد بصورت گرد/گاز در هوا	glycoprotein remodeling	تغییر مدل گلیکوپروتئین
gap repair	تعمیر بینابینی	locomotion	تغییر مکان
gene repair	تعمیر ژن (ی)	residue	تفاله
substitution, switching	تعویض	diversity	تفاوت
sequencing	تعیین توالی	protocol	تفاهم نامه
determinant, epitope	تعیین کننده	gene subtraction	تفریق ژن (ی)
antigenic determinant, epitope	تعیین کننده پادگنی	annotation	تفسیر
trophic	تغذیه ای	differentiation, dissociation, segregation (= separation)	تفکیک
malnutrition	تغذیه ناقص	hollow fiber separation	تفکیک فیبر میان تهی
conversion, mutation, variation	تغییر	independent assortment	تفکیک مستقل
climate change	تغییر آب و هوا	gun	تفنگ (ی)
antigenic switching	تغییر آنتی ژن	demand	تقاضا
somaclonal variation	تغییرات همسانه بدنی/سوماکلونال	test cross	تقاطع آزمونی
		reciprocal crosses	تقاطع دوطرفه
idiovariation	تغییر ارثی	crossing-over	تقاطع کروموزومی
metaplasia	تغییر بافت	crossing-over	تقاطع و تبادل
labile	تغییر پذیر	crossing of sibling	تقاطع هم نیا ئی
variance	تغییر پذیری	distribution	تقسیم
post-translational modification of protein	تغییر پروتئین پساترجمه	cleave	تقسیم اول/دوم سلول تخم
post-translational modification	تغییر پسا ترجمه ای	nascent cleavage	تقسیم در حال شکل گیری
		mentation	تقسیم دی.ان.ای.
somatic variant	تغییر تنی	mitosis	تقسیم رشتمانی
compensating variation	تغییر جبرانی	premature centromere division	تقسیم سانترومر نابهنگام/زود هنگام
frameshift	تغییر چارچوب		
genetically modified	تغییر داده ژنتیکی	cell division	تقسیم سلولی
modifier	تغییر دهنده	breakdown	تقسیم شدن/کردن
biological response modifier	تغییر دهنده واکنش بیولوژیکی	meiosis, reduction division	تقسیم کاهشی
		cell division	تقسیم یاخته
genetic modification	تغییر ژنتیکی	fragmentation	تقطیع
transformation	تغییر شکل	denaturation, inversion	تقلیب
bacterial transformation	تغییر شکل باکتریائی	mimetics	تقلیدی
transformant	تغییر شکل دهنده	amplification, enhancement	تقویت
biotransformation	تغییر شکل زیستی	potentiates	تقویت اثر
glycoprotein remodeling	تغییر طراحی گلیکو پروتئین	immuno-enhancing	تقویت ایمنی
chromatin modification	تغییر کروماتینی	rapid cycle DNA amplification (RCDA)	تقویت چرخه سریع دی.ان.ای
thermal denaturation	تغییر ماهیت حرارتی	cycloserine enrichment	تقویت سایکلوسرین
chromatin remodeling	تغییر مدل کروماتین	rapid amplification of cDNA ends	تقویت سریع پایانه های سی.دی.ان.ای.

amplify	تقویت کردن	monoculture	تک کشت
chloramphenicol	تقویت کلرآمفنیکل	monoploid	تک لاد
amplification		haplophase	تک لاد چهر
promoter	تقویت کننده	haplotype	تک لاد مونه
enhancer	تقویت گر	lipid monolayer	تک لایه لیپید
univalent	تک ارزشی	monocotyledon	تک لپه ای
evolution	تکامل	monogenic	تک منشاء
co-evolution	تکامل توام	complementation	تکمیل
directed evolution	تکامل جهت دار	technology	تکنولوژی
in-vitro evolution	تکامل در محیط کشت	antisense	تکنولوژی آنتی سنس/ضد
molecular evolution	تکامل مولکولی	technology)anti-sense	سنس/غیر کد کننده/غیر
phyletic evolution	تکامل نژادی	technology(	قابل ترجمه
directed evolution	تکامل هدایت شده	recombinant DNA	تکنولوژی دی.ان.ای. نوترکیب
monoclonal	تک بنیانی	technology	
monomer	تک پار	biosensor technology	تکنولوژی زیست حسگر
monoecious	تک پایه	biotechnology	تکنولوژی زیستی
mono zygote	تک تخمکی	radioimmunotechnique	تکنیک ایمنی پرتویی
amplification, replication	تکثیر	stopped-flow technique	تکنیک جریان متوقف
gene conversion	تکثیر بیش از انتظار ژن (ی)	rDNA techniques	تکنیکهای آر.دی.ان.ای. ئی
plasmid amplification	تکثیر پلاسمیدی	pathogenesis	تکوین بیماری
gene amplification	تکثیر ژن (ی)	stuffer fragment	تکه استافر
amplify	تکثیر کردن	Okazaki fragment	تکه اوکازاکی
virus replication	تکثیر ویروسی	fragmentation	تکه تکه شدن
monoclonal	تک دودمانی	mentation	تکه سازی
replication	تکرار	monocyte	تک هسته
terminal repeat	تکرار پایانه	agranulocyte, mononuclear	تک هسته ای
long terminal repeat	تکرار پایانه بلند	uracil fragments	تکه های اوراسیل
inverted terminal repeat	تکرار پایانه معکوس	random fragments	تکه های تصادفی
trinucleotide repeats	تکرار تری نوکلئوتیدی	protoplast	تک یاخته
simple sequence repeat	تکرار توالی ساده	attenuation	تکیدگی
tandem repeats	تکرار توام/متوالی	affinity tag	تگ افینیتی
direct repeat	تکرار مستقیم	affinity tag	تگ خویشاوندی/قرابت
short tandem repeats	تکرارهای اتصال کوتاه	fragmentation, lysis	تلاشی
CA repeats (repeats of	تکرار های سی.ای.	backcross	تلاقی برگشتی
cytosine and adenine)		wide cross	تلاقی دور
single strand	تک رشته	integrated	تلفیقی
monogenic	تک زا	in vitro fertilization	تلقیح این ویترو
monogenic	تک زاد	artificial insemination	تلقیح مصنوعی
monogenic	تک ژن	artificial insemination by	تلقیح مصنوعی توسط دهنده
unicellular	تک سلولی	donor	
monosaccharide	تک قندی	sodium pump	تلمبه سدیمی

English	Persian
telocentric	تلوسنتریک
telophase	تلوفاز
telomere	تلومر
telomerase	تلومراز
ion trap	تله یونی
quadrupole ion trap	تله یونی چهارقطبی/کوادروپول
teliospore	تلیواسپور
exposure	تماس
casual contact	تماس اتفاقی
dermal contact	تماس پوستی
dermal contact	تماس جلدی
casual contact	تماس روزمره
casual contact	تماس عادی
casual contact	تماس غیر دائم
indirect contact	تماس غیر مستقیم
casual contact	تماس غیر مستمر
diploid	تمام دانه
total cell DNA	تمام دی.ان.ای. سلولی
differentiation	تمایز
cell differentiation	تمایز سلولی
differentiation	تمایز یابی
affinity	تمایل
codon bias	تمایل کدون
concentration	تمرکز
mechanical cleaning	تمیز کردن مکانیکی
lymph	تنابه
lymphatic	تنابه ای
periodicity	تناوب
virulent	تند
fitness	تندرستی
tachypnea	تند نفسی
blast cell	تنده یاخته
tachypnea	تندی نفس
depression	تنزل
strain	تنش
abiotic stress	تنش ابیوتیک
oxidative stress	تنش اکسایشی
biotic stress	تنش زیستی
abiotic stress	تنش عوامل فیزیکی محیط
formulation	تنظیم
cystic fibrosis	تنظیم کننده رسانائی بین

English	Persian
transmembrane conductance regulator	غشائی فیبروز سیستی
inhalation, respiration	تنفس
aerobic respiration	تنفس ایروبیک
tachypnea	تنفس سریع
cell respiration	تنفس سلولی
respirometer	تنفس سنج
physiology	تنکار شناسی
microbial physiology	تنکار شناسی میکروبی
vasoconstrictor	تنگ کننده عروق
secondary constriction	تنگای ثانویه
diversity, variance, variation, variety	تنوع
somatic variant	تنوع بدنی
improved variety	تنوع بهینه
somaclonal variation	تنوع تن تاگی
domestic animal diversity	تنوع حیوانات اهلی
biological diversity	تنوع زیست شناختی
agricultural biological diversity	تنوع زیست شناختی کشاورزی
agrobiodiversity	تنوع زیست-کشاورزی
biodiversity	تنوع زیستی
domestic biodiversity	تنوع زیستی حیوانات اهلی
agricultural biodiversity	تنوع زیستی کشاورزی
genetic diversity	تنوع ژنتیکی
cultural diversity	تنوع کشت
species diversity	تنوع گونه
cultural diversity	تنوع مزروعی
molecular diversity	تنوع ملکولی
isolation	تنها سازی
refractile body	تنه شکننده/منکسر کننده نور
somatic	تنی
heredity	توارث
heritability	توارث پذیری
quantitative inheritance	توارث صفات کمی
maternal inheritance	توارث مادری
matroclinal inheritance	توارث متمایل به مادر
Mendelian inheritance	توارث مندلی
mendelian inheritance in man	توارث مندلی در انسان
cytogenetics	توارث یاخته

convention, protocol	توافق نامه	short interfering RNA	توالی های مزاحم کوتاه آر.ان.ای.
advanced informed agreement	توافقنامه ادوانس اینفورم	small interfering RNA	توالی های مزاحم کوچک آر.ان.ای
advanced informed agreement	توافقنامه پیش آگاه	targeting sequence	توالی هدفگیر
cascade, sequence	توالی	autonomous(ly) replicating sequence	توالی همانند ساز خودمختار
address sequence	توالی آدرس	protein sequencer	توالی یابی پروتئین
alu sequence	توالی آلو در ملکول دی.ان.ای.	massively parallel signature sequencing	توالی یابی حجیم موازی امضائی
shotgun sequencing	توال یابی شات گانی	DNA sequencing	توالی یابی دی.ان.ای.
amino acid sequence	توالی اسید آمینه	whole-genome shotgun sequencing	توالی یابی شات گانی همه ژنوم
control sequence	توالی بازبینی/کنترل/نظارت/کاربری/تنظیم/بازرسی	concomitant, diploid	توام
base sequence	توالی بازها	aptitude, asset	توان
ecological succession	توالی بوم شناختی	competency	توانائی
terminator sequence	توالی پایان دهنده	heritability	توانائی ارثی
LI repeat sequence	توالی تکراری ال.آی	establishment potential	توان استقرار
regulatory sequence	توالی تنظیمی	capacity	توانایی
consensus sequence	توالی توافقی	phosphorylation potential	توان فسفات افزایی
autonomous consensus sequence	توالی توافقی خودمختار	potentiates	توانمند سازی
flanking sequence	توالی جانبی	totipotency	توانمندی
recognition sequence (site)	توالی (جایگاه) شناسائی	tuberculin	توبرکولین
palindromic sequence	توالی جناس قلب	tubulin	توبولین
insertion sequence	توالی درون جای دهی	particle cannon/gun	توپ ذره ای
trailer segment/sequence	توالی دنباله رو	solid	توپر
DNA sequence	توالی دی.ان.ای.	topoisomerase	توپوایزومراز
regulatory DNA sequence	توالی دی.ان.ای. تنظیمی	endothelium	توپوش
leader sequence	توالی راهنما	vascular endothelium	توپوش آوندی
coding sequence	توالی رمز گذار	agglomeration, aggregation, bulk	توده
Shine-Dalgarno sequence	توالی شاین-دالگارنو		
signal sequence	توالی علامتی	bulk	توده ای
upstream activator sequence (UAS)	توالی فعال گر بالادست	mole	توده (درون رحم)
flanking sequence	توالی کناری	stacked gene	توده ژنی
kozak sequence	توالی کوزاک	stacking	توده کردن
spinning cup protein sequencer	توالی گر پروتئین پیمانه گردان	niche	تورفتگی
palindromic sequence	توالی مساوی الطرفین	tumor	تورم
marker sequence	توالی نشانگر	taurocholate	توروکولات
intervening sequences	توالی های مداخله گر	taurine	تورین
		dispensing, distribution	توزیع
		sustainable development	توسعه پایدار

61

English	Persian	English	Persian
sustainable development	توسعه قابل دوام	growth, neoplasm, tumor	تومور
annotation	توضیح	oncogene	تومورزا
rational expectation	توقع عقلانی	cellular oncogene	تومور زایی یاخته ای
quick-stop	توقف فوری	oncology	تومورشناسی
quick-stop	توقف کوتاه	myeloma	تومور مغز استخوان
toxemia	توکسمی	wilms tumor	تومور ویلمز
tetanus toxoid	توکسوئید کزاز	hallucination	توهم
toxin	توکسین	invagination	توی خود برگشتی
marine toxin	توکسین دریایی	invagination	توی هم رفتگی
labile toxin	توکسین لابایل	benthos	ته
cholera toxin	توکسین وبا	invasive	تهاجمی
tocotrienols	توکوترینول ها	biological threat	تهدید زیست شناختی
tocopherols	توکوفرول	credible threat	تهدید قابل قبول
togavirus	توگاویروس	credible threat	تهدید موثق
togaviridae	توگاویروس ها	benthos	ته زی
toxoplasma	توگزوپلاسما	chlorine residual	ته مانده کلر
tularemia	تولارمی	settling	ته نشانی
ulceroglandular tularemia	تولارمی زخمی-غده ای	settling	ته نشستن
typhoidal tularemia	تولارمی شبه تیفوئید	residue	ته نشین
parturition	تولد	settling	ته نشین شدن
tolerance	تولرانس	settle	ته نشین کردن/شدن
acquired tolerance	تولرانس اکتسابی	settling	ته نشینی
generation, product	تولید	aeration	تهویه
suppuration	تولید چرک	DNA profiling (DNA fingerprinting)	تهیه نما/طرح دی.ان.ای.
abiogenesis	تولید خود به خود	T (= thymine)	تی.
spontaneous generation	تولید فوری/خودبخودی	TR (= terminal repeat)	تی.آر.
phyto-manufacturing	تولید فیتو	TRF (= telomere repeat factor), TRF (= thyroid releasing factor)	تی.آر.اف.
generating	تولید کننده		
generation	تولید مثل		
inbreeding	تولید مثل بین خودی	tRNA (= transfer RNA)	تی. آر.ان.ای.
sexual reproduction	تولید مثل جنسی	isoacceptor tRNA	تی.آر.ان.ای. ایزو آکسپتور
abiogenesis	تولید مثل خود به خودی	TRA (= T cell receptor alpha)	تی.آر.ای.
captive breeding	تولید مثل در اسارت		
asexual reproduction	تولید مثل غیر جنسی	TRP (= tricho-rhino-phalangeal)	تی.آر.پی.
breed	تولید مثل کردن		
vegetative reproduction	تولید مثل گیاهی	TIL (= tumor-infiltrating lymphocyte)	تی.آی.ال.
xenogenesis	تولید مثل متناوب		
replication	تولید مجدد	THF (= tetro-hydro-folate)	تی.اچ.اف.
no-tillage crop production	تولید محصول کشاورزی بدون زراعت	TS (= thymidylate synthase)	تی.اس.
clone	تولید مصنوعی	TSH (= thyroid stimulating	تی.اس.اچ.

hormone)		pyrophosphate)	
TSD (= Tay-Sachs disease)	تی.اس.دی.	titer	تیتر
TF (= transcription factor)	تی.اف.	TT (= tetanus toxoid)	تی.تی.
TF II (= transcription factor for control of RNA polymerase II)	تی.اف.2	T-T (= thymine-thymine dimer)	تی.-تی.
TFN (= transferrin)	تی.اف.ان.	TGF-β (= transforming growth factor-beta)	تی.جی.اف.بتا
Tm (= melting temperature)	تی.ام.	TDF (= testis determining factor)	تی.دی.اف.
tmRNA (= transfer-messenger RNA)	تی.ام.آر.ان.ای.	T-DNA	تی.دی.ان.ای.
TMV (= tobacco mosaic virus)	تی.ام.وی.	TDA (= thymus-dependent area)	تی.دی.ای.
thiamine	تیامین	TTD (= trucho-thio-dystrophy)	تی.دی.دی.
TNF (= tumor necrosis factor)	تی.ان.اف.	fog, turbidity	تیرگی
Tm (= melting temperature)	تی. اندیس ام.	turbidimetry	تیرگی سنجی
TOF (= tetralogy of Fallot)	تی.او.اف.	tyrosine	تیروزین
TACF (= telomer-associated chromosome fracionation)	تی.ای.اف.سی.	protein tyrosine kinase	تیروزین کیناز پروتئینی
		receptor tyrosine kinase	تیروزین کیناز گیرنده
		solanaceae	تیره سیب زمینی
TAP (= transporter associated with antigen presentation)	تی.ای.پی.	vulgaris	تیره گسترده لوبیاهای خوراکی
		TCR (= T-cell receptor)	تی.سی.آر.
		TCRA (= T-cell receptor)	تی.سی.آر.ای.
TAP tagging (= tandem affinity purification tagging)	تی.ای.پی. تگینگ	TCA (= tri-carboxyllic acid cycle)	تی.سی.ای.
		slide cover	تیغک
TAPVR (= total anomalous pulmonary venous return)	تی.ای.پی.وی.آر.	slide	تیغه
		TK (= thymidine kinase), Tk (= thymidine kinase)	تی.کی.
TBP (= TATA-binding protein)	تی.بی.پی.	thylakoid	تیلاکوئید
TBG (= thyroxine-binding-globuline)	تی.بی.جی.	thale cress	تیل کرس
		typhimurium (= salmonella tymphiumurium)	تیمفیموریوم
idiotype	تیپ ایده آل		
wild type	تیپ وحشی	salmonella typhimurium	تیمفیموریوم سالمونلا
TP (= thymidine phosphorylase)	تی.پی.	salmonella tymphiumurium	تیمفیوموریوم سالمونلا
		thymoleptics	تیمولپتیک
TPA (= tissue plasminogen activator), tPA (= tissue-type plasminogen activator)	تی.پی.ای.	thymidylate synthase	تیمیدیلات سینتاز
		thymidine kinase	تیمیدین کیناز
		thymine	تیمین
		tyndallization	تیندالیزاسیون
TPP (= thiamine	تی.پی.پی.	thioesterase	تیواستراز

thioredoxin	تیوردوکسین

ث

COOH (= carboxyl group)	ث.او.او.هاش.
consistency, stability	ثبات
genetic equilibrium	ثبات ژنتیکی
patent	ثبت شده
asset	ثروت

ج

switching	جابجائی
reading frame shift	جابجائی چارچوب قرائت
transboundary movement	جابجائی فرا مرزی/سرحدی
unintended transboundary movement	جابجائی ناخواسته بین مرزی/سرحدی
Robertsonian translocation	جابجاشدگی رابرتسونی
transposition	جا به جا شدگی
locomotion	جا به جایی
translocation	جابه جایی
gene translocation	جابه جایی ژن (ی)
chromosomal translocation	جا به جایی کروموزومی
embedding	جا دادن
attractant	جاذب
whiskers	جاروی کوچک
embedding	جاسازی کردن
solid	جامد
community	جامعه
climax community	جامعه اوج
climax community	جامعه بالیست
sociobiology	جامعه-زیست شناسی
organism	جاندار
biotype	جانداران هم نژاد
biota	جانداران یک پهنه
pathogenic organism	جاندار بیماری زا
organisms with novel traits	جاندار دارای نشانویژه های جدید
nontarget organism	جاندار غیر هدف
microorganism	جاندار میکروسکوپی
model organism	جاندار نمونه

substitution	جانشین سازی
base substitution	جانشین سازی باز
substitution	جانشینی
junctional sliding	جانکشنال اسلایدینگ
fauna	جانوران
parasite	جانور انگلی
transgenic animal	جانور تراژنی
gnotobiotic animal	جانور گنوتوبیوتیک
enzootic	جانورگیر
transgenic animal	جانور واریخته
topotaxis	جای آرایی
eschar, sequela	جای زخم
locus	جایگاه
A,P site	جایگاه ای.،پی.
peptidyle site (P-site)	جایگاه پپتیل
binding site	جایگاه پیوند
ribosome-binding site	جایگاه پیوند ریبوزوم
transcription factor binding site	جایگاه پیوند عامل رونویسی
CAP site	جایگاه سی.ای.پی.
recognition site	جایگاه شناسائی
catalytic site	جایگاه فروکافتی
Cos site	جایگاه کاس
active site	جایگاه کنش ور
sequence tagged site	جایگاه نشاندار شده با توالی
target site (= recognition site)	جایگاه هدف
isosemantic substitution	جایگزینی ایزوسمانتیکی
base substitution	جایگزینی باز
transversion	جایگزینی ناهمجنس
compensation	جبران
dosage compensation	جبران دوز(ی)
progenitor	جد
separation	جدائی
endometrium	جدار زهدان/رحم
apoenzyme	جدازیما
excision	جدا سازی
isolation	جداسازی
high-throughput screening	جدا سازی با ظرفیت بالا
transgressive segregation	جداسازی ترانسگرسیو
single cell isolation	جداسازی تک سلولی

English	Persian
membrane affinity seperation	جداسازی تمایل غشائی
mass screening	جداسازی توده ای
gel filtration	جداسازی ژلی
magnetic cell sorting	جدا سازی سلول مغناطیسی
uncoupling	جدا کردن
whirlpool separator	جداکننده گردابی
differentiation, dissociation, isolation	جدایش
cell differentiation	جدایش یاخته ای
dissociation, isolation	جدایی
absorption, affinity, assimilation, uptake	جذب
bulking	جذب آب در روده
dermal absorption	جذب از راه پوست
biological uptake	جذب بیولوژیکی
dermal absorption	جذب جلدی
biologic uptake	جذب زیستی
absorption, adsorption	جذب سطحی
dermal adsorption	جذب سطحی پوستی
absorbance	جذب کنندگی
anabolism, assimilation	جذب و ساخت
genetic assimilation	جذب و ساخت ژنتیکی
anabolism	جذب و هضم
exotic germplasm	جرم پلاست نابومی/بیگانه/خارجی
germplasm	جرم پلاسم
elite germplasm	جرم پلاسم نخبه/ممتاز/الیت/برگزیده
residue weight	جرم رسوبات
protoplast	جرم زنده
molecular mass	جرم ملکولی
drift, flow, flux	جریان
random genetic drift	جریان ژنتیکی تصادفی
gene flow	جریان ژن (ی)
gating current	جریان مهار کننده
N-segment	جزء ان.
fragmentation	جزء به جزء کردن
trailer segment/sequence	جزء دنباله رو
slow component	جزء کند
secretory component	جزء مترشح
CG island	جزیره سی.جی.
probe (= DNA probe)	جستجو
solid	جسم
analyte	جسم تجزیه ای
inclusion body	جسم درون بسته/گنجیده
chloroplast	جسم سبزینه ای
perikaryon	جسم سلولی
supernatant	جسم شناور
solid	جسم صلب
protein inclusion body	جسم میانبار پروتئینی
Pribnow box	جعبه پریبناو
TATA box (= Cis-acting sequence in RNA polymerase II promoters)	جعبه تی.ای.تی.ای.
GC box	جعبه جی.سی.
CAAT box	جعبه سی.ای.ای.تی.
Goldberg-Hogness box	جعبه گلدبرگ-هاگنس
Hogness box	جعبه هاگنس
biogeography	جغرافیای زیستی
placenta	جفت
base pair, bP (= base Pair), Bp = base pair	جفت باز
tertiary base pair	جفت باز ترشیاری/سه گانه
conjugation	جفت شدگی
gene linkage	جفت شدگی ژن (ی)
conjugate	جفت شدن
induced fit	جفت شدن القایی
interamolecular association	جفت شدن بین مولکولی
kilobase pair	جفت کیلو باز
breeding, mating	جفت گیری
panmixia	جفت گیری تصادفی
sibmating (= crossing of sibling)	جفتگیری هم نیائی
megabase pairs	جفت مگاباز
liver	جگر
liver	جگر سیاه
algicide	جلبک کش
algae	جلبک ها
cyanobacteria	جلبکهای سبز-آبی
case, skin	جلد

English	Persian	English	Persian
enhancer	جلوبرنده	surplus embryo	جنین اضافی
prevention	جلوگیری	embryology	جنین شناسی
immunocontraception	جلوگیری از بارداری از لحاظ ایمنی	surplus embryo	جنین مازاد
		barley	جو
gene silencing	جلوگیری نمایانی ژن (ی)	granulation tissue	جوانه بافتی
solid	جماد	germinate	جوانه زدن
company	جمع	gem	جواهر
additive	جمع پذیر	incubation	جوجه کشی
chiasmata	جمع کیاسما	genus, homogeneous, species	جور
community, population	جمعیت		
demography	جمعیت شناسی	variation, xenogenesis	جوراجوری
receptor population	جمعیت گیرنده	homograft	جور پیوند
vagility	جنبش	homozygote, homozygous	جور تخم
enzyme kinetics	جنبش شناسی آنزیمی	homothalism	جور ریسگی
kinetic	جنبشی	isozyme	جور زیما
paravertebral	جنب مهره ای	isozyme	جور زیمایه
vagile	جنبنده	strain	جوره
genus	جنس	disabled strain	جوره خنثی شده
gynandromorph	جنسیت آمیخته	inbred strain	جوره درون آمیخته
homogametic sex	جنسیت جور زامه ای	homokaryon	جور هسته
heterogametic sex	جنسیت ناجور زامه ای/هتروگامتی	Abelson strain of murine	جوره (مربوط به جوندگان) آبلسونی
homogametic sex	جنسیت یک جور/هوموگامتی	live vaccine strain	جوره واکسن زنده
warfare	جنگ	homozygous	جور یوغ
munition, weapon	جنگ افزار	rash	جوش
mass-casualty biological weapon	جنگ افزار بیولوژیکی کشتار انبوه	meteorology	جو شناسی
		oxygen deficient atmosphere	جو کم اکسیژن
multi-agent munition	جنگ افزار چند عاملی		
submunition	جنگ افزار خوشه ای/فرعی	tannin	جوهر دباغی
bioweapon	جنگ افزار میکروبی	caffeine	جوهر قهوه
biological warfare	جنگ بیولوژیکی	citric acid	جوهر لیمو
chemical warfare	جنگ شیمیائی	tannin	جوهر مازو
primary forest	جنگل اولیه	palmitic acid	جوهر نخل
silviculture	جنگل پروری	atmospheric	جوی
secondary forest	جنگل ثانویه	genitourinary tract	جهاز تناسلی-ادراری
agroforestry	جنگل داری زراعی	mutation	جهش
natural forest	جنگل طبیعی	loss-of-function mutation	جهش از بین برنده کارآئی
silviculture	جنگل کاری	acquired mutation	جهش اکتسابی
biological warfare, biowarfare	جنگ میکروبی	amber mutation	جهش امبر
		suppressor mutation	جهش بازدارنده
embryo	جنین	gain of function mutation	جهش بدست آوردن کاربری

English	Persian
leaky mutation	جهش بی صدا
nonsense mutation	جهش بی معنی
back mutation	جهش پسرو
forward mutation	جهش پیش رو
frameshift mutation	جهش تغییر چارچوب
single-site mutation	جهش تک جایگاهی
deletion mutation	جهش حذفی
temperature-sensitive mutation	جهش حساس به دما
silent mutation	جهش خاموش
leaky mutation	جهش خفیف
mutate	جهش دادن
down promoter mutation	جهش داون پروموتر
insertion mutation	جهش دخولی
frameshift mutation	جهش دگر قالب
insertion mutation	جهش رخنه ای
mutagen	جهش زا
mutagenesis, mutagenicity	جهش زایی
site-directed mutagenesis	جهش زایی جهت یافته
directed mutagenesis	جهش زایی هدایت شده
genetic mutation	جهش ژنتیکی
gene mutation	جهش ژن (ی)
silent mutation	جهش ساکت
samesense mutation	جهش ساکت/هم معنی
leaky mutation, silent mutation	جهش ساکن
somatic mutation	جهش سوماتیکی/بدنی
polar mutation	جهش قطبی
polarity mutation	جهش قطبیتی
mutate	جهش کردن
lethal mutation	جهش کشنده
down promoter mutation	جهش کم کننده نسخه برداری
back mutation	جهش معکوس
missense mutation	جهش نادرست
point mutation	جهش نقطه ای
constitutive mutation	جهش نهادی
homeotic mutation	جهش هومئوتیک
mutate	جهش یافتن
mutant	جهش یافته
idiotrophic mutant	جهش یافته ایدیو تروفیک/گزین پرور

English	Persian
chain terminating mutant	جهش یافته پایان بخش زنجیره
lethal mutant	جهش یافته کشنده
conditional lethal mutant	جهش یافته کشنده شرطی
leaky mutant	جهش یافته ناقص
G (= Gibbs free energy), G (= guanine)	جی.
G1 (= G1 phase of the cell cycle)	جی.1
G1P (= glucose-1-phosphate)	جی.1 پی.
G2 (= G2 phase of the cell cycle)	جی.2
G3P (= glyceraldehyde-3-phosphate)	جی.3پی.
G6P (= glucose-6-phosphate)	جی.6پی.
G6PD (= glucose-6-phosphate dehydrogenase)	جی.6پی.دی.
GE (= gene expression)	جی.ئی.
GH (= growth hormone)	جی.اچ.
GS (= glutamine synthetase)	جی.اس.
GSH (= glutathione (reduced form))	جی.اس.اچ.
GSSG (= glutathione disulfide (oxidized form))	جی.اس.اس.جی.
GSD (= glutathione synthetase deficiency)	جی.اس.دی.
GLC (= gas-liquid chromatography)	جی.ال.سی.
GM-CSF (= granulocyte-macrophage colony-stimulating factor)	جی.ام.-اس.اف.
GMF (= genetically modified food)	جی.ام.اف.
GMP (= guanosine mono-phosphate)	جی.ام.پی.
GMT (= geometric mean titer)	جی.ام.تی.

English	Persian
GNRP (= guanine nucleotide releasing protein)	جی.ان.آر.پی.
GABA (= gamma-aminobutyric acid)	جی.ای.بی.ای.
GAP (= GTPase activating protein)	جی.ای.پی.
gibberella zeae	جیبرلا زیا
gibberellin	جیبرلین
GBY (= gonado blastoma Y)	جی.بی.وای.
GPI (= glucose phosphate isomerase)	جی.پی.آی.
GT (= gene therapy)	جی.تی.
GTF (= general transcription factor), GTF (= glocosyl transferase factor)	جی.تی.اف.
GTP (= guanosine tri-phosphate)	جی.تی.پی.
GTPase (= guanosine tri-phosphatase)	جی.تی.پی.آز
JGS (= juvenile galactosia lidosis)	جی.جی.اس.
G-DNA	جی.- دی.ان.ای.
GDP (= guanosine di-phosphate)	جی.دی.پی.
GC-MS (= combined gas chromatography-mass spectrometre)	جی.سی.-ام.اس.
GCP (= good clinical practice)	جی.سی.پی.
GCPS (= Grieg's cephalo polysyndactyly syndrome)	جی.سی.پی.اس.
g+	جی مثبت
mercury	جیوه

چ

English	Persian
machine	چارا

English	Persian
open reading frame	چار چوب باز خوانی باز
reading frame	چارچوب قرائت
coordinated framework for regulation of biotechnology	چارچوب هماهنگ برای تنظیم زیست فن شناسی
tetraploid	چارلا
primer	چاشنی
fat(s)	چاق
glottis	چاکنای
chalone	چالون
triticale	چاودم
levorotary	چپ بر
levorotary	چپ گرد
levorotary	چپ گردان
levorotary	چپ گردانی
levorotary	چپ گردش
adipose, fat(s)	چرب
fat(s), lipid (= fat)	چربی
monounsaturated fat	چربی تک اشباع نشده
liposome	چربی تن
lipophilic	چربی خواه
adipose	چربی دار
lipophilic	چربی دوست
lipogenesis	چربی زایی
lipolysis	چربی کافت (ی)
sebum	چربی مترشحه
medium chain saturated fats	چربی های اشباع شده زنجیره محیط
spinning	چرخاندن
machine	چرخ کردن
nitrogen cycle	چرخه ازت
tricarboxylic acid cycle	چرخه اسید تری کاربوکسیل
tri-carboxyllic acid cycle	چرخه اسید تری-کربوکسیلیک
citric acid cycle	چرخه اسید سیتریک
urea cycle	چرخه اوره
cyclic	چرخه ای
open circular	چرخه ای باز
interphase cycle	چرخه اینترفاز
futile cycle	چرخه بی ثمر
multiplication cycle	چرخه تکثیر
periodicity	چرخه تولید مثل

life cycle	چرخه زندگی	multiplex, multivalent	چند تایی
cell cycle	چرخه سلولی	pleiotropic	چند جانبه
futile cycle	چرخه عبث	multipotent	چند خاصیتی
lytic infection cycle	چرخه عفونت کافتی	pleiotropic	چند رخ
lysogenic infection cycle	چرخه عفونت لیزوژنیک	pleiotropy	چند رخی
phosphatidylinositol cycle	چرخه فسفاتیدیلینوزیتول	polyribosome	چند رناتن
lytic cycle	چرخه کافتی	polymorphism	چند ریختی
Calvin cycle	چرخه کالوین	single-nucleotide	چند ریختی تک نوکلئوتید
krebs cycle	چرخه کربس	polymorphism	
Cori cycle	چرخه کوری	single nucleotide	چند ریختی تک نوکلئوتیدی
nitrogen cycle	چرخه نیتروژن	polymorphism	
cell cycle	چرخه یاخته ای	DNA polymorphism	چند ریختی شدن دی.ان.ای.
pus	چرک	polygenic	چند زاد
pyuria	چرک شاشی	multigenic	چند ژنی
sepsis	چرکی شدگی	polygenic	چند ژنی
flaccid	چروکیده	multiple sclerosis	چند سختینگی
adhesive	چسب	polycistronic	چند سیسترونی
collagen	چسب زا	multiplex, polymorphism	چند شکلی
colloidal	چسب سان	DNA polymorphism	چند شکلی شدن دی.ان.ای.
colloid	چسب مانند	simple sequence length	چندشکلی طولی توالی ساده
agglutination	چسبندگی	polymorphism	
viscosity	چسبندگی	restriction fragment length	چند شکلی طولی قطعات
colloidal	چسبی	polymorphism (RFLP)	محدود کننده
port	چشم	multivalent	چند ظرفیتی
line-source	چشمه خطی	multifactorial	چند عاملی
droplet	چکه	multiplex	چند عضوی
encapsulated	چکیده	polysaccharide	چند قنده
condenser	چگالنده	polysaccharide	چند قندی
density	چگالی	non-starch polysaccharide	چند قندی بدون نشاسته/فاقد
densitometer	چگالی سنج		نشاسته
buoyant density	چگالی شناوری	pluripotent	چند قوه زا
cruciferae	چلیپائیان	polyploid	چندگان
clubbing	چماقی شدن	multiplex	چندگانه
lawn	چمن	pleiotropic	چند گرا
coiled coil	چنبر (ه) پیچیده	pleiotropy	چند گرایی
plectonemic coiling	چنبره پیچیده رشته	polymorphism	چند گونگی
random coil	چنبره تصادفی	polyploid	چند لاد
pleiotropic	چند اثر	granulocyte	چند هسته ای
polyvalent	چند ارزشی	cancer	چنگار
multiplex	چند بخشی	chelation	چنگک سازی
oligomer, telomere	چند پار	wand	چوب جادوئی

English	Persian
pungi stick	چوب خاردار/پانجی
saponin	چوبک
lignin	چوب مایه
lignin	چوبینه
overlapping reading frame	چهارچوب بازخوانی همپوش
reading frame	چهارچوب خواندن
tetraploid	چهار لا
protein folding	چین خوردگی پروتئین
crop	چینه دان

ح

English	Persian
event	حادثه
annotation	حاشیه
fertilization	حاصلخیزسازی
fecundity, fertility	حاصلخیزی
transition state	حالت انتقال
parasitism	حالت انگلی
ground state	حالت حداقل انرژی
transition state	حالت عبور
emergency	حالت فوق العاده
transition state	حالت گذار
transition state	حالت میانی
transition state	حالت واسطه
Holliday mode	حالت هالیدی
carrier, transfer, vector, vehicle	حامل
baculovirus expression vector (BEV)	حامل اکسپرشن باکولو ویروس
electron carrier	حامل الکترون
electron carrier	حامل الکترونی
bifunctional vector	حامل دو کاربردی/بایفانکشنال
DNA vector	حامل دی.ان.ای.
gestation	حاملگی
molecular vehicle	حامل ملکولی
cyst	حباب
vesicle	حبابچه
vacuoles	حباب ها
volume	حجم
minute volume	حجم در دقیقه
specific volume	حجم مشخص

English	Persian
bulk	حجیم
level	حد
safe minimum standard	حداقل استاندارد ایمنی
minimal risk level	حداقل سطح ریسک
minimum tillage	حداقل کشت و زرع
ceiling value	حد اکثر ارزش
maximum sustainable yield	حداکثر برداشت پایدار
maximum sustainable yield	حداکثر برداشت معقول
maximum residue level	حداکثر پس مانده مجاز
maximum permissible concentration	حداکثر تجمع مجاز
maximum enzyme velocity	حداکثر سرعت آنزیم
maximum contaminant level	حد اکثر سطح آلاینده
maximum permissible concentration	حداکثر غلظت مجاز
stringency	حدت
virulent	حدت دار
nontranscribed spacer	حد فاصل نسخه برداری نشده
ceiling limit	حد مجاز
standard	حد مطلوب
ceiling value	حد نهائی ارزش
deletion, excision	حذف
allelic exclusion	حذف آللی
intercalary deletion	حذف اینترکالاری
intercalary deletion	حذف تداخلی
intercalary deletion	حذف میان بافتی
partial monosomy	حذف نسبی
temperature	حرارت بدن
optimum temperature	حرارت بهینه
thermal	حرارتی
character	حرف
locomotion	حرکت
sense	حس
temperature-sensitive	حساس به دما
sensitization	حساس سازی
allergy, sensitivity	حساسیت
photosensitivity	حساسیت به نور
desensitization	حساسیت زدائی
genetic sensitivity	حساسیت ژنتیکی
quorum sensing	حس حد نصاب

immunosensor	حسگر ایمنی
lipid sensor	حسگر چربی
genosensor	حسگر ژنی
lipid sensor	حسگر لیپید
microbial sensor	حسگر میکروبی
sense	حسی
insect	حشره
Asian corn borer	حشره آفت ذرت آسیائی
corn borer	حشره ذرت
Asian corn borer	حشره ذرت آسیائی
European Corn Borer, pyralis	حشره ذرت اروپایی
coffee berry borer	حشره سنجد تلخ
insecticide, pesticide	حشره کش
nadir	حضیض
protection of human health and environment	حفاظت از بهداشت و محیط زیست انسان
conservation of biodiversity	حفاظت تنوع زیستی
ex-situ conservation of farm animal genetic diversity	حفاظت تنوع ژنتیکی دام و طیور در خارج از محل
in-situ conservation of farm animal genetic diversity	حفاظت تنوع ژنتیکی دام و طیور در محیط طبیعی
ex-situ conservation	حفاظت در خارج از محل
in-situ conservation	حفاظت در محل
conservation of farm animal genetic resource	حفاظت ژنتیکی دام و طیور
environmental protection	حفاظت محیط زیست
conservation	حفاظت منابع طبیعی/محیط زیست
biological shield	حفاظ زیستی
cell, lumen	حفره
ozone hole	حفره ازن
peritoneal cavity	حفره صفاقی
vacuoles	حفره ها
conserved	حفظ شده
patent	حق انحصاری اختراع
breeder's rights	حقوق اصلاحگر/پرورنده
central dogma	حکم اساسی
chakrabarty decision	حکم چاکرابارتی
solvation	حلال پوشی
solvated	حلال پوشیده

solvalysis	حلال کافت
soluble	حل شدنی
nasopharynx	حلق-بینی
nasopharynx	حلق و بینی
displacement loop	حلقه جابجائی
rolling circle	حلقه رولینگ
z-ring	حلقه زد
hairpin loop	حلقه سنجاق سری
truck	حمل کردن
vehicle	حمل کننده
accession, onset	حمله
seizure	حمله بیماری
biological attack	حمله بیولوژیکی
chemical attack	حمله شیمیائی
seizure	حمله صرعی
domain	حوزه
area of release	حوزه آزادسازی
B-domain	حوزه بی.
aeration basin	حوزه تهویه
breakpoint cluster region	حوزه خوشه شکننده
apple domains	حوزه (های) اپل/سیبی
aeration basin	حوزه هوادهی
metabolic pool	حوضچه متابولیکی
biocide	حیات کش
biotic	حیاتی
garden	حیاط
rDNA small laboratory animals	حیوانات کوچک آزمایشگاهی آر.دی.ان.ای. ئی
hybrid	حیوان دورگه

خ

in-vitro	خارج بدنی
extracellularly	خارج سلول(ی)
extracellular	خارج سلولی
knockout	خارج کردن
acanthocephalus	خار سر
acanthocephala	خار سران
acanthor	خار سرچه
acanthella	خار سرک
itching	خارش

English	Persian	English	Persian
itching, pruritic	خارش دار	external cost	خسارت ظاهری
itching, pruritic	خارشی	fatigue	خستگی
itching	خاریدن	desiccator	خشکانه
BOD (= biochemical oxygen demand)	خاز	lyophilization	خشکانیدن انجمادی
		powdered	خشک شده
origin	خاستگاه	drum dryer	خشک کننده بشکه ای/طبلی
COD (= chemical oxygen demand)	خاش	trait	خصلت
		periodicity	خصلت تناوبی
colligative property	خاصیت جمعیتی	characterization assay	خصلت سنجی
landfill	خاکچال	personal, proprietary	خصوصی
pica	خاک خوری	trait	خصوصیت
geoponics	خاک کشت	novel trait	خصیصه جدید
mole	خال	characterization of animal genetic resource	خصیصه یابی منابع ژنتیکی جانوری
homozygote, homozygous	خالص		
enantiopure	خالص انانتیومری	lane	خط
mole	خال گوشتی	tolerance	خطای مجاز
extinct	خاموش	standard error	خطای معیار
extinguisher	خاموش کننده	hazard, risk	خطر
dormancy	خاموشی	residual risk	خطر پسماندی
kindred, lineage	خاندان	biohazard	خطر زیستی
domestication	خانگی کردن	malignant	خطرناک
alu family	خانواده آلو	health hazard	خطرناک (برای سلامتی)
multigene family	خانواده چند ژنی	cell line	خط سلولی
gene family	خانواده ژن (ی)	animal cell line	خط سلولی حیوانی
terminator	ختم کننده	black-lined	خط سیاه (ذرت)
fouling, knockout	خراب کردن	aspirate	خفه شدن
primer	خرج	upstream	خلاف جریان آب
particle	خرد	germinate	خلق کردن
microphage	خردخوار	grade	خلوص
telomere-associated chromosome fractionation	خردشدگی کروموزومی وابسته به تلومر	dentifrice	خمیر دندان
		yeast	خمیر مایه
		neutrophil	خنثی خواه
particle	خردک	disposal	خنثی سازی
DNA fragmentation	خرد کردن دی.ان.ای.	incubation	خواباندن
juncea	خردل صحرایی	narcosis	خواب رفتگی
juncea	خردل وحشی	carotid artery	خوابرگ
microenvironment	خرد محیط	incubation	خوابیدن
particle	خرده	biochemical oxygen demand (BOD)	خواست اکسیژن زیست شیمیائی (خاز)
harvesting	خرمن کردن		
gene pool	خزانه ژن (ی)	chemical oxygen demand (COD)	خواست اکسیژن شیمیائی (خاش)
algae	خزه ها		

heme	خون
plasma, serum	خونابه
hematocrit	خون بهر
hematogenous	خون زا
erythropoiesis	خونسازی
hematocrit	خون سنجه
hemolymph	خون-لنف
hematochezia	خون مدفوعی
consanguineous	خویش
inbreeding	خویش آمیزی
kindred	خویشاوندان
callus	خیز
malignant edema	خیز مهلک

د

tRNA deacylase	دآسیلاز تی.آر.ان.ای.
deaminase	دآمیناز
deamination	دآمیناسیون
alcohol dehydrogenase	دئیدروژناز الکل
double minute	دابل ماینیوت
data mining	داتا ماینینگ
dot blot, dot-blot	دات بلات
reverse dot blot	دات بلات معکوس
dot blotting	دات بلاتینگ
intracellular	داخل سلولی
illegal traffic	داد و ستد غیر قانونی
medical informatics	داده شناسی پزشکی
bioinformatics	داده شناسی زیستی
data mining	دادهکاوی
data	داده (ها)
baseline data	داده های خط مبنا
baseline data	داده های معیار
baseline data	داده های مقدماتی
vagile	دارای تحرک
transgenic	دارای ژنهای پیوندی
symbiotic	دارای همزیستی
scaffold	داربست
pharmacokinetics	دارو جنبش شناسی
chemotherapy	دارو درمانی
pharmacodynamics	دارو دینامیک شناختی

acclimatization	خوپذیری
person	خود
auto-correlation	خود ارتباطی
homeostasis	خود ایستایی
autoimmunity	خود ایمنی
self-fertilization	خودباروری
self-assembly	خود بر پائی
homeostasis	خود پایداری
autoradiography	خود پرتو نگاری
autopolyploid	خود پر/چند لاد
autotroph, auxotroph, prototroph	خود پرور
autograft	خود پیوند
homeostasis	خود تعادلی
autosome	خودتن
chemo-autotroph	خود خوار شیمیائی
autotroph, prototroph	خود غذا
auxotroph	خودغذا
autophosphorylation	خود فسفات افزائی
directed self-assembly	خود گردایش هدایت شده
self-pollination	خود گرده افشانی
autogamous	خود گشن
self-fertilization	خود گشنی
autogamy, self-pollination	خودگشنی
autoimmunity	خود مصونی
auto-correlation	خود همبستگی
autologous	خودی
feeder	خور-
trophic	خوراکی
endosperm	خورش
feeder	خورنده
aromatic	خوشبو
benign	خوش خیم
cluster	خوشه
cluster	خوشه ای
clustering	خوشه ای شدن
cluster of differentiation	خوشه جدایش
gene cluster	خوشه ژن (ی)
staphylococcus	خوشه گوییزه
acclimatization	خوگیری
cold acclimation	خوگیری با سرما

73

pharmacophore	دارو زا
pharmacogenetics	دارو زاد شناسی
pharmacogenomics	دارو ژن شناسی
pharmacology	دارو شناسی
chemopharmacology	دارو شناسی شیمیائی
placebo	دارونما
pharmacovigilance	دارو هشیاری
placebo	داروی بی اثر
investigational new drug	داروی جدید تحقیقی
treatment investigational new drug	داروی جدید درمان تحقیقی
psoralen	داروی داءالصدف
toxin	داروی سمی
orphan drug	داروی غیر اقتصادی
placebo	داروی کاذب
site-specific drug	داروی مختص به جایگاه
dendrite, dentrite	دارینه
dendritic	دارینه ای
duster	داستر
Wright dust feeder	داست فیدر رایت
uricotelic	دافع اوره
insecticide	دافع حشرات
daffodil rice	دافودیل رایس
deoxy	داکسی
deoxy adenosine	داکسی آدنوزین
deoxy thymidine tri phosphate	داکسی تیرامیدین تری فسفات
deoxy thymidine di phosphate	داکسی تیرامیدین دی فسفات
deoxy thymidine mono phosphate	داکسی تیرامیدین مونو فسفات
deoxyribonuclease	داکسی ریبونوکلئاز
deoxyribovirus	داکسی ریبوویروس
deoxy cytosine triphosphate	داکسی سیتوزین تری فسفات
deoxynuclotid	د اکسی نوکلوتید
Dalton	دالتون
test-range	دامنه آزمایش
domain of protein	دامین (دومین) پروتئین
pharmacognosy	دانش داروبابی
technology	دانش فنی
pyrenoid	دانک

daunorubicin	دانوروبیسین
daunomycin	دانومایسین
particle	دانه
sizing	دانه بندی
castor bean	دانه سمی کرچک
recalcitrant seed	دانه مقاوم
rash	دانه های پوستی
magnetic bead	دانه های جادویی
downstream	داون استریم
dyad	دایاد
daidzen	دایدزن
daidzein, daidzin	دایدزین
nicked circle	دایره شکافته/شکسته
disomy	دایزومی
dynein	داینئین
Dynafog	داینافاگ
dynorphin	داینورفین
deinococcus radiodurans	داینوکوکوس رادیودورانس
WF (= von Willebrand factor)	دبلیو.اف.
WAP (= whey acidic protein)	دبلیو.ای.پی.
WPW (= Wolff-Parkinson-White)	دبلیو.پی.دبلیو.
deproteinization	دپروتئینیزاسیون
determinant	دترمینان
detoxication	دتوکسیکیشن
daughter	دختر
daughter	دخترانه
daughter	دختری
insertion, introgression	دخول
gene insertion	دخول ژن (ی)
absorbance	درآشامندگی
absorption	درآشامی
income	درآمد
upstream	در بالای رودخانه
encapsulated	در بر گرفته
capping	درپوش گذاشتن
in-situ	در جا
in-vivo	در جاندار
in-situ	درجای خود

in-situ	درجای طبیعی	intravenous therapy	درمان تزریقی
grade	درجه	biologic response modifier therapy	درمان توسط تغییر دهنده پاسخ بیولوژیکی
expected progeny difference	درجه بندی عددی ژنتیک والدین دام	gene replacement therapy	درمان توسط جایگزینی ژن (ی)
grade	درجه بندی کردن	intravenous therapy	درمان داخل وریدی
expressivity	درجه تظاهر	gene therapy	درمان ژن (ی)
temperature	درجه حرارت	adoptive cellular therapy	درمان سلولی انتخابی
turbidity	درجه کدر بودن	clinic	درمانگاه
temperature	درجه گرما	clinical	درمانگاهی
sustainable	در حد معقول (مصرف)	curative	درمانی
aerobe	در (حضور) هوا	in-situ	در محل
dendrimer	درختبار	in-situ	در محل طبیعی
neem tree	درخت زیتون تلخ	in-vitro	در محیط کشت
dendritic	درختواره	exposure	در معرض چیزی قرارگرفتن
arthralgias	دردهای مفاصل	in-vivo	در موجود زنده
congenital dislocation of the hip	در رفتگی مفصل ران مادرزادی	in-vivo	در نسج زنده
mischarging	در رفتن اشتباهی	portal	دروازه ای
in-vivo	در زنده	harvesting	دروکردن
holotype, type specimen	درست مونه	inbreeding	درون آمیزی
acromegaly	درشت پایانکی	endocytosis	درون بری
mega base pair	درشت جفت باز	injector	درون پاش
macromutation	درشت جهش	endoderm	درون پوست
macrophage	درشت خوار	endodermal, endodermic, intradermal	درون پوستی
alveolar macrophage	درشت خوار حفره ای	endothelium	درون پوش
macronutrient	درشت خوراک	vascular endothelium	درون پوشه آوندی
in-vitro	در شیشه	lymphatic endothelium	درون پوشه لنفاوی
standard	درفش	endosome	درون تن
quarantine	در قرنطینه نگهداشتن/بودن	systeomics	درون تنی
absorption	درکشی	insertion	درون جای دهی
absorbance	درکشیدگی	insertion or duplication	درون جای دهی یا نسخه برداری
in-vitro	در لوله آزمایش	intake	درون جذب
antidote, curative, remediation	درمان	acceptable daily intake	درون جذب قابل قبول روزانه
anaerobic treatment	درمان آن ایروبیک/غیر هوازی	adequate intake	درون جذب کافی
therapy of antisense mRNA	درمان ام.آر.ان.ای. بی معنی	chronic intake	درون جذب مزمن/حاد/جدی
aerobic treatment	درمان ایروبیکی	absorption	درون جذبی
immunotherapy	درمان ایمنی	endosperm	درون دانه
antibiotic therapy	درمان با آنتی بیوتیک	endophyte, endophytic	درون رست
immunosuppressive therapy	درمان بازدارنده ایمنی	endophytism	درون رستی
curative	درمان بخش	endophyte	درون روی

endocrinology	درون ریز شناسی
inbreeding	درون زاد گیری
endotoxin	درون زهر
endotoxin	درون زهرابه
delta endotoxin	درون زهرابه دلتا
intracellular	درون سلولی
endocardium	درون شامه قلب
incubation	درون کمون
incubation period	درون نهفتگی
invagination	درون نیامی
endospore	درون هاگ
inbreeding	درون همسری
endocytosis, intracellular	درون یاختگی
receptor-mediated endocytosis	درون یاختگی به واسطه گیرنده
protoplasm	درون یاخته
port	دریچه
desferroxamine manganese	دزفرواکسامین منگنز
deoxyribose	دزوکسی ریبوز
deoxyribonucleotide	دزوکسی ریبو نوکلئوتید
deoxyribonucleic acid (DNA)	دزوکسی ریبونوکلئیک اسید
deoxynivalenol	دزوکسی نیوا لنول
degeracy	دژراسی
degeneration	دژنراسیون
desaturase	دساچیوراز
hand	دست
product	دستاورد
disposal	دسترسی
genetic manipulation	دستکاری ژنتیکی
gene manipulation, gene modification	دستکاری ژن (ی)
device, equipment, machine, organism, unit	دستگاه
immune system	دستگاه ایمنی بدن
humidifier	دستگاه بخور/رطوبت ساز/مرطوب کننده
sprayer	دستگاه پاشیدن
dispenser	دستگاه پخش/توزیع (کننده)
feeder	دستگاه تغذیه اتوماتیک
chromatin remodeling	دستگاه تغییر آرایه کروماتین
machine	
genitourinary tract	دستگاه تناسلی-ادراری
immune system	دستگاه دفاعی بدن
roller bottle apparatus	دستگاه رولر باتل
central nervous system	دستگاه عصبی مرکزی
blood pressure gauge	دستگاه فشار خون
Golgi apparatus, Golgi's apparatus	دستگاه گلژی
centrifuge	دستگاه مرکز گریز
molecular machine	دستگاه ملکولی
self-assembling molecular machine	دستگاه مولکولی خود تجمع/برپا ساز
microchannel fluidic devices (= microfluidics)	دستگاههای ریز لوله سیالاتی
cluster	دسته
flux	دسته اشعه
cell sorting	دسته بندی سلولی
expression profiling	دسته بندی نمایانی/تظاهر/هویدائی
hand, hand-held	دستی
desensitization	دسنسیتیزاسیون
desulfovibrio	دسولفو ویبریو
tumor	دشبل
plastid	دشتاره
plastidome	دشتاره مونه
mitochondrion	دشته تن
mitochondria	دشته تن ها
plasmolysis	دشته کافتی
surface plasmon	دشته مونه سطحی
nucleoplasm	دشته هسته
plasmocyte	دشته یاخته
plasmid	دشتیزه
multi-copy plasmid	دشتیزه چند نسخه ای
stringent plasmid	دشتیزه سخت
plasma	دشتینه
blood plasma	دشتینه خون
antagonist	دشمن
environmentalism	دفاع از محیط زیست
biological defense	دفاع بیولوژیکی
disposal	دفع
graft rejection	دفع پیوند

hyperacute rejection	دفع حاد عضو پیوندی
defensins	دفنسین ها
decarboxylase	دکربوکسیلاز
glutamic acid decarboxylase	دکربوکسیلاز اسید گلوتامیک
dextran	دکستران
cyclodextrin	دکسترین چرخه ای
limit dexterin	دکسترین معیار
allelopathy	دگر آسیبی
outcrossing	دگر آمیزی
allopatric	دگر بوم
heterotroph	دگر پرور
heterotrophic	دگر پروری
heterotrophic	دگر خواری
metaplasia	دگر دشتاری
metastasis, polymorphism	دگر دیسی
heterochromatin	دگر رنگینه
allosterism	دگر ریختاری
heteromorphism	دگر ریختی
heterologous	دگر ساخت
heterology	دگر ساختی
transformation	دگر سازی
heterologous	دگر سان
allotrope, allotropy	دگرشکل
heterotroph	دگر غذا
heterotrophic	دگر غذایی
frameshift	دگر قالب
metabolism	دگرگشت
metabolite	دگرگشته
metabolic	دگرگشتی
allogamous, open pollination	دگر گشن
allogamy	دگر گشنی
dissimilation	دگرگون سازی
metabolism, mutation, transformation	دگرگونی
biotransformation	دگرگونی زیستی
intermediary metabolism	دگر گوهرش میانجی
nitrogen metabolism	دگر گوهرش نیتروژن
metabolite	دگر گوهره
metabolic product	دگرگوهره

allopatric	دگر نیاک
strain	دگروشی
allele	دگره
dominant allele	دگره بارز
heterokaryosis	دگر هستگی
heterokaryon	دگر هسته
heterosis	دگرینگی
import	دلالت داشتن بر
placebo	دل خوشکنک
scab	دلمه
agglomeration	دلمه شدن
bucket	دلو
temperature	دما
room temprature	دمای اتاق
boiling point	دمای جوش
melting temperature	دمای ذوب
DNA melting temperature	دمای ذوب دی.ان.ای.
optimum temperature	دمای مطلوب
expiration	دم بر آوری
respiration	دم زنی
spirometer	دم سنج
spirometery	دم سنجی
respiration	دمش
moribund	دم مرگ
blower	دمنده
denaturant gradient gel electrophoresis	دناتورانت گرادیان ژل الکتروفوروز
tailing	دنبال کردن/رفتن
sequela	دنباله
dentogram	دنتوگرام
dentinogenesis imperfecta	دنتینوژنسیس ایمپرفکتا
dentifrice	دندان شوی
dendrite	دندانه
dendrimer	دندرایمر
dendrite	دندریت
dendritic	دندریتی
bivalent	دو ارزشی
stability	دوام
bipolar	دوانتهایی
cascade	دوانه
duplex	دو بخشی

English	Persian	English	Persian
diploid	دو برابر	gestation period	دوره آبستنی
two-dimensional	دوبعدی	cyclic	دوره ای
epimer	دوپار	periodicity	دوره ای بودن
dopastin	دوپازتین	lag phase	دوره تاخیر/تاخر/لنگی
duplication	دوپلیکاسیون	gestation	دوره تکوین
gene duplication	دوپلیکاسیون ژن (ی)	gestation	دوره جنینی
bivalent	دو تایی	incubation period	دوره جوجه کشی
diploid, duplex	دوتایی	quaternary period	دوره چهارم
hermaphrodite	دوجنسی	growth phase	دوره رشد
smoke	دود	tertiary period	دوره سوم/سه گانه
smoke	دود دادن	incubation	دوره شکل گیری
sodium dodecyl sulfate	دودسیل سولفات سدیم	krebs cycle	دوره کربس
smoke	دود کردن	incubation, incubation period	دوره کمون
puffer	دودکن		
lineage, pedigree, strain	دودمان	convalescence	دوره نقاهت
cloning	دودمان سازی	latency period	دوره نهفتگی
megabase cloning	دودمان سازی کلان باز	communicable period	دوره واگیر
molecular cloning	دودمان سازی ملکولی	dose	دوز
positional cloning	دودمان سازی موقعیتی	heterogametic	دوزامه ای
clone	دودمان سلولی	cumulative radiation dose	دوز پرتوئی تراکمی
pedigree	دودمانه	gene dosage	دوز ژن (ی)
smog	دودمه	chlorine dose	دوز کلر
distal	دور	dosimetry	دوزیمتری
periodicity	دوران	bifurcate	دو شاخه شدن
quaternary period	دوران کواترنری	replication fork	دوشاخه همانند سازی
futile cycle	دور بی حاصل	showering	دوش گرفتن
revolutions per minute	دور در دقیقه	bivalent	دو ظرفیتی
waste	دورریز	bipolar	دوقطبی
disposal	دور ریزی	duplex, twins	دوقلو
hybrid	دورگه	monozygotic twins	دوقلوهای
interspecific hybrid	دورگه بین گونه ای		یکسان/مونوزیگوت/تک
interspecific hybrid	دورگه دوگونه		تخمکی
colony hybridization	دورگه سازی کولونی	disaccharide	دوقندی
hybridization	دورگه شدن	doxycycline	دوکسی سیکلین
annealing	دورگه شدن زنجیره دی.ان.ای.	spindle	دوکی شکل
	یا آر.ان.ای	heteroduplex	دوگانه ناجور
fluorescence in situ hybridization (FISH)	دورگه گیری به روش پرتوافشانی در محل	central dogma	دوگم اصلی/بنیادی/مرکزی
		facultative	دو گونه زی
southern hybridization	دورگه گیری به روش ساترن	diploid	دولا
first filial hybrid	دورگه نسل اول	diploid	دولاد
lifetime	دوره	diplophase	دولاد چهر

lipid bilayer	دولایه چربی	Dhh (= Desert Hedgehog Protein)	دی.اچ.اچ.
exogenous	دو لپه	DHF (= di-hydro folate)	دی.اچ.اف.
biphasic	دو مرحله ای	DHFR (= di-hydro folate reductase)	دی.اچ.اف.آر.
second meiotic anaphase	دومین آنافاز میوتیک		
second meiotic metaphase	دومین متافاز میوتی	DHAP (= di hydoxy acetone phosphate)	دی.اچ.ای.پی.
devilbiss	دویل بیس		
nozzle	دهانه	DHPR (= di-hydro pteridine reductase)	دی.اچ.پی.آر.
Micronair spray nozzle	دهانه اسپری مایکرون ایر		
oropharyngeal	دهانی-حلقی	DHT (= di-hydro testosterone)	دی. اچ. تی.
oral-facial-digital	دهانی-صورتی-انگشتی		
left atrium	دهلیز چپ	DHTR (= di-hydro testosterone receptor)	دی.اچ.تی.آر.
donor	دهنده		
glucose-6-phosphate dehydrogenase	دهیدروژناز فسفات-6-گلوکز	diadzein (= daidzein)	دیادزین
		amylase	دیاستاز
formaldehyde dehydrogenase	دهیدروژناز فورمالدهید	diastereoisomer	دیاستریوایزومر
		diacylglycerol (DAG)	دیاسیلگلیسرول
lactate de-hydrogenase	دهیدروژناز لاکتات	carbon dioxide	دی اکسید کربن
lactate dehydrogenase	دهیدروژناز لاکتاز	supercritical carbon dioxide	دی اکسید کربن فوق حیاتی
malate dehydrogenase	دهیدروژناز مالات		
flavin-linked dehydrogenase	دهیدروژناز وابسته به فلاوین	diakinesis	دیاکینزیس
		diagnostis	دیاگنوستیز
dehydrogenases	دهیدروژنازها	dielectrophoresis	دی الکتروفورز
histidinol dehydrogenase	دهیدروژناز هیستیدینول	dialysis	دیالیز
dehydrogenation	دهیدروژناسیون	DM (= myotonic dystrophy)	دی.ام.
dehydroxyphenylalanine	د هیدروکسی فنیل آلانین	DMAD (= diabetes mellitus optic atrophy deafness)	دی.ام.ای.دی.
D (= Dalton)	دی.		
DRPA (= dentato rubral-pallidolusyian atrophy)	دی.آر.پی.ای.	DMD (= Duchenne muscular dystrophy)	دی.ام.دی.
		DNase (= deoxyribonuclease)	دی. ان.آز
DE (= DNA extraction and quality control)	دی.ئی.		
		DNASe (= deoxyribonuclease)	دی.ان.آز
DEAE (= diethylamino-D ethanol)	دی.ئی.ای.ئی.		
		DNA (= deoxyribo nucleic acid)	دی.ان.ای.
DEB (= mutagen dipoxybutane)	دی.ئی.بی.		
		junk DNA (= selfish DNA)	دی.ان.ای.آشغال
diauxy	دیا اوکسی	selfish DNA	دی.ان.ای.آشغال/خودخواه
diabetes	دیابت	a-DNA	دی.ان.ای.-آلفا/آ.
atopic diathesis	دیاتز آتوپیک	mtDNA	دی.ان.ای. ام.تی.
diethylamino-D ethanol	دی اتیل آمینو-دی. اتانول	blunt-end DNA	دی.ان.ای. انتها صاف
DHEA (= di-hydro epiandrosterone)	دی.اچ.ئی.ای.	blunt-end DNA	دی.ان.ای.انتها کور
		naked DNA	دی.ان.ای برهنه

English	Persian
BDNA, b-DNA	دی. ان. ای. بی
biotinylated-DNA	دی.ان.ای بیوتینی شده
DNA polymerase	دی.ان.ای. پلیمراز
DNA polymerase I	دی.ان.ای. پلیمراز 1
DNA polymerase III	دی.ان.ای. پلیمراز3
denatured DNA	دی.ان.ای. تغییر یافته/تقلیب شده
denaturation DNA	دی.ان.ای. تقلیب
repetetive DNA	دی.ان.ای.تکراری
single-stranded DNA	دی.ان.ای. تک رشته ای
single-copy DNA	دی.ان.ای.تک نسخه
telomeric DNA	دی.ان.ای.تلومریک
total cell DNA	دی.ان.ای.تمام سلولی
DNA topoisomerase	دی.ان.ای.-توپوایزومراز
tumor DNA	دی.ان.ای. توموری
random amplified polymorphic DNA	دی.ان.ای. چند ریخت تکثیر یافته تصادفی
transfer DNA	دی.ان.ای. حامل
spacer DNA	دی.ان.ای. حد فاصل
transferred DNA	دی.ان.ای. حمل شده
disarmed DNA	دی.ان.ای. خنثی شده
zigzag DNA	دی.ان.ای دارای ساختمان زیگزاگی
nascent DNA	دی.ان.ای. در حال شکل گیری
heterologous DNA	دی.ان.ای. دگر ساخت
denaturation DNA	دی.ان.ای. دناتوره
duplex DNA	دی.ان.ای دوبخشی
heteroduplex DNA	دی.ان.ای. دوگانه ناجور
donor DNA	دی.ان.ای. دهنده/دونر
linker DNA	دی.ان.ای.رابط
ribosomal DNA	دی.ان.ای. ریبوزومی
microsatellite DNA	دی.ان.ای. ریز قمر
Z-DNA (= zigzag DNA)	دی.ان.ای. زد
DNA gyrase	دی.ان.ای. ژیراز
cytoplasmic DNA	دی.ان.ای. سیتوپلاسمی
ccc DNA	دی.ان.ای. سی.سی.سی.
anti-messenger DNA	دی.ان.ای.ضد پیک/پیامبر
non-coding DNA	دی.ان.ای. غیر رمزگذار
DNA fingerprint	دی.ان.ای. فینگرپرینت
DNA fingerprinting (DNA profiling)	دی.ان.ای. فینگرپرینتینگ
catenated DNA	دی. ان.ای. کاتنیتد/متصل
concatermeric DNA (concatemeric DNA)	دی.ان.ای. کنکاترمریک
concatemeric DNA	دی.ان.ای. کنکاتمریک
DNA glycosylase	دی.ان.ای. گلیکوسیلاز
DNA ligase	دی.ان.ای. لیگاز
satellite DNA	دی.ان.ای.ماهواره ای
DNA methylase	دی.ان.ای. متیلاز
sense DNA	دی.ان.ای. معنی دار
complementary DNA	دی.ان.ای. مکمل
double-stranded complementary DNA (dscDNA)	دی.ان.ای. مکمل دو رشته ای
mitochondrial DNA (mtDNA)	دی.ان.ای. میتوکندریایی
foreign DNA	دی.ان.ای. نا بومی/بیگانه
heteroduplex DNA	دی.ان.ای. ناجور دوتایی
heterologous DNA	دی.ان.ای ناهمساخت
recombinant DNA	دی.ان.ای. نوترکیب
A form DNA	دی.ان.ای نوع آ.
I-DNA	دی.ان.ای.نوع آی.
beta-DNA	دی.ان.ای. نوع بتا
alpha DNA	دی.ان.ای. نوع/شکل آلفا
semiconservative DNA	دی.ان.ای.نیمه محافظه کار
Watson-Crick DNA	دی.ان.ای. واتسون-کریک
denatured DNA	دی.ان.ای. واسرشته
promiscous DNA	دی.ان.ای.ولگرد
target DNA	دی.ان.ای. هدف
core DNA	دی.ان.ای. هسته/اصلی
nuclear DNA	دی.ان.ای.هسته ای
DNA helicase (gyrase)	دی.ان.ای. هلیکاز
DNP (= dinitrophenol)	دی.ان.پی.
DOPA (= dihydroxyphenylalanine)	دی.او.پی.ای.
DOGS (= dioctadectylamidoglycyls permine tetra fluoroacetic acid)	دی.او.جی.اس.
DA (= Dalton)	دی.ای.
DASH (= dynamic allele-specific	دی.ای.اس.اچ.

hybridization)	
DAG (= diacylglycerol)	دی.ای.جی.
DB (= dot blot)	دی.بی.
DBP (= DNA-binding-protein)	دی.بی.پی.
DBC (= double-stranded DNA binding site)	دی.بی.سی.
muramyl di-peptide	دی پپتید مورامیل
dipeptidyl-peptidase	دی پپتیدیل-پپتیداز
dip-pen lithography	دیپ-پن لیتوگرافی
dip-pen nanolithography	دیپ-پن نانولیتوگرافی
deprotection	دیپروتکشن
Dipel	دیپل
diploid	دیپلوئید
diplotene	دیپلوتن
diplophase	دیپلوفاز
mutagen dipoxybutane	دیپوکسی بوتان جهش زا
mutagen dipoxybutane	دیپوکسی بوتان موتاژن
DPI	دی.پی.آی.
DPD (= dihydro pyrimidine dehydrogenase)	دی.پی.دی.
DTH (= delayed-type hypersensitivity reaction)	دی.تی.اچ.
DGI1 (= dentinogenesis imperfecta)	دی.جی.آی. 1
dideoxy	دیداکسی
dideoxynucleotide	دی داکسی نوکلئوتید
dideoxynucleoside triphosphate (dd NTP)	دیداکسی نوکلئوزید تریفسفات
hemophilia	دیربندآمدن خون
derepression	دی ریپرس کردن
enzyme derepression	دی ریپرس کردن زیمایه/دیاستاز/آنزیم
derepression	دی ریپرشن
enzyme derepression	دی ریپرشن زیمایه/دیاستاز/آنزیم
paleontology	دیرین شناسی
asphyxiating thoracic dysplacia	دیزپلازی توراسیک آسفیکسیتینگ
uniparental disomy (UPD)	دیزومی تک والدی
disaccharide	دی ساکارید

ectrodactyly-ectodermal dysplasia-clefting	دیسپلازی-شکاف اکتروداکتیلی-اکتودرمال
displacement loop	دیسپلیسمنت لوپ
trucho-thio-dystrophy	دیستروفی تروکو-تیو
dystrophication	دیستروفی شدن
limb-girdle muscular dystrophy type 2A	دیستروفی عضلانی کمر وپائین تنه نوع آ2.
Duchenne muscular dystrophy (DMD)	دیستروفی عضلانی نوع دوشن
dystrophication	دیستروفیکیشن
meta chromatic leano dystrophy	دیستروفی متاکروماتیک لینو
myotonic dystrophy	دیستروفی میوتونیک
disjunction	دیسجانکشن
dysgenic	دیسژنیک
dissimilation	دیسسیمیلیشن
expressive dysphasia	دیسفیژیای آشکار/بارز.
disk	دیسک
biological disk	دیسک بیولوژیکی
super-oxide dismutase	دیسموتاز ابر-اکسیدی
dicentric	دیسنتریک
glutathione disulfide (oxidized form)	دیسولفیدگلوتاتیون (اکسید شده)
plastid	دیسه
DCR (= dominant control regions)	دی. سی.آر.
diptheria toxin	دیفتریا توکسین
guanosine di-phosphate	دی-فسفات گوانوزین
uridine di-phosphate	دی-فسفات یوریدین
defensins	دیفنزین ها
circular dichroism	دیکروئیزم دایره ای
diglyceride	دی گلیسرید
dimer	دیمر
primidine dimmer	دیمر پریمیدین
thymine-thymine dimer	دیمر تیمین-تیمین
cyclobutyl dimer	دیمر سیکلو بوتیل
dynamics	دینامیک
flavin adenine dinucleotide	دی نوکلئوتید فلاوین آدنین
reduced flavin adenine dinucleotide	دینوکلئوتید فلاوین آدنین تضعیف شده
dinitrophenol	دینیتروفنول

81

English	Persian
denaturing gradient gel electrophoresis	دینیچرینگ گریدینت ژل الکتروفورسیز
cell membrane (cell wall)	دیواره سلول
cell wall	دیواره سلولی
cell wall	دیواره یاخته
D.V (= DNA vaccine)	دی.وی.
di-hydro pteridine reductase	دی.- هایدرو پتریدین رداکتاز
dihydro pyrimidine dehydrogenase	دی هایدروپیریمیدین دهیدروژناز
di-hydro testosterone	دی.-هایدروتستوسترون
di-hydro folate	دی.- هایدروفولات
dihydrofolate reductase	دی هایدروفولات رداکتاز
di-hydro folate reductase	دی.- هایدروفولات رداکتاز
dihybrid	دی هیبرید
di-hydro epiandrosterone	دی-هیدرواپیآندوسترون
di hydoxy acetone phosphate	دی هیدروکسی استون فسفات
dihydroxyphenylalanine	دی هیدروکسی فنیل آلانین

ذ

English	Persian
epidemic pneumonia	ذات الریه همه گیر
resource	ذخائر
extractive reserve	ذخیره استخراج
biosphere reserve	ذخیره زیست کره
microparticle	ذرات بسیار ریز
suspended particulate matter (SPM)	ذرات معلق
corn	ذرت
waxy corn	ذرت روغنی
particle	ذره
alpha particle	ذره آلفا
proteinaceous infectious particle	ذره آلوده پروتئیناسیوز
aerosol particle	ذره اسپری شده
aerosol particle	ذره افشانه ای
aerosol, atomizer, sprayer	ذره پاش
aerosol	ذره پاشی
aerosolize(d)	ذره پاشیدن
aerosol particle	ذره جامد/مایع معلق در هوا
signal recognition particle	ذره شناسائی پیام

English	Persian
magnetic particle	ذره مغناطیسی
virion	ذره ویروس
melting	ذوب سازی
melting	ذوب شدن
flux	ذوب کننده
localized melting	ذوب منطقه ای
ablation	ذوب یخ

ر

English	Persian
reagin	راژین
aerobic reactor	راکتور ایروبیک
packed bed reactor	راکتور بستر فشرده
bio reactor, bioreactor	راکتور زیستی
hollow fibre reactor	راکتور فیبر میان تهی
reovirus	رئوویروس
reoviridae	رئوویروس (ها)
rhabdoviridae	رابدوویروس (ها)
liaison	رابط
liaison	رابطه
rapamycin	راپامایسین
relaxed	راحت
free radical	رادیکال آزاد
radioactive	رادیواکتیو
radioisotope	رادیو ایزوتوپ
radiobiology	رادیوبیولوژی
radiotherapy	رادیوتراپی
deinococcus radiodurans	رادیو دورانس د اینوکوک
ras (= ras oncogene)	راس
dextrorotary	راست بر
dextrorotary	راست گرد
racemate	راسمات
racemase	راسماز
racemic	راسمیک
hypophosphataemic ricket	راشیتیسم مقاوم به ویتامین دی.
raffinose	رافینوز
rocket	راکت
mitochondrion	راکیزه
mitochondria	راکیزه ها
domestication	رام کردن

rhamnose	رامنوز	grade	رده بندی کردن
ethnobiology	رامه شناسی	probe (= DNA probe),	ردیاب
lane. pathway	راه	tracer	
promoter	راه انداز	nucleic acid probe	ردیاب اسید نوکلئیک
core promoter	راه انداز اصلی	dissolved oxygen probe	ردیاب اکسیژن محلول
project leader	راهبر پروژه	oligonucleotide probe	ردیاب اولیگونوکلئوتید
respiratory tract	راه تنفسی	multi-locus probe	ردیاب چند نقطه ای
leader	راهنما	DNA probe	ردیاب دی.ان.ای
infiltration	راه یابی	bioprobe	ردیاب زیستی
consultation	رایزنی	genetic probe	ردیاب ژنتیکی
rice blast	رایس بلاست	gene probe	ردیاب ژن (ی)
ricin	رایسین	fluorogenic probe	ردیاب فلوئوروژنیک/فلوئورسانس
rhinovirus	راینو ویروس		زا/شب
liaison	ربط		نمایی/پرتوافشانی/شارندگ
repressor	ریرسور		ی زا
replisome	رپلیزوم	RNA probes	ردیاب های آر.ان.ای.
replicase	رپلیکاز	hybridization probe	ردیاب هیبریداسیون
DNA replication	رپلیکاسیون دی.ان.ای.	biological monitoring	ردیابی بیولوژیکی
replicon	رپلیکون	contact tracing	ردیابی تماسها
retrovirus	رترو ویروس	biologic monitoring	ردیابی زیستی
amphotropic retrovirus	رتروویروس آمفوتروپیک	microbial source tracking	ردیابی منشاء میکروبی
endogenous retro-virus	رتروویروس آندوژنوس	resveratrol	رزوراترول
reticulocyte	رتیکولوسیت	electron paramagnetic	رزونانس پارامغناطیسی
rough endoplasmic	رتیکولوم اندوپلاسمیک زبر	resonance	الکترونی
reticulum		nuclear magnetic resonance	رزونانس مغناطیسی هسته
retinitis pigmentosa	رتینایتیس پیگمنتوزا	regiospecific	رژیواسپسیفیک
retinene	رتینن	regioselective	رژیوسلکتیو
retinoid	رتینوئید	expressivity	رسایی
retino blastoma	رتینوبلاستوما	habitat	رستنگاه
(retinoblastoma)		plant	رستنی
retinol	رتینول	hypostasis, residue, settling	رسوب
reversion	رجوع	chlorine residual	رسوبات کلر
phosphorus	رخشا	immune percipitate	رسوب ایمنی
phenotype	رخ مانه	settling	رسوب گذاری
phenotype	رخ مون	infiltration	رسوخ
infiltration, insertion	رخنه	infiltrate	رسوخ دادن/کردن
tracer	ردگیر	mitosis	رشتمان
redox	ردوکس	sequence, strand	رشته
taxonomy	رده بندی شناسی	antisense strand	رشته آنتی سنس/غیر کد کننده
biological warfare agent	رده بندی عامل جنگ میکروبی	anticoding strand	رشته آنتی کدون
classification		template strand	رشته الگو

lagging strand	رشته پسرو
leading strand	رشته پیشرو
fibroblast	رشته تنده
intermediary filament	رشته حد وسط
coding strand	رشته رمز گزار
intermediary filament	رشته سیتواسکلتی واسطه
Y long arm	رشته طولانی وای.
anticoding strand	رشته غیر رمز کننده
non-coding strand	رشته غیر رمز گذار
codogenic strand	رشته کدوژنیک/کد ساز
plus strand	رشته مثبت
sense strand	رشته معنی دار
negative strand (= antisense strand)	رشته مکمل مثبت
minus strand, negative strand (= antisense strand)	رشته منفی
anticoding strand	رشته نامفهوم
evolution, growth	رشد
anaerobic growth	رشد آن-ایروبیک/غیر هوازی
auxotroph	رشد پرور
early development	رشد زودهنگام
heterosis	رشد غیر مترقبه
neoplastic growth	رشد مربوط به نوسازی
neoplastic growth	رشد نئوپلاستی
aerobic growth	رشد هوازی
decontamination	رفع آلودگی
derepression	رفع سرکوب
enzyme derepression	رفع سرکوب
	زیمایه/دیاستاز/آنزیم
antagonism, competition	رقابت
dilution	رقت
character, grade, strain	رقم
disabled strain	رقم خنثی شده
inbred strain	رقم درون آمیخته
cultivar	رقم زراعی
cultivar	رقم کشته شده
landrace	رقم محلی
Abelson strain of murine	رقم (مربوط به جوندگان) آبلسونی
live vaccine strain	رقم واکسن زنده

antagonist	رقیب
dilution	رقیق سازی
dilution	رقیق شدگی
dilution	رقیق کردن
recombination	رکمبیناسیون
recombinant	رکمبینان
DNA backbone	رکن دی.ان.ای.
recombinase	رکومبیناز
additive recombination	رکومبیناسیون افزایشی
genetic recombination	رکومبیناسیون ژنتیکی
homologous recombination	رکومبیناسیون هومولوگ
aneurysm	رگ آماس
angiogenesis	رگ زایی
vasodilator	رگ گشا
regulon	رگولون
line	رگه
inbred line	رگه خویش آمیز
code, codon	رمز
initiation codon	رمز آغازین
exon	رمزآور
termination codon, terminator sequence	رمز اختتام
termination codon	رمز انقطاع
amber codon, nonsense codon	رمز بی معنی
amber codon	رمز پایان دهنده
stop codon	رمز پایانی
genetic code	رمز ژنتیکی
triplet	رمز سه حرفی
protein encoding	رمزگذاری پروتئین
miscoding	رمزگذاری غلط
genetic code	رمز وراثتی
ribosome	رناتن
burn-through range	رنج برن ترو
paint	رنگ
chromosome painting	رنگ آمیزی کروموزومها
high resolution chromosome banding	رنگ آمیزی کروموزوم ها با دقت بالا
gram stain	رنگ آمیزی گرام
bleaching	رنگبری
acid-fast stain	رنگ ثابت شده با اسید

84

chromatophore	رنگ دار	rhodanese	رودانیز
blue cone pigment	رنگدانه مخروط آبی	rhodospin	رودواسپین
chromoplast	رنگ دیسه	bowel	روده
paint	رنگ زدن	bowel	روده ای
quencher dye	رنگ سیر/اشباع کننده	enterotoxin	روده زهرابه
paint	رنگ کردن	staphylococcal enterotoxin	روده زهرابه استافیلو کوکی
gram stain	رنگ گرام	streptococcal enterotoxin	روده زهرابه استرپتوکوکی
phytochrome	رنگ گیاهی	epiphyte, epiphytic	رورست
acid-fast stain	رنگ مقاوم به اسید	epiphytism	رورستی
chromatography	رنگ نگاری	cancer epigenetics	روزاد شناسی سرطان
ion-exchange chromatography	رنگ نگاری تبادل یونی	lumen	روزن
		nozzle, port, vent	روزنه
gas-liquid chromatography	رنگ نگاری گاز-مایع	agar-diffusion method	روش آگار دیفیوژن
gas chromatography	رنگ نگاری گازی	traditional breeding method	روش اصلاح نژاد سنتی
high-pressure/high performance liquid chromatography	رنگ نگاری مایع فشار و کارکرد بالا	explosion method	روش انفجاری
		independent cutting method	روش انقطاع مستقل
pigmentation	رنگیزه دارشدن	BESS method (= base excision sequence scanning)	روش بس
pigmentation	رنگی شدن		
chromosome	رنگین تن	BESS method (= base excision sequence scanning)	روش بی.ئی.اس.اس.
chromatin	رنگینه		
flourescent dye	رنگینه فلوئورسان		
fluorescent dye	رنگینه فلوئوروسنت	BESS t-scan method	روش تی. اسکن بس
chymosin, renin, v	رنین	chromosome walking	روش راهبری کروموزومی
rho	رو	Sanger's method	روش سانگر
prevalence	رواج	Sanger-Coulson method	روش سانگر کولسون
tolerance	رواداری	Symba process	روش سیمبا
flux	روان ساز	biological warfare agent identification method	روش شناسایی سلاحهای میکروبی
tolerance	روایی		
rubratoxin	روبراتوکسین	biological warfare agent identification method	روش شناسایی عامل جنگ میکروبی
moribund	رو به مرگ		
moribund	رو به نزع	technology	روش فنی
rubitecan	روبیتکان	systematics	روش مند
epiderm	روپوست	Roche Molecular Biochemicals	روش مولکولار بیوکمیکالز
epidermal, epidermic	روپوستی		
episome	روتن	patent	روشن
rotenone	روتنون	supernatant	رو شناور
rotor fermentor	روتور فرمنتور	luminescence	روشنایی
rotating biological contractor	روتیتینگ بیولوژیکال کنتراکتور	shotgun cloning method	روش همسانه سازی شات گانی
ghost	روح	step aeration method	روش هوادهی پله ای

castor oil	روغن کرچک	riboflavin	ریبوفلاوین
mid-oleic vegetable oils	روغن گیاهی میان-اولئیکی	ribulose	ریبولوز
adipose	روغنی	ribonuclease	ریبونوکلئاز
cover slip	روکش	ribonucleoprotein,	ریبونوکلئوپروتئین
plating	روکش دادن	ribo-nucleo-protein	
epiphysis	رومغزی	small nuclear	ریبونوکلئوپروتئین کوچک هسته
down processing	روند کاهشی	ribonucleoprotein	
adsorption	رونشینی	(snRNP)	
dermal adsorption	رو نشینی پوستی	ribonucleotide	ریبونوکلئوتید
transcript	رونوشت	ribonucleoside	ریبونوکلئوزید
RNA transcript	رونوشت آر.ان.ای.	repiration	ریپایریشن
gene transcript	رونوشت ژن (ی)	enzyme repression	ریپرس کردن
transcription	رونویسی		زیمایه/دیاستاز/آنزیم
embryo	رویان	enzyme repression	ریپرشن زیمایه/دیاستاز/آنزیم
culture	رویاندن	mold	ریخت
embryology	رویان شناسی	morphogenetic	ریخت زا
event, incidence, occurrence	رویداد	morphology	ریخت شکل شناسی
biological incident	رویداد بیولوژیکی	morphology	ریخت شناسی
genetic event	رویداد ژنتیکی	colonial morphology	ریخت شناسی دست جمعی
growth	رویش	safety-pin morphology	ریخت شناسی سیفتی پین
germinate	رویش جرم	genotype	ریخته ارثی
somatic	رویشی	phenotype	ریخته ظاهری
precautionary approach	رویکرد اخطارانه	rare cutter	ریرکاتر
buffy coat	رویه بافی	microarray	ریز آرایه
germinate	روییدن	protein microarray	ریز آرایه پروتئینی
apomixis	رها آمیزی	DNA microarray	ریز آرایه دی.ان.ای.
transboundary release	رهائی فرا مرزی/سرحدی	cDNA microarray	ریز آرایه زنجیره واحد دی.ان.ای
apoenzyme	رهازیما		
covert release	رها سازی پنهان/مخفی	mini preparation	ریز آماده سازی
intended release	رهاسازی دلخواه	micropropagation	ریز ازدیادی
planned release	رها سازی عمدی	microtome	ریز بر
unintended release	رها سازی نا خواسته	nanocrystals	ریز بلور
uncoupling	رها کردن	microscopy	ریزبینی
delivery	رهایی	electron microscopy	ریزبینی الکترونی
accidental release	رهایی تصادفی	scanning tunneling electron	ریزبینی الکترونی تونل زنی
precautionary approach	رهیافت اخطارانه ای	microscopy	پویشی
ribose	ریبوز	scanning tunneling	ریزبینی تونل زنی پویشی
ribose-5-phosphate	ریبوز-5-فسفات	microscopy	
ribosome	ریبوزوم	confocal microscopy	ریزبینی دوچشمی
ribozyme	ریبوزیم	confocal microscopy	ریزبینی دو کانونی
riboswitch	ریبوسوئیچ	confocal microscopy	ریزبینی کنفوکال

atomic force microscopy	ریزبینی نیروی هسته ای
nanobiology	ریز بیو لوژی
acromicria	ریزپایانکی
micropipette	ریز پیپت
electropermeabilization	ریزتراواسازی
nanocomposites	ریز ترکیبها
microinjection	ریز تزریق
micromodification	ریز تغییر
nanotechnology	ریز تکنولوژی
microsequencing	ریز توالی گری
micro sensor	ریز حسگر
microphage	ریزخوار
nanoscience	ریز دانش
ribosome	ریزدانه پالاد
metachromatic granule	ریزدانه دگر رنگ
nanoparticle	ریز ذره
superparamagnetic nanoparticle	ریز ذره ابر فرا مغناطیسی
nanobot	ریز روبات
microsphere	ریز سپهر
mini cell	ریز سلول
micrometer	ریز سنج
nanofluidics	ریز سیالات
microfluidics	ریز سیال شناسی
nanowire	ریز سیم
continuous perfusion	ریزش پیوسته
laser capture microdissection	ریزشکافی لیزری
droplet	ریز قطره
minisatellite	ریز قمر
particle	ریزک
DNA fragmentation	ریز کردن دی.ان.ای.
vesicle	ریز کیسه
lipid vesicle	ریزکیسه چربی
cytoplasmic vesicle	ریزکیسه سیتوپلاسمی
nanoshells	ریز گلوله
microtubule, nanotube	ریز لوله
peptide nanotube	ریز لوله پپتیدی
microtubule	ریز لوله چه
single-walled carbon nanotube	ریز لوله کربن تک دیواره

nanolithography	ریز لیتوگرافی
micromachining	ریز ماشینکاری
nanopore	ریز منفذ
microfilament	ریز میله
rhizosphere	ریزواسفر
rhizobium	ریزوبیوم
rhizobia	ریزوبیوم ها
particle	ریزه
phagocytosis	ریزه خواری
receptor fitting (RF)	ریسپتور فیتینگ
ristocetin	ریستوستین
risk	ریسک
recurrent risk	ریسک عود کننده/برگشتی
individual risk	ریسک فردی
spinning	ریسندگی
free radical	ریشه آزاد
radicular, systeomics	ریشه ای
mandrake root	ریشه تمیس
rhizosphere	ریشه سپهر
hyphae	ریشه قارچها
rhizosphere	ریشه گاه
rhizospheric	ریشه گاهی
mandrake root	ریشه مهر گیاه
rifampicin	ریفآمپیسین
rifampin	ریفآمپین
refractile body	ریفرکتایل بادی
vitamin d-dependency rickets	ریکتزی متکی به ویتامین دی.
X-linked hypophosphatemic rickets	ریکتزی هیپوفسفاتمیکی مربوط به کروموزوم اکس.
hypophosphataemic ricket	ریکت هایپوفسفاتمیکی
high-frequency recombinant	ریکومبینان پر بسآمد
relaxin	ریلاکسین

ن

zein	زئین
redundancy	زائد بودن
gene	زاد
generation, parturition	زادآوری
genome	زادان

genecology	زادبوم شناسی	gap period	زمان بینابینی
structural gene	زادژن	chronobiochemistry	زمان-بیوشیمی
genetics	زادشناسی	generation time	زمان تولید
generation	زادگان	latency period	زمان درنگ
breeding	زادگیری	generation time	زمان زادآوری
outcrossing	زادگیری چلیپایی	mean lifetime	زمان میانگین عمر
breed	زادگیری کردن	generation time	زمان نسل
genotype	زادمون	activated partial	زمان نیمه ترومبوپلاستین فعال
generation	زادو ولد	thromboplastin time	شده
offspring	زاده ها	fallow	زمین آیش
gamete	زامه	fat(s)	زمین حاصلخیز
xanthopterin	زانتوپترین	geomicrobiology	زمین زیست شناسی میکروبی
xanthosine	زانتوزین	geomicrobiology	زمین زیوه شناسی
xanthoxin	زانتوکسین	geotropic	زمین گرا
bond angle	زاویه پیوند	geotropism	زمین گرایی
fecundity, fertility	زایایی	wasteland	زمین مرده
generation	زایش	geomicrobiology	زمین میکروبشناسی
fauna	زایگان	domain	زمینه
generator	زایگر	sequence	زنجیره
xylan	زایلن	A chain	زنجیره آ.
xylose	زایلوز	fibrinogen, alpha chain	زنجیره آلفا فیبرینوژن
xylulose	زایلولز	electron transport chain	زنجیره انتقال الکترون
xylitol	زایلیتول	control sequence	زنجیره
xylene	زایلین		بازبینی/کنترل/نظارت/کاربر
delivery, parturition	زایمان		ی/تنظیم/بازرسی
germinate	زاییده شدن	J chain	زنجیره جی.
ultrasonic	زبر صوتی	concatemer	زنجیره چند واحدی
necrotic ulcer	زخم بافت مرده	light chain	زنجیره سبک
necrotic ulcer	زخم نکروزی	cessation cassette	زنجیره قطع کننده/جلوگیری
con-till (= conservation	زراعت حفاظتی		کننده
tillage)		concatemer	زنجیره کنکاتامر
agricultural	زراعی	autonomous replicating	زنجیره همانند ساز اتونوموس
gall, bile	زرداب	segment (ARS) element	
lutein	زرده	parasitism	زندگی انگلی
xanthophyll	زردینه	parasitism	زندگی طفیلی
xanthophyll	زردینه برگ	resuscitation	زنده کردن
xeroderma pigmentosum	زرودرما پیگمنتوسوم	rust	زنگ
activated charcoal	زغال جذب	rust	زنگ آهن
activated charcoal	زغال چوب اشتار	rust	زنگار
activated charcoal	زغال فعال	rust	زنگ خوردگی
activated charcoal	زغال فعال شده	rust	زنگ زدگی

English	Persian	English	Persian
rust	زنگ زدن	biome	زی بوم
smut	زنگ سیاه	leucine zipper	زیپ لوسین
smut	زنگ گیاهی	zearalenone	زیرالنون
gynandromorph	زن-مرد ریخت	hypostasis	زیر ایستایی
xenoantibody	زنوآنتی بادی	substrate	زیربستر
xenoantigen	زنوآنتی ژن	chromogenic substrate	زیر بستر رنگ زا
xenoantiserum	زنوآنتی سروم	bed	زیر بنا
xenobiochemistry	زنوبیوشیمی	submetacentric	زیرپس میانپار
xenopus laevis	زنوپوس لائه ویس	hypoderm	زیرپوست
xenograft	زنوگرافت	hypodermal, hypodermic	زیرپوستی
zoaster	زوآستر	showering	زیر دوش شستن
zooplankton	زوئوپلانکتون	substrate	زیر ساخت
zoo blot	زو بلات	subcellular	زیر سلولی
grade	زوج	subculture	زیر-کشت
base pair, bP (= base Pair)	زوج باز	subspecies	زیرگونه ای
diploid	زوجی	hypophysis	زیرمغزی
zoogloea	زوگلوئیا	hypothalamus	زیر نهنج
zoonoses	زونوز	monomer, subunit	زیر واحد
zoonotic	زونوزی	bio	زیست
gall, toxin	زهر	bioleaching	زیست آبشویی
virulent	زهرآگین	bioassay	زیست آزمون
toxicity	زهر آگینی	biopesticide	زیست آفت کش
virulence	زهرآگینی	biosynthesis	زیست آمایی
malignant	زهرآلود	biorythm	زیست آهنگ
mycotoxin, toxin, vomitoxin	زهرابه	bionics	زیستار شناسی
		bioluminescence	زیست افروزش
pathogen toxin	زهرابه بیماری زا	bioregion	زیست اقلیم
endotoxin	زهرابه داخلی	bioaccumulant	زیست انباشت
marine toxin	زهرابه دریایی	bioaccumulate	زیست انباشتن
lethal toxin	زهرابه مهلک	bioaccumulation	زیست انباشتی
epsilon toxin	زهر اپسیلون	bioinformatics	زیست انفورماتیک
diptheria toxin	زهر دیفتری	biosafety	زیست ایمنی
toxoid	زهر ماند	viability	زیستایی
bile	زهره	bio-bar code	زیست بارکد
zeatin	زیاتین	biorecovery	زیست بازیافت
redundancy	زیادی	biome	زیست بوم
zeaxanthin	زیازانتین	viability	زیست پذیری
zootoxin	زیا زهر	biopolymer	زیست پرپار
zeaxanthin	زیاکسانتین	bioluminescence	زیست تابی
hazard	زیان	biotransformation	زیست تبدیلی
mercury	زیبق	biodegradable	زیست تجزیه پذیر

biodeteroriation	زیست تخریبی	bioastronautics	زیست فضانوردی
biochip	زیست تراشه	biotechnology	زیست فناوری
protein biochip	زیست تراشه پروتئین	environmental biotechnology	زیست فناوری زیست محیطی
biotechnology	زیست تکنولوژی	agrobiotechnology	زیست فناوری کشاورزی
bioregulator	زیست تنظیم گر	biotechnology	زیست فن شناسی
biomass	زیست توده	agrobiotechnology	زیست فن شناسی کشاورزی
active biomass	زیست توده فعال/کنش ور		
biotope	زیست جا		
biotope	زیست جای	biophysics	زیست فیزیک
biosensor	زیست حسگر	bioenergy	زیست کارمایه
niche	زیستخوان	biosphere	زیست کره
biopharmaceutical	زیست دارویی	biocide	زیست کش
bioremediation	زیست درمانی	biotope, habitat	زیستگاه
bioavailability	زیست دستیابی/دسترسی	biocoenosis	زیستگاه نیادی
biotransformation	زیست دگرگونی	biocoenosis	زیستگاه نیمه طبیعی
biomonitoring	زیست ردیابی	biotype	زیست گروه
biogenesis	زیست زایی	biogeography	زیست گیتا شناسی
biosynthesis	زیست ساختی	bioreceptor	زیست گیرنده
bioassay	زیست سنجش	protein bioreceptor	زیست گیرنده پروتئین
bioassay	زیست سنجی	biofilm	زیست لایه
biological, biologics	زیست شناختی	bionics	زیست لگام شناسی
neurobiologist	زیست شناس اعصاب	biomotor	زیست محرکه
biology	زیست شناسی	biomotor	زیست موتور
neurobiology	زیست شناسی اعصاب	bioengineering	زیست مهندسی
combinatorial biology	زیست شناسی ترکیبی	bioregion	زیست ناحیه
sociobiology	زیست شناسی جامعه	biorythm	زیست نواخت
structural biology	زیست شناسی ساختاری	bio reactor, bioreactor	زیست واکنشگر
cryobiology	زیست شناسی سرما	biological, biotic	زیستی
exobiology	زیست شناسی کیهانی	biologicals	زیستی ها
computational biology	زیست شناسی محاسبه ای	biocatalyst	زی فروکافنده
immunobiology	زیست شناسی مصونیت	sigma (= σ)	زیگما
molecular biology	زیست شناسی ملکولی	zygospore	زیگواسپور
biochemistry	زیست شیمی	zygote	زیگوت
biochemical	زیست شیمیائی	enzyme	زیمایه
biochemist	زیست شیمیدان	extremozyme	زیمایه غائی/فرینه/نهائی
activated biofilter	زیست صافی آکتیو شده	zymurgy	زیمورژی
bioscience	زیست علم	zymogen	زیموژن
bioassay	زیست عیاریابی	biotype	زیمون
bioinorganic	زیست غیر آلی	grade	زینه
biodegradation	زیست فروزینگی	biocide	زیوا کش
bioaugmentation	زیست فزونی	biota	زیوگان

microbe	زیوه	BRCA 1 gene	ژن بی.آر.سی.ای. نوع 1
microbiology	زیوه شناسی	BRCA 2 gene	ژن بی.آر.سی.ای. نوع 2
		BXN gene	ژن بی.اکس.ان.
ژ		BLA gene	ژن بی.ال.ای.
		BAR gene	ژن بی.ای.آر.
gene	ژانه	bcr-abl gene	ژن بی.سی.آر.-ای.بی.ال.
gel	ژل	gap gene	ژن بینابینی/گپ
acrylamide gel	ژل آکریلامید	jumping gene	ژن پرشی
agarose gel	ژل آگاروز	contiguous gene	ژن پیاپی
gelatinase	ژلاتیناز	PAT gene	ژن پی.ای.تی.
gelatinization	ژلاتینیزاسیون	early gene	ژن پیشین
gelation	ژلاسیون	linked gene/marker	ژن پیوسته
sizing gel	ژل اندازه گیری	gene targeting	ژن تارگتینگ
sequencing gel	ژل تعیین توالی	stacked gene	ژن تجمع یافته
orthogonal field alternation gel	ژل تغییر قائمی زمینه	modifying gene	ژن تغییر دهنده
spacer gel	ژل حد فاصل	regulatory gene	ژن تنظیم کننده
gelsolin	ژلسولین	immortalizing oncogene	ژن توموری پایدار
gel	ژله	transforming oncogene	ژن توموری تغییر یابنده
gem	ژم	cellular oncogene	ژن توموری سلولی
gene	ژن	oncogene	ژن توموری غده زا
architectural gene	ژن آرکیتکچرال	dominant(-acting) oncogene	ژن توموری (-فعال) غالب
AcuronTM gene	ژن آکورون	recessive oncogene	ژن توموری نهفته
antimutator gene	ژن آنتی موتاتور	tra gene	ژن تی.آر.ای.
HAP gene (= highly available phosphorous gene)	ژن اچ.ای.پی.	genetics	ژنتیک
		immunogenetics	ژنتیک ایمنی
		genetic targeting	ژنتیک تارگتینگ
gene stacking	ژن استکینگ	radiation genetics	ژنتیک تشعشعی
FAD gene (= flavin adenine dinucleotide)	ژن اف.ای.دی.	population genetics	ژنتیک جمعیت
		pharmacoenvirogenetics	ژنتیک داروشناختی محیطی
genecology	ژن اکولوژی	cytogenetics	ژنتیک سلولی
NARK gene	ژن ان.ای.آر.کی.	molecular genetics	ژنتیک ملکولی
MYC gene/oncogene	ژن/انکوژن ام.وای.سی.	cytogenetics	ژنتیک یاخته ای
immediate early gene	ژن اولیه فوری	jumping gene, roving gene	ژن جا به جا شدنی
ALS gene (= acetolactate synthase gene)	ژن ای.ال.اس.	mutator gene	ژن جهشگر
		GO gene	ژن جی.او.
BAR gene	ژن بار	multiple aleurone layer (MAL) gene	ژن چند لایه آلورونی
suppressor gene	ژن بازدارنده		
naked gene	ژن برهنه	phosphate transporter gene	ژن حامل فسفات
gene + chromosome	ژن بعلاوه کروموزوم	extrachromosomal gene	ژن خارج کروموزومی
BRCA gene	ژن بی.آر.سی.ای.	extranuclear gene	ژن خارج هسته ای
		selectable marker gene	ژن دارای نشانه قابل انتخاب

English	Persian	English	Persian
gene therapy	ژن درمانی	homologue gene	ژن مشابه
somatic cell gene therapy	ژن درمانی سلول جسمی	artificial gene	ژن مصنوعی
germ cell gene therapy	ژن درمانی سلول جنسی	architectural gene	ژن معماری
germ cell gene therapy	ژن درمانی یاخته تناسلی	heat shock gene	ژن (مقاوم به) شوک گرمائی
generator	ژنراتور	fusion gene	ژن ممزوج
ras gene	ژن راس	selector gene	ژن منتخب
redement napole gene	ژن ردمنت ناپول	napole gene	ژن ناپول
generic	ژنریک	nod gene	ژن ناد
structural gene	ژن ساختاری	immortalizing gene	ژن نامیرا سازی
architectural gene	ژن ساختمانی	regulatory gene	ژن نظم دهنده
maker gene	ژن سازنده	lux gene	ژن نور
genestein	ژنستئین	constitutive gene	ژن نهادی
ras oncogene	ژن سرطان زای موشی	recessive gene	ژن نهفته
breast cancer gene	ژن سرطان سینه	nif gene	ژن نیف
tumor-suppressor gene	ژن سرکوبگر تومور	hemizygous gene	ژن نیم جور تخم
roving gene	ژن سرگردان	sublethal gene	ژن نیمه-کشنده
toxicogenomics	ژن سم شناسی	pseudogene	ژن واره
citrate synthase gene	ژن سنتاز سیترات	genotoxic	ژنوتوکسیک
switch gene	ژن سوئیچ/کلید	genotype	ژنوتیپ
gene switching	ژن سوئیچینگ	genosensor	ژنوسنسور
extranuclear gene	ژن سیتوپلاسمی	genophore	ژنوفور
acetolactate synthase gene	ژن سینتاز استو لاکتیت	genome	ژنوم
indicator gene	ژن شاخص	functional genomics	ژنوم شناسی
genomics	ژن شناسی		کارکردی/عملی/کاربردی
structural genomics	ژن شناسی ساختاری	genomic blotting	ژنومیک بلاتینگ
mass-applied genomics	ژن شناسی کاربرد توده ای	genonema	ژنونما
reverse genetics	ژن شناسی معکوس	housekeeping genes	ژنهای اداره کننده
antimutator gene	ژن ضد جهشگر	orthologous genes	ژنهای ارتولوگ
bcr-abl gene	ژن عامل سرطان خون انسان	additive genes	ژن های افزایشی
dominant gene	ژن غالب	reiterated genes	ژنهای برگشتی
polygene	ژن فرعی	orphan genes	ژن های بیکاره
highly available phosphorous gene	ژن فسفر شدیدا در دسترس	late gene	ژن های تازه
		housekeeping genes	ژنهای خانه داری
mutable gene	ژن قابل جهش	proto-oncogene	ژنهای سلولی سرطان زا
pseudogene	ژن کاذب	syntenic genes	ژن های سینتنیک
lethal gene	ژن کشنده	additive genes	ژن های فزاینده
reporter gene	ژن گزارشگر	plantigens	ژن های گیاهی
lux gene	ژن لاکس	housekeeping genes	ژنهای هاوس کیپینگ
jumping gene	ژن متحرک	overlap genes, overlapping genes	ژن های هم پوشان
analog gene	ژن متشابه		
sex-linked gene	ژن متصل به جنسیت	hometic genes	ژنهای هومتیکی

target gene	ژن هدف	supramolecular assembly	ساختار مولکولی بسیار بزرگ
analog gene, homologue gene	ژن همانند	primary structure	ساختار نخستین
		Holliday structure	ساختار هالیدی
contiguous gene	ژن هم جوار	structure	ساختمان
fusion gene	ژن همجوش	native structure	ساختمان بومی
allele	ژن همردیف	cell membrane structure	ساختمان دیواره سلولی
analog gene, v	ژن همسان	tertiary structure	ساختمان سومین
hemizygous gene	ژن همیزایگوس	cell membrane structure	ساختمان غشاء سلولی
holandric gene	ژن هولاندریک (پدر به پسر)	structure	ساختن
genistein	ژنیستئین	sarafotoxin	سارافوتوکسین
genistin	ژنیستین	sarcosine	سارکوزین
syntenic genes	ژنهای هم مکان	sarcolemma	سارکولما
gibberella zeae	ژیبرلا زیا	sarcoma	سارکوما
gibberellin	ژیبرلین	sarkomycin	سارکومایسین
gyrase (= DNA gyrase)	ژیراز	Kaposi's sarcoma	سارکومای کاپوزی
		acclimatization	سازش با محیط
س		cold acclimatization	سازش با محیط سرد
		staggered conformation	سازش پله ای
submetacentric	سابمتاسنتریک	adaptor	سازش دهنده
saprozoic	ساپروزوئیک	native conformation	سازش طبیعی
saprophagy	ساپروفاژی	adaptation, compensation, consistency	سازگاری
saprophyte	ساپروفیت		
saponin, saponnin (=saponin)	ساپونین	histocompatibility	سازگاری بافتی
		histocompatibility	سازگاری نسجی
satratoxin	ساتراتوکسین	ecosystem	سازگان بوم شناختی
southern blot/transfer	ساترن بلات/ترانسفر	multienzyme system	سازگان چند آنزیمی
southern blotting	ساترن بلاتینگ	structure	سازمان دادن
southwestern blot	سات وسترن بلات	organism	سازواره
conformation, construct, structure, synthesis	ساخت	T-cell independent mechanism	سازوکار مستقل سلول تی.
structure	ساختار	T-cell dependent mechanism	سازوکار وابسته سلول تی.
protein structure	ساختار پروتئین		
unipartite structure (chromosome)	ساختار تک قسمتی/منفرد	pathway feedback mechanisms	سازوکار های پس خور مسیر (شیمیائی)
secondary structure of protein	ساختار ثانوی پروتئین	hour	ساعت
		saccharase	ساکاراز
quaternary structure	ساختار چهارم	saccharomyces cereviciae (= baker's yeast)	ساکارومایسز سرویسیا
tertiary structure	ساختار سومین		
tertiary structure of protein	ساختار سومین پروتئین	saccharification	ساکاریفیکاسیون
cap structure	ساختار کلاهکی	saxitoxin	ساکسیتوکسین
lock-washer structure	ساختار لاک-واشر	electrostatic	ساکن الکتریسیته ای

93

English	Persian	English	Persian
salting out	سالتینگ آوت	chloroplast	سبزدیسه
salting in	سالتینگ این	germinate	سبز کردن
senescence	سالخوردگی	legume	سبزی
worried well	سالم نگران	chlorophyll	سبزینه
salmonella	سالمونلا	whiskers	سبیل (در گربه و حیوانات)
salvarsan	سالوارسان	biological shield	سپر زیستی
organism	سامانه	acropachia	ستبر استخوانی
centrosome	سانتروزوم	acropachyderma	ستبر پایانکی
centromere	سانترومر	pachytene	ستبر نوار
centrifuge	سانتریفوژ	sterile	سترون
centrifugation	سانتریفوژاسیون	sterilization	سترون سازی
active-enzyme centrifugation	سانتریفوژاسیون آنزیم فعال شده	sizing column	ستون اندازه گیری
differential centrifugation	سانتریفوژاسیون دیفرانسیلی	packed column	ستون فشرده
density gradient centrifugation	سانتریفوژ با شیب چگالی	cetylpyridinium	ستیل پیریدینیوم
		dura mater	سخت شامه
zonal centrifugation	سانتریفوژ ناحیه ای	atherosclerosis	سخت شدگی سرخرگها
centrioles	سانتریول(ها)	induration	سخت شدن همراه با تورم
celsius	سانتیگراد	hardening	سخت گردانی
centimorgan	سانتی مورگان	sclerotium	سختینه
emergency	سانحه	permeability barrier	سد تراوایی
sangivamycin	سانگیوا مایسین	genetic block	سد ژنتیکی
cybrid	سایبرید	blood-brain barrier	سد مغزی-خونی
allosteric site	سایت آلوستریک	sodium	سدیم
acceptor junction site	سایت اتصال دریافت کننده	contagion, infection, infectiousness, infestation, transmission	سرایت
acceptor site	سایت گیرنده/اکسپتور		
sequence-tagged site (STS)	سایت نشاندار شده با توالی	indirect transmission	سرایت غیر مستقیم
cytidine diphosphate	سایتیدین دیفسفات	contaminate	سرایت کردن
cytidine monophosphate	سایتیدین مونوفسفات	cerebrose	سربروز
sizing	سایزینگ	oversight	سرپرستی
ablation	سایش	cap structure	سرپوش
cyclosporin	سایکلو اسپورین	serpin	سرپین
cyclooxygenase	سایکلو اکسیژناز	proband	سر تبار
cycloserine	سایکلوسرین	origin	سرچشمه
cyclophosphamide	سایکلوفسفامید	replication origin	سرچشمه همانند سازی
cycloheximide	سایکلوهگزیمید	carotid artery	سرخرگ کاروتید
crown gall	سایدگی تاج	redness	سرخی
powder	ساییدن	fog	سردرگمی
powdered	ساییده	proband	سر دودمان
etiology	سبب شناسی	decanting	سرریزکردن
chloroplast	سبز دش	decanter	سرریزکن

94

decantation	سرریزکنی	serology	سرم شناسی
cancer	سرطان	seronegative	سرم منفی
oncogene	سرطان زا	immune sera	سرم های ایمنی
genotoxic carcinogen	سرطان زای ژنوتوکسیک	environmental fate	سرنوشت محیط زیست
carcinogenicity, oncogenesis	سرطان زایی	serotonin	سروتونین
oncology	سرطان شناسی	serotype	سروتیپ
cancerocidal	سرطان کش	seroconversion	سروکانورژن
hereditary non-polyposis colorectal cancer	سرطان کولو-رکتال ارثی غیر پولیپوزی	asymptomatic seroconversion	سروکانورژن آسیمپتوماتیک
inhibition, quelling, repression, silencing	سرکوب	maternal serum AFP	سروم مادری ای.اف.پی.
feedback inhibition	سرکوب پسخور	cascade	سری
uncompetetive inhibition	سرکوب غیر رقابتی	sericin	سریسین
catabolite repression	سرکوب فرو گوهره	serine	سرین
cumulative feedback inhibition	سرکوب فیدبک تراکمی	flaccid, labile	سست
enzyme repression	سرکوب کردن زیمایه/دیاستاز/آنزیم	cessation cassette	سسیشن کاست
		level	سطح
inhibitor, silencer, suppressor	سرکوبگر	spread plate	سطح پخش
		exposure	سطح تماس
amber suppressor	سرکوبگر امبر/عنبری	containment level	سطح تنگداشت
chain terminating suppressor	سرکوبگر پایان بخش زنجیره	physical containment level	سطح حصر فیزیکی و ایمنی
pancreatic trypsin inhibitor	سرکوبگر تریپسین پانکراتیکی	acceptable level of risk	سطح ریسک قابل قبول
inhibitor of sister chromatid separation	سرکوبگرجداسازی کروماتیدهای خواهر	no-observed effects level	سطح عدم مشاهده اثر
		no-observed adverse effects level	سطح عدم مشاهده اثر سوء
crossover suppressor	سرکوبگر کراس اوور	trophic level	سطح غذایی
lactose repressor	سرکوبگر لاکتوز	theoretical plate	سطح فرضیه ای
gene silencing	سرکوب نمایانی ژن (ی)	containment level	سطح محدودیت
down processing, down regulation	سرکوب واکنش عادی	hybridization surface	سطح هیبریدیزاسیون/هم تیرگی
down regulating	سرکوب واکنش عادی یک اندام یا سیستم	bucket	سطل
genetic drift	سرگردانی ژنتیکی	biosafety levels	سطوح زیست ایمنی
serum	سرم	cefazolin	سفازولین
serum albumin	سرم آلبومین	cephalosporin	سفالواسپورین
psychrophile	سرما خواه	ceftriaxone	سفتریاکزون
psychrophile	سرما دوست	gel	سفت شدن
blood plasma, blood serum	سرم خون	leukocytosis	سفید گویچه تبسی
serologist	سرم شناس	leukopenia	سفید گویچه کاستی
		leukocyte	سفید گویچه/یاخته
		achromotrichia	سفید مویی مادرزاد
		albumin	سفیده تخم مرغ
		ectromelia	سقط اندام

English	Persian
ceiling limit	سقف مجاز
coronary thrombosis, infarction	سکته قلبی
secretase	سکرتاز
secretagogue	سکرتاگوگ
secretor	سکرتور
secretogogue	سکرتوگوگ
secretin	سکرتین
cecropin	سکروپین
cecropin a	سکروپین ای.
cecrophins	سکروفین ها
sexduction	سکسداکشن
population	سکنه
sequon	سکوئون
tuberculosis	سل
weapon	سلاح
biological weapon, bioweapon	سلاح بیولوژیکی
weapon of mass destruction	سلاح تخریب گروهی
toxin weapon	سلاح توکسینی
toxin weapon	سلاح زهرابه ای
chemical weapon	سلاح شیمیائی
weapon of mass destruction	سلاح کشتار جمعی
incapacitate	سلب صلاحیت کردن
cell turnover	سل ترن اوور
lupus	سل جلدی
phylum	سلسله
isogenic lines	سلسله های ایزوژنیکی
celsius	سلسیوس
selectin	سلکتین
cell line	سل لاین
selenoprotein	سلنوپروتئین
selenocysteine	سلنوسیستئین
selenomethionine	سلنومتیونین
cell	سلول
Hfr cell	سلول اچ.اف.آر.
facultative cell	سلول اختیاری
antigen presenting cell, antigen-presenting cell (APC)	سلول ارائه دهنده آنتی ژن
cellular adhesion receptor	سلولار ادهیژن ریسپتور

English	Persian
cellular response	سلولار ریسپانس
cellulase	سلولاز
Spot-Formi cell	سلول اسپات-فرمی
F-cell	سلول اف.
F'	سلول اف. پرایم/پریم
F+-cell	سلول اف. مثبت
LAK cell (= lymphokin activated killer cell)	سلول ال.ای.کی.
initial cell, primary cell	سلول اولیه
mesenchymal adult stem cell	سلول بنیادی بالغ میان آگنه ای
fetal calf stem	سلول بنیادی جنین گوساله
hemopoietic stem cells	سلول بنیادی خون ساز
hemopoietic stem cells	سلول بنیادی هموپوئتیک
phagocyte	سلول بیگانه خوار
stem cell	سلول پایه
adult stem cell	سلول پایه بالغ
totipotent stem cell	سلول پایه پر توان
embryonic stem cell	سلول پایه جنینی
human embryonic stem cell	سلول پایه جنینی انسان
pluripotent stem cell	سلول پایه چند قوه زا
germ cell, oocyte, zygote	سلول تخم
antibody-secreting cell	سلول تراوش کننده آنتی بادی
haploid cell	سلول تک لاد
mononuclear cell	سلول تک هسته ای
gamete	سلول تناسلی
somatic cell	سلول تنی
T-cell	سلول تی.
killer T cell (=cytotoxic cell)	سلول تی. کشنده
accessory cell	سلول جانبی
somatic cell	سلول جسمی
sex cell	سلول جنسی
adipocyte	سلول چربی
adipocyte	سلول چربی(حیوانی)
dendritic cell	سلول دندرایتی/دندریتی
diploid cell	سلول دوتایی/دیپلوئید
pluripotent stem cell	سلول دودمانی چند قوه زا
cell hybrid	سلول دو رگه
cellulose	سلولز
gamete	سلول زایشی

sector cell	سلول سکتوری/ترشحی	toxin	سم
soma cell	سلول سوما	epsilon toxin	سم اپسیلون
cytotoxic cell	سلول سیتوتوکسیک	ion-channel-binding toxin	سم اتصال مجرای یون
cytology	سلول شناسی	staphylococcal toxin	سم استافیلو کوکی
accessory cell	سلول فرعی	fusion toxin	سم امتزاج/همجوشی
flow cell	سلول فلو/جریان	bacterial toxin	سم باکتری
cultured cell	سلول کشت شده	brevetoxin	سم بروه
cultured cell	سلول کشتی	botulin toxin	سم بوتولین
killer cell	سلول کشنده	botulinum toxin	سم بوتولینوم
lymphokin activated killer cell	سلول کشنده فعال شده توسط لنفوکین	pathogen toxin	سم بیماری زا
		duster	سم پاش
zygote	سلول گشنیده	Chapin plant and rose powder duster	سم پاش چاپین
vegetative cell	سلول گیاهی		
recipient cell	سلول گیرنده	London Fog foggers	سمپاش لندن فاگ
langerhans cell	سلول لانگرهانس	dust	سم پاشیدن
dendritic langerhans cell	سلول لانگر هانس دندریتی/دندرایتی	Pertussis toxin	سم پرتوسیس
		protein toxin	سم پروتئینی
Leydig cell	سلول لایدیگ	nadir	سمت القدم
sperm mother cell	سلول مادر اسپرم	pyrogenic toxin	سم تب آور/زا
mast cell	سلول ماست	zootoxin	سم جانوری
anchorage-dependent cell	سلول متکی به لنگرگاه/آنکوریج	algal toxin	سم جلبکی
Bursa dependent cells	سلول متکی/وابسته به بورسا	exotoxin	سم خارجی
compound heterozygote	سلول مرکب	marine toxin	سم دریایی
competent cell	سلول مستعد/توانمند	pesticide	سم دفع آفات
host cell	سلول میزبان	diptheria toxin	سم دیفتری
phagocyte	سلول میکروب خوار	detoxication	سم زدائی
microbiological cell	سلول میکروبیولوژیکی	cytotoxic	سم سلولی
rod cell	سلول میله ای	shiga toxin	سم شیگا
haploid cell	سلول هاپلوئید	neurotoxin	سم عصب(ی)
long-term culture-initiating cells	سلولهای آغاز گر کشت بلند مدت	toxoid	سم غیرفعال
		paralytic shellfish toxin	سم فلج کننده صدف(ها)
ES cells (= embryonic stem cells)	سلول(های) ئی.اس.	paralytic cobra toxin	سم فلج کننده کبرا
		von Willebrand's toxin	سم فون ویلبراند
beta cells	سلول های بتا	fungal toxin	سم قارچی
peripheral blood mono-nuclear cells	سلولهای تک هسته ای فرعی خون	tetanus toxin	سم کزاز
		lethal toxin	سم کشنده
naive T cells	سلولهای تی. ساده	clostridium perfringens toxin	سم کلوستریدیوم پرفرینژن
entrapped cells	سلول های در تله/گیر افتاده		
hela cell	سلول هلا	toxoid	سم گونه
eukaryotic cell	سلول یوکاریوتیک	phytotoxin, plant toxin	سم گیاهی
cell hybrid	سل هیبرید	labile toxin	سم لابایل

metabolic poison	سم متابولیکی
cholera toxin	سم وبا
malignant, virulent	سمی
toxicity	سمیت
aluminum toxicity	سمیت آلومینیم
acute toxicity	سمیت حاد
synaptobrevin	سناپتوبروین
cluster	سنبله
synthase	سنتاز
citrate synthase	سنتاز سیترات
ligase (DNA), synthetase	سنتتاز
fatty acid synthetase	سنتتاز اسید چرب
centrin	سنترین
synthesis	سنتز
DNA synthesis	سنتز دی. ان.ای.
unscheduled DNA synthesis	سنتز دی.ان.ای. خارج از برنامه
gene synthesis	سنتز ژن (ی)
solid-phase synthesis	سنتز فاز جامد
synthesizer	سنتز کننده
de novo synthesis	سنتز نوپدید/دینوو
photosynthesis	سنتز نوری
synthesizer	سنتسایزر
meter	—سنج
biotype	سنج زیستی
assay, assessment, measurement	سنجش
enzyme assay	سنجش آنزیم
linked enzyme assay	سنجش آنزیم وابسته
enzyme immunoassay	سنجش ایمنی آنزیم
enzyme-linked immunoassay	سنجش ایمنی وابسته به آنزیم
enzyme-linked immunosorbent assay	سنجش ایمونوسوربنت متصل به آنزیم
enzyme-linked-immunosorbent assay	سنجش ایمونوسوربنت وابسته به آنزیم
biological measurement	سنجش بیولوژیکی
audiometrician	سنجشگر شنوایی
multiplex assay	سنجش مرکب
luminescent assay	سنجش نورافشانی/نورزائی
meter, standard	سنجه
Alport syndrome	سندرم آلپورت
alkaptonuria	سندرم ادرار چای/کوکاکولا
Hunter's syndrome	سندرم هانتر
syndrome	سندروم
Angelman syndrome	سندروم آنجلمان
Edward syndrome	سندروم ادوارد
cerebro-oculo-facial skeletal syndrome	سندروم اسکلتی سربرو-اوکلو-فاسیال
fragile X syndrome	سندروم اکس شکننده
Ehlers-Danlos syndrome	سندروم اهلرز-دانلوس
Beckwith-Weideman syndrome	سندروم بکویت-ویدمان
Prader-Willi syndrome	سندروم پرادر-ویلی
turner syndrome	سندروم ترنر
trisomy-18 syndrome (Edward syndrome)	سندروم تریزومی 18
Chediak-Higashi syndrome	سندروم چدیاک-هیگاشی
oculo-dental-digital syndrome	سندروم چشم-دهان-انگشت
Down's syndrome	سندروم داون
Grieg's cephalo polysyndactyly syndrome	سندروم سفالو پلی سینداکتیلی گریگ
Wolf-Hirosch horn syndrome	سندروم شاخ ولف-هیروش
prune belly syndrome	سندروم شکم آلوئی
testicular feminization syndrome (FTS)	سندروم فمینیزه شدن بیضه ای
Acquired Immune Deficiency Syndrome (AIDS)	سندروم کاهش ایمنی اکتسابی (ایدز)
Klinefelter syndrome	سندروم کلاین فلتر
Cockayne syndrome	سندروم کوکین
Lesch Nyhan syndrome	سندروم لش نایهان
Marfan syndrome	سندروم مارفان
myelo-dysplastic syndrome	سندروم مایلو-دیسپلاستیک
nephrotis syndrome	سندروم نفروتیس
congenital nephrotic syndrome	سندروم نفروتیک مادر زاد
congenital nephrotic syndrome of other types	سندروم نفروتیک مادر زاد نوع دیگر
congenital nephrotic syndrome of Finnish type	سندروم نفروتیک مادر زاد نوع فنلاندی

English	Persian	English	Persian
Wardenberg syndrome	سندروم واردنبرگ	sulfosate	سولفوسات
diabetes mellitus optic atrophy deafness	سندروم ولفرام	ethyl methane sulfonate	سولفونات اتیل متان
		sodium sulfite	سولفیت سدیم
Wiskott Aldrich syndrome	سندروم ویسکوت آلدریچ	hydrogen sulfide	سولفید هیدروژن
antigenic switching	سوئیچ آنتی ژنی	solenoid	سولنوئید
switching	سوئیچ کردن	somatacrin	سوماتاکرین
antigenic switching	سوئیچ کردن آنتی ژنها	somatostatin	سوماتواستاتین
gene switching	سوئیچ کردن ژن (ی)	somatoplasm	سوماتوپلاسم
Swingfog TM	سوئینگ فاگ	somatotropin	سوماتوتروپین
swarming	سوارمینگ	bovine somatotropin	سوماتوتروپین گاوی
subtilisin	سوبتیلیزین	somatotrophin (= somatotropin)	سوماتوتروفین
subtilin	سوبتیلین		
superoxide dismutase	سوپر اکسید دیس موتاز	somatoliberin	سوماتولیبرین
human superoxide dismutase	سوپر اکسید دیسموتاز انسانی	somatomammotropin	سوماتوماموتروپین
		somatomedin	سوماتو مدین
catabolism	سوخت	Somalia	سومالی
biofuel	سوخت بیولوژیکی	soman	سومان
injector	سوخت پاش	sonography	سونوگرافی
metabolism	سوخت و ساز	sense	سوهش
nitrogen metabolism	سوخت و ساز ازت	sense	سوهشی
intermediary metabolism	سوخت و ساز واسطه ای	malnutrition	سوء تغذیه
port	سوراخ	soya bean	سویا
vent	سوراخ	strain	سویه
sorbose	سوربوز	transgressive variation	سویه ترانسگرسیو
sorbitol	سوربیتول	disabled strain	سویه خنثی شده
sorcin	سورسین	inbred strain	سویه درون آمیخته
surfactant	سورفکتانت	somatic variant	سویه سوماتیکی
sociobiology	سوسیوبیولوژی	Abelson strain of murine	سویه (مربوط به جوندگان) آبلسونی
sucrase	سوکراز		
sucrose	سوکروز	live vaccine strain	سویه واکسن زنده
succus entericus	سوکوس انتریکوس	splice variants	سویه های بهم تابیده
solanine	سولانین	triploid	سه بخشی
alpha-solanine	سولانین آلفا	solid	سه بعدی
keratan sulfate	سولفات کراتان	triplet	سه تائی
magnesium sulfate	سولفات منیزیم	triploid	سه جزئی
mucoitin sulphate	سولفات موکوئیتین	triplet	سه حرفی
sulfatide lipidosis	سولفاتید لیپیدوز	hybrid	سه رگه
sulphobromo phtahalein	سولفوبروموفتالئین	triploid	سه قسمتی
sulphonation	سولفودار شدن/کردن	triplet	سه قلو
sulphonated	سولفودار شده	triploid	سه گان
sulforaphane	سولفورافین	triploid	سه گانه

99

triploid	سه لاد
moiety	سهم
oversight	سهو
C (= cytosine)	سی.
CR (= complement receptor)	سی.آر.
CRM (= cross reacting material)	سی.آر ام.
CRM (= chromatin remodeling machine)	سی.آر.ام.
CRP (= catabolite activator protein)	سی.آر.پی.
cis	سی.آی.اس.
CHEF (= contour-clamped homogeneous electric fields)	سی.چ.ئی.اف.
CHS (= Chediak-Higashi syndrome)	سی.چ.اس.
CHD (= congenital heart disease)	سی.چ.دی.
CS (= Cockayne syndrome)	سی.اس.
CF (= cystic fibrosis)	سی.اف.
CFE (= capital femoral epiphysis)	سی.اف.ئی.
CFTR (= cystic fibrosis transmembrane conductance regulator)	سی.اف.تی.آر
CFU (= colony forming units)	سی.اف.یو.
flux	سیال
CLL (= chronic lymphocytic leukemia)	سی.ال.ال.
fluidized	سیال شده
sialoadhesin	سیالو ادهسین
sialidase	سیالیداز
CM (= centimorgan)	سی.ام.
CMI (= cell mediated immunity)	سی.ام.آی.
CML (= chronic myelocytic leukemia)	سی.ام.ال.
CMP (= cytidine	سی.ام.پی.

monophosphate)	
CMD (= congenital myeloperoxidase deficiency)	سی.ام.دی.
CMV (= cytomegalovirus)	سی.ام.وی.
CNS (= central nervous system), CNS (= congenital nephrotic syndrome)	سی.ان.اس.
CNF (= congenital nephrotic syndrome of Finnish type)	سی.ان.اف.
CNO (= congenital nephrotic syndrome of other types)	سی.ان.او.
cyanobacteria	سیانوباکتری
cyanosis	سیانوز
cyanotic	سیانوزی
cyanogen	سیانوژن
COFS (= cerebro-oculo-facial skeletal syndrome)	سی.او.اف.اس.
COPD (= chronic-obstructive-pulmonary disease)	سی.او.پی.دی.
portal vein	سیاهرگ باب
jugular vein	سیاهرگ گردن
cutaneous anthrax	سیاه زخم پوستی
smut	سیاهک
CAH (= congenital adernal hyperplasia)	سی.ای.اچ.
CASP (= CTD-associated SR-like protein)	سی.ای.اس.پی.
CAAT (= CAAT box)	سی.ای.ای.تی
CAP (= catabolite activator protein)	سی.ای.پی.
C-banding	سی. باندینگ
sibriomotsin	سیبریوموتزین
sebum	سیبوم
CPSF (= cleavage and polyadenylation	سی.پی.اس.اف.

specificity factor)		chromosome)	
citreoviridin	سیترئوویریدین	mannuronic acid	سیدواسترپتومایسین اسید مانورونیک
cytoplast	سیتوپلاست	sidostreptomycin	
cytoplasm	سیتوپلاسم	cidofovir	سیدوفوویر
cortical cytoplasm	سیتوپلاسم پوسته ای	CD (= circular dichroism)	سی. دی.
cortical cytoplasm	سیتوپلاسم کورتیکی	CDH (= congenital	سی.دی.اچ.
apurinic cellular citotoxicity	سیتوتوکسیته سلولی آپورینیک	dislocation of the hip)	
antibody dependant cellular	سیتوتوکسیته سلولی متکی به	c-DNA (= complementary	سی دی. ان. ای.
cytotoxicity	آنتی بادی	DNA)	
cytotoxicity	سیتوتوکسیسیته	cDNA (= complementary	سی دی.ان.ای
cytotoxic	سیتوتوکسیک	DNA)	
cytotoxic	سیتوتوگزیک	CDA (= congenital	سی.دی.ای.
cytosol	سیتوزول	dyserythropoietic	
cytosine	سیتوزین	anaemia)	
cytosine arabinoside	سیتوزین آرابینوزید	CDP (= cytidine	سی.دی.پی.
cytogenetics	سیتوژنتیک	diphosphate)	
sitostanol	سیتوستانول	Cdk (= cyclin-dependent	سی.دی.کی.
beta sitostanol	سیتوستانول بتا	kinase)	
b sitostanol	سیتوستانول بی.	sirtuin	سیرتوئین
sitosterol	سیتوسترول	quenching	سیر کردن
beta sitosterol	سیتوسترول بتا	sirenin	سیرنین
cytochemistry	سیتوشیمی	cis	سیز
cytochalasin	سیتوکالازین	cisplatin	سیزپلاتین
cytochrome	سیتوکروم	sisomicin	سیزومیسین
cytochrome C	سیتوکروم سی.	cyst	سیست
cytokinetics	سیتوکینتیک	cysteine	سیستئین
cytokinesis	سیتوکینزیس	cystatin	سیستاتین
cytokines	سیتوکین ها	CystX	سیست ایکس
cytolysin	سیتولایزین	cistron	سیسترون
cytology	سیتولوژی	gene array system	سیستم آرایه ژن (ی)
cytolysis	سیتولیز	human leucocyte antigen	سیستم آنتی ژن لوکوسیتی انسانی
cytolysin	سیتولیزین	system	
flow cytometry	سیتومتری فلو/جریان	systematics	سیستماتیک
cell cytometry	سیتومتری یاخته ای	hospital information system	سیستم اطلاعات بیمارستان
CTL (= cytotoxic T	سی.تی.ال.	line-source delivery system	سیستم انتقال چشمه خطی
lymphocytes)		innate immune system	سیستم ایمنی سرشتی
CTP (= cytidine	سی.تی.پی.	micro total analysis system	سیستم تحلیل تظاهر ژن
triphosphate)		micro total analytical	سیستم تحلیلی تظاهر ژن
CGD (= chronic	سی.جی.دی.	system	
granulomatous disease)		restriction-modification	سیستم تحول محدودیت
CGY (= cell growth Y	سی.جی.وای	system	

101

English	Persian
SOS repair system	سیستم ترمیم اس.او.اس.
cell-free gene expression system	سیستم تظاهر ژنی بدون سلول
baculovirus expression vector system	سیستم حامل اکسپرشن باکولو ویروس
treatment system	سیستم درمان
two-hybrid system	سیستم دورگه
bacterial two-hybrid system	سیستم دو رگه باکتریایی
yeast two-hybrid system	سیستم دو رگه مخمری
micro-electromechanical system	سیستم ریز-الکترومکانیکی
zyme system	سیستم زایم
long-range biological standoff detection system	سیستم شناسائی ابر بیولوژیکی دوربرد
biological integrated detection system	سیستم شناسایی منسجم بیولوژیکی
micro-electromechanical system	سیستم میکرو الکترومکانیکی
host vector system	سیستم ناقل-میزبان
insertional knockout system	سیستم ناک اوت دخولی
nitrogenase system	سیستم نیتروژناز
sustainable intensification of animal production systems	سیستمهای افزایش پایدار تولید حیوانات
bio microelectromechanichal systems	سیستمهای بیو میکروالکترومکانیکی
systeomics	سیستمیک
cystitis	سیستیت
cystic fibrosis	سیستیک فیبروز
cystine	سیستین
cystinuria	سیستینوری
cystinosis	سیستینوز
krebs cycle	سیکل کربس
cyclodextrin	سیکلودکسترین
cycloheximide	سیکلوهگزیمید
cyclin	سیکلین
CKI (= cyclin-dependent kinase inhibitor)	سی.کی.آی.
flux	سیلان
cilia	سیلیا

English	Persian
ciliary neurotrophic factor (CNTF)	سیلیاری نروتروفیک فاکتور
silica	سیلیس
silica	سیلیکا
mercury	سیماب
symbiant	سیمبیانت
symport	سیمپورت
quantum wire	سیم کوانتومی
synaptotagmin	سیناپتوتگمین
synapse	سیناپس
aminoacyl-tRNA synthetase	سینتاز آمینوآسیل-تی.آر.ان.ای.
acetolactate synthase	سینتاز استو لاکتیت
endothelial nitric oxide synthase	سینتاز اکسید نیتریک اندوتلیال
syntrophsysteminy	سینتروفیستمینی
syntrophism	سینتروفیسم
syntomycin	سینتومایسین
syndactyly	سینداکتیلی
syndesis	سیندز
syndesine	سیندزین
syndecan	سیندکان
synergy	سینرژی
synergid	سینرژید
syncytium	سینسیتیوم
synkaryon	سینکاریون
synexin	سینکسین
syngamy	سینگامی
syngraft	سینگرافت
CVS (= chorionic villus sampling)	سی.وی.اس.
acromelanism	سیه پایانکی

ش

English	Persian
template	شابلون
chaperone	شاپرون
chaperonins	شاپرونین ها
indicator	شاخص
paternity index	شاخص پدری
quantitative character	شاخصه کمی
branch, phylum	شاخه

English	Persian
long arm of a human chromosome	شاخه بلند کروموزوم انسانی
cladistics	شاخه بندی
p arm (= short arm of a human chromosome)	شاخه پی.
spacer arm	شاخه حد فاصل
dendritic	شاخه دار
branch	شاخه شاخه شدن
codogenic strand	شاخه کدوژنیک/کد ساز
short arm of chromosome	شاخه کوتاه کروموزوم
short arm of a human chromosome	شاخه کوتاه کروموزوم انسانی
q arm (= long arm of a human chromosome)	شاخه کیو
keratin	شاخینه
flux	شار
fluidized	شاراینده شده
DNA shuffling	شافل دی.ان.ای.
chaconine	شاکونین
membrane	شامه
meningitis	شامه آماس
meninge	شامه گان
cell membrane (cell wall)	شامه یاخته
competency	شایستگی
endemic	شایع/رایج در محیطی خاص
affinity	شباهت
bioluminescence	شب تابی
ghost	شبح
berseem clover	شبدر برسیم/مصری
fluorescence	شب رنگی
sewreage system	شبکه پساب/فاضلاب
rough endoplasmic reticulum	شبکه درون دشته ای زبر
endoplasmic reticulum (ER)	شبکه درونی سیتوپلاسم
food web	شبکه غذایی
retinoid	شبکیه مانند
fluorescence	شب نمایی
pseudoallele	شبه آل
terpenoid	شبه ترپن
pararetrovirus	شبه رتروویروس
pseudo-xanthoma elasticum	شبه زانتوما الاستیکوم

English	Persian
pseudogene	شبه ژن
conventional pseudogene	شبه ژن قراردادی
toxoid	شبه سم
retinoid	شبه شبکیه
pseudo-pseudo-hypo-parathyroidism	شبه شبه هیپو پاراتیروئیدیسم
pseudodominance	شبه غلبه
alkaloid	شبه قلیایی
carotenoid	شبه کاروتن
pseudomonas	شبه موناس
viroid	شبه ویروس
procaryotes, prokaryotes	شبه هسته داران
pseudohemophilia	شبه هموفیلی
pseudo-hypo-parathyroidism	شبه هیپو پاراتیروئیدیزم
analogous	شبیه
simulation	شبیه سازی
pedigree	شجره
pedigree	شجره نامه
dendrite	شجری
dendritic	شجری
person	شخص
dispenser	شخص پخش/توزیع (کننده)
affected party	شخص مبتلا/متاثر
personal	شخصی
character	شخصیت
sex-conditioned character	شخصیت شرطی شده توسط جنسیت
sex-influenced character	شخصیت متاثر از جنسیت
sex-limited character	شخصیت محدود به جنس
stringency, virulence	شدت
syrup of ipecac	شربت اپیکا
syrup of ipecac	شربت الیون کوکی
annotation	شرح
company	شرکت
initiation	شروع
promoter, starter	شروع کننده
showering	شست وشو
molecular beacon	شعاع نور ملکولی
sense	شعور
predator	شکارچی

English	Persian
predator	شکارگر (حیوان)
niche, nick, vent	شکاف
hydrolytic cleavage	شکافتگی آبکافتی
orthophosphate cleavage	شکافتگی ارتوفسفات
hydrofluoric acid cleavage	شکافتگی اسید هیدروفلئوریک.
pyrophosphate cleavage	شکافتگی پیروفسفات
centric fission	شکافت مرکزی
synaptic cleft	شکاف هموری/سیناپسی
fatigue, nick	شکستگی
refraction	شکست (نور)
configuration, conformation	شکل
phenotype	شکل ظاهری
germinate	شکل گرفتن
replicative form of M13	شکل همانند ساز ام.13.
bowel	شکم
bowel, rumen	شکمبه
gastric	شکمی
dissociation	شکند
flaccid	شل
brassica	شلغم بیابانی (کلم پیچ و خانواده کلم)
hose	شلنگ
unit	شمار
meter	-شمار
bacteria count	شمارش باکتری ها
liquid-scintillation counting	شمارش درخشش-مایع
viable count	شمارش زیستی
viable count	شمار قابل رویش/زیست
meter	شمارگر
sizing	شماره بندی
MIM code number (= mendelian inheritance in man)	شماره کد ام.آی.ام.
copy number	شماره نسخه (ای)
familiarity	شناخت
recon	شناسائی
DNA typing	شناسائی از روی تیپ دی.ان.ای.
gene sequencing	شناسائی توالی ژن (ی)
rapid microbial detection	شناسائی سریع میکروب
sentinel surveillance	شناسائی سنتینلی/پاسداری

English	Persian
cellular pathway mapping	شناسائی مسیر سلولی
indicator	شناساگر
high-throughput identification	شناسایی با ظرفیت پذیرش بالا
de novo sequencing	شناسایی فی البداهه/بدون پیش آگاهی/اطلاع قبلی
cell recognition	شناسایی یاخته
indicator	شناسه
pathognomonic	شناسه بیماری
biomarker	شناسه زیستی
raft, supernatant	شناور
plankton	شناوران
phytoplankton	شناوران گیاهی
lipid raft	شناور چربی
floatation	شناورسازی
flotation	شناوری
audiometrist	شنوایی آزما
audiometer	شنوایی سنج
audiometry	شنوایی سنجی
audiologist	شنوایی شناس
audiology	شنوایی شناسی
audiogram	شنوایی نگاره
showdomycin	شودومایسین
denitrification	شوره برداری
halophile	شوره خواه
nitrification	شوره سازی
nitrification	شوره گذاری
shock	شوک
septic shock	شوک عفونی
fame	شهرت
regulated article	شیئی تحت کنترل
nozzle	شیپوره
photon	شیدپار
emulsion	شیرمایه
glycolysis	شیرین کافت (ی)
gluconeogenesis	شیرین نوزایی
schizogone	شیزوگون
schizogony	شیزوگونی
shikimate	شیکیمات
Shigella	شیگلا
chimera	شیمر

chemicalization	شیمیایی کردن/سازی
combinatorial chemistry	شیمی ترکیبی/تلفیقی
chemopharmacology	شیمی داروشناسی
chemotherapy	شیمی درمانی
biogeochemistry	شیمی زیست زمینی
cytochemistry	شیمی سلولی
coordination chemistry	شیمی کو ئوردیناسیون/هم آهنگی
chemotaxis	شیمی گرایی
accession, outbreak, prevalence	شیوع
nosocomial spread	شیوع بیمارستانی
disease outbreak	شیوع بیماری
morbidity	شیوع مرض
current good manufacturing practices	شیوه های عمل مناسب کنونی

ص

saponification	صابونی شدن/کردن
export	صادرات
export	صادر کردن
exporter	صادر کننده
level	صاف
percolating filter	صافی پرکولاتوری
high-efficiency particulate air filter	صافی تصفیه ذرات هوا با کارآیی بالا
trickle filter	صافی چکه ای
gel filtration	صافی کردن ژلی
export	صدور
express	صریح
inheritance	صفات ارثی
plant functional attributes	صفات عملکردی گیاه
peritoneal	صفاقی
trait	صفت
disk	صفحه
disk	صفحه گرد
bile	صفرا
solid	صلب
guild	صنف
precursor	صورت ابتدایی

conformation	صورت بندی
boat conformation	صورت بندی قایقی
bronchi	صورت جمع برونکوس به معنی نایژه
safe safety	ضامن مطمئن
neurologic sequelae	ضایعه عصب شناختی
contact zone thickness	ضخامت منطقه تماس
hydrophobic	ضد آب
antioxidant	ضد اکسنده
antioxidant	ضد اکسیدان
anti-interferon	ضد انتر فرون
anticoagulant	ضد انعقاد
antibacterial	ضد باکتری
fungicide	ضد باکتری/قارچ
antiplatelet	ضد پلاکت
antipyretic	ضد تب
counterterrorism	ضد تروریزم
anticonvulsant	ضد تشنج
anticooperativity	ضد تعاونی
anti-oncogene	ضد تومور زا
antihistamine	ضد حساسیت
antibiosis	ضد حیات
antibiosis	ضد حیاتی
analgesic	ضد درد
antiproliferative	ضد رشد و تولد
antiangiogenesis	ضد رگ زایی
anticodon	ضد رمزه
antibiotic	ضدزندگی
antitoxin	ضد زهرابه
anti-oncogene	ضد سرطان زا
antisera	ضد سرم ها
antitoxin, immunotoxin	ضد سم
antisense	ضد سنس
anti-infective	ضد عفونت
decontamination, disinfection, sterilization	ضد عفونی
aseptic, sterile	ضد عفونی شده
antifolate	ضد فولات
antifungal	ضد قارچ
antifungal	ضد قارچی
antiplatelet	ضد گروه خون

English	Persian
anticomplementary	ضد مکمل
antiviral	ضد ویروس
shock	ضربه
shock	ضربه روحی
safety factor	ضریب اطمینان
attenuation coefficient	ضریب انحطاط
coefficent of coincidence	ضریب انطباق
safety factor	ضریب ایمنی
partition coefficient	ضریب تجزیه
coefficent of coincidence	ضریب تصادف
attenuation coefficient	ضریب تضعیف
respiratory quotient	ضریب تنفسی
sedimentation coefficient	ضریب ته نشینی
partition coefficient	ضریب جدا سازی
attenuation coefficient	ضریب خاموشی
risk ratio	ضریب ریسک
Faraday constant	ضریب فاراده
activity coefficient	ضریب فعالیت
viscosity	ضریب گران روی/چسبندگی
correlation coefficient	ضریب همبستگی
coefficent of coincidence	ضریب هم رخدادی
depression	ضعف
attenuation	ضعیف سازی
attenuate	ضعیف شدن
attenuated	ضعیف شده
attenuate	ضعیف کردن
attenuator	ضعیف کننده
affinity tag	ضمیمه افینیتی
affinity tag	ضمیمه خویشاوندی
affinity tag	ضمیمه قرابت/خویشاوندی

ط

English	Persian
plague	طاعون
secondary septicemic plague	طاعون خونی ثانوی
secondary pneumonic plague	طاعون ریوی ثانوی
bubonic plague	طاعون میمونی
tolerance	طاقت
alternative medicine, complementary and alternative medicine	طب جایگزین
alternative medicine, complementary and alternative medicine	طب غیر سنتی
bed, grade	طبقه
taxonomy	طبقه بندی
FAB classification (= French-American-British Cooperative Group)	طبقه بندی اف.ای.بی.
biological warfare agent classification	طبقه بندی سلاحهای میکروبی
alternative medicine, complementary medicine	طب مکمل
modelling	طراحی
tissue engineering	طراحی بافت
protein engineering	طراحی پروتئین
computer-assisted drug design	طراحی دارو به کمک کامپیوتر
metabolic engineering	طراحی متابولیکی
rational drug design	طراحی معقول دارو
drug design	طرح دارو
phylogenetic profiling	طرح دودمانی
functional plan	طرح عملی
functional plan	طرح کاربردی
competitive exclusion	طرد رقابتی
party of transit	طرف ترانزیت
proponent	طرفدار
party concerned	طرف درگیر/مربوطه
receiving party	طرف دریافت کننده
party of export	طرف صادر کننده
affected party	طرف مبتلا/متاثر
party of origin	طرف مبدا
party of import	طرف وارد کننده
predator	طعمه گیر
parasite, parasitic	طفیلی
lifetime	طول عمر/زندگی
serum lifetime	طول عمر سرم
spectrum	طیف
burn-through range	طیف برن ترو
spectrometer	طیف سنج

combined gas chromatography-mass spectrometre	طیف سنج انبوه رنگ نگاری گازی ترکیب شده
mass spectrometer	طیف سنج جرمی
wide spectrum	طیف گسترده
raman optical activity spectroscopy	طیف نمائی فعالیت نوری رامان
ultraviolet spectroscopy	طیف نمایی فرا بنفش
near-infrared spectroscopy	طیف نمایی نزدیک به مادون قرمز
spectrophotometer	طیف نور سنج
broad spectrum	طیف وسیع

ظ

express	ظاهر کردن
petri dish	ظرف پتری
capacity, volume	ظرفیت
immunological tolerance	ظرفیت ایمونولوژیکی
carrying capacity	ظرفیت برد
carrying capacity	ظرفیت حمل
carrying capacity	ظرفیت زیست محیطی
capacity building	ظرفیت سازی
serum-trypsin-inhibitory capacity	ظرفیت سرکوبگری سروم تریپسین

ع

incapacitate	عاجز کردن
syndrome	عارضه
cerebro-oculo-facial skeletal syndrome	عارضه اسکلتی سربرو-اوکلو-فاسیال
sequela	عارضه باقیمانده
endocardial cushion defect	عارضه بالشتک آندوکاریال
side effect	عارضه جانبی
major histocompatibility complex	عارضه عمده بافت سازگاری
Acquired Immune Deficiency Syndrome (AIDS)	عارضه کاهش ایمنی اکتسابی (ایدز)

Cockayne syndrome	عارضه کوکین
congenital nephrotic syndrome	عارضه نفروتیک مادر زاد
congenital nephrotic syndrome of other types	عارضه نفروتیک مادر زاد نوع دیگر
congenital nephrotic syndrome of Finnish type	عارضه نفروتیک مادر زاد نوع فنلاندی
diabetes mellitus optic atrophy deafness	عارضه ولفرام
adventitious	عارضی
generic	عام
agent, operator	عامل
Rh factor	عامل آر.اچ.
initiation factor 2	عامل آغازین 2
initiation factor (IF)	عامل آغازین/اولیه
contamination	عامل آلودگی
anticrop agent	عامل آنتی کراپ/ضد محصول
antimateriel agent	عامل آنتی ماتریال/ضد ماده
capture agent	عامل اتصال
edaphic factor	عامل ادافیک
fusogenic agent	عامل ادغام/امتزاج/همجوشی زائی
oxidant	عامل اکسایش
oxidizing agent	عامل اکسندگی/اکسید کنندگی
platelet-derived wound healing factor	عامل التیام زخم مشتق از پلاکت
transfer factor	عامل انتقال
cartilage-inducing factor	عامل ایجاد غضروف
chelating agent	عامل ایجاد کننده کلات
safety factor	عامل ایمنی
fertility factor	عامل باروری
bacteriostat	عامل بازدارنده رشد باکتری
bacterial agent	عامل باکتریایی
azurophil-derived bactericidal factor (ADBF)	عامل باکتری کش مشتق از آزوروفیل
colony stimulating factor	عامل برانگیزنده کولونی
granulocyte colony stimulating factor	عامل برانگیزنده کولونی گویچه سفید دانه ای
agent, pathogen, pathogenic	عامل بیماری زا

107

agent	عامل
airborne pathogen	عامل بیماری زای هوایی
chaotropic agent	عامل بی نظمی دوست
biologic agent	عامل بیولوژیکی
biological agent	عامل بیولوژیکی
aerosolized biological agent	عامل بیولوژیکی افشانه شده
antimicrobial agent	عامل پاد زیوه ای
response element	عامل پاسخ
serum response factor	عامل پاسخ سروم
pyrogen	عامل تب آور
cleavage stimulation factor	عامل تحریک کلیواژ
megakaryocyte stimulating factor	عامل تحریک یاخته کلان هسته ای
cleavage and polyadenylation specificity factor	عامل تخصیص پلی آدنیلاسیون و کلیواژ
intercalating agent	عامل تداخلی
elongation factor	عامل تداوم
bioconcentration factor	عامل تراکم زیستی
glocosyl transferase factor	عامل ترانسفرازگلوکوسیل
bacterial agent	عامل ترکیزه ای
thrombolytic agent	عامل ترومبولیتیک لخته خون
testis determining factor	عامل تعیین بیضه
modifying factor	عامل تغییر دهنده
transforming growth factor-beta	عامل تغییر رشد بتا
chromatin remodeling element	عامل تغییر مدل کروماتینی
telomere repeat factor	عامل تکرار تلومر
nonmass casualty agent	عامل تلفات غیر انبوه
integration factor	عامل تلفیق
bioconcentration factor	عامل تمرکز زیستی
regulatory element	عامل تنظیم کننده
gene	عامل توارث
competence factor	عامل توانش
toxin agent	عامل توکسین/زهرابه
biological threat agent	عامل تهدید زیست شناختی
partitioning agent	عامل جداسازی
chemotactics factor	عامل جذب شیمیائی
sex factor	عامل جنسیت
biological warfare agent	عامل جنگ بیولوژیکی
chemical warfare agent	عامل جنگ شیمیائی
biological warfare agent	عامل جنگ میکروبی
tabun	عامل جی.ای.
thrombolytic agent	عامل حل کننده لخته خون
peptide elongation factor	عامل دراز شدگی پپتید
elongation factor 1	عامل دراز شدن 1
curing agent	عامل درمان کننده
growth factor	عامل رشد
angiogenic growth factor	عامل رشد آنژیوژنیک
human growth factor	عامل رشد انسانی
insulin growth factor	عامل رشد انسولین
hematologic growth factor	عامل رشد خون شناختی
fibroblast growth factor	عامل رشد رشته تنده
acidic fibroblast growth factor	عامل رشد رشته تنده اسیدی
epidermal growth factor	عامل رشد رو پوستی
platelet-derived wound growth factor	عامل رشد زخم مشتق از پلاکت
stem cell growth factor	عامل رشد سلول پایه
T-cell growth factor	عامل رشد سلول تی.
nerve growth factor	عامل رشد عصب
acidic fibroblast growth factor	عامل رشد فایبرو پلاست اسیدی
basic fibroblast growth factor (= fibroplast growth factor)	عامل رشد فیبروبلاست بازی
fibroplast growth factor (= fibroblast growth factor)	عامل رشد فیبروپلاست
hematopietic growth factor	عامل رشدگلبولهای قرمز خون
platelet-derived growth factor	عامل رشد مشتق از پلاکت
stem cell growth factor	عامل رشد یاخته دودمانی
angiogenesis factor	عامل رگ زایی
rho factor	عامل رو
transcription factor (TF)	عامل رونویسی
release factor	عامل رها سازی
polymerase I and transcript release factor	عامل رها سازی پلیمراز 1 و رو نوشت
thyroid releasing factor	عامل رها سازی تیروئید
risk factor	عامل ریسک

English	Persian
biologic agent, biological agent	عامل زیست شناختی
environmental factor	عامل زیست محیطی
etiological agent	عامل سبب شناختی
environmental etiological agent	عامل سبب شناختی محیطی
causative agent	عامل سببی
leukemia inhibitory factor	عامل سرکوب لوکمی
migration inhibition factor	عامل سرکوب مهاجرت
cytostatic agent	عامل سیتواستاتیکی
ciliary neurotrophic factor (CNTF)	عامل سیلیاری نروتروفیک
competence factor	عامل شایستگی
catalyst	عامل شتاب دهنده
chemical agent	عامل شیمیائی
anti-infective agent	عامل ضد عفونت
antimicrobial agent	عامل ضد میکروبی
antivirial agent	عامل ضد ویروس
antihemophilic factor	عامل ضد هموفیلی
uncertainty factor	عامل عدم قطعیت
ciliary neurotrophic factor (CNTF)	عامل عصب گرای مژکی
infectious agent	عامل عفونت زا/آلوده کننده
α- trans - inducing factor	عامل فرا تحریک آلفا
biological stressor	عامل فشار بیولوژیکی/زیستی
platelet activating factor	عامل فعال کننده پلاکت
von Willebrand factor	عامل فون ویلبراند
fibrinolytic agent	عامل فیبرین کافت
fibrinolytic agent	عامل فیبرینولایتیک
stringent factor	عامل قطعیت
chaotropic agent	عامل کائوتروپ
cachectic factor	عامل کاشکتیک
nonmass casualty agent	عامل کشتار غیر انبوه
lethal factor	عامل کشنده
co-factor	عامل کمکی
chemotactics factor	عامل کیموتاکتیک
clotting factor VIII	عامل لخته کننده نوع 8
clotting factor IX	عامل لخته کننده نوع 9
movable genetic element	عامل متحرک ژنتیکی
bacteriostat	عامل متوقف کننده رشد باکتری
replication licensing factor	عامل مجوز همانند سازی

English	Persian
limiting factor	عامل محدودکننده
activator	عامل محرک
granulocyte-macrophage colony-stimulating factor	عامل محرک کولونی گرانولوسیت ماکروفاژ
co-factor	عامل مشترک
contact zone element	عامل منطقه تماس
pyrogen	عامل مولد تب
biological control agent	عامل مهار/کنترل بیولوژیکی
host factor	عامل میزبانی
aerosolized biological agent	عامل میکروبی افشانه ای
atrial natriuretic factor	عامل ناتری-اورتیک دهلیزی
incapacitating agent	عامل ناتوان کننده
transcription factor for control of RNA polymerase II	عامل نسخه برداری برای مهار پلیمراز 2 آر.ان.ای.
ligand-activated transcription factor	عامل نسخه برداری فعال شده توسط لیگاند
general transcription factor	عامل نسخه برداری کلی
tumor necrosis factor	عامل نکروز دهنده تومور
unit factor	عامل واحد
hormone response element	عامل واکنش هورمون
viral agent	عامل ویروسی
replication factor C	عامل همانندسازی سی.
isolation	عایق بندی
turnover number	عدد بازیابی
turnover number	عدد تبدیل
turnover number	عدد روگشت
turnover number	عدد نوسازی
coordination number	عدد هم آرایی
heterogeneity	عدم تجانس
handedness	عدم تقارن (در دستها)
genetic imbalance	عدم توازن ژنتیکی
nonproliferation	عدم فزونی/تکثیر
dormancy, inactivation	عدم فعالیت
x-inactivation	عدم فعالیت اکس.
nonproliferation	عدم گسترش
nondisjunction	عدم گسستگی
linewidth	عرض خط
pancreatic juice	عصاره پانکراتیکی
neural	عصبی
neuropsychiatric	عصبی-روانی

109

English	Persian	English	Persian
myocard)ium(	عضله قلب	Jimson weed	علف جیمسون
organ	عضو	herbicide	علف کش
xenogenic organ	عضو نا همسرشت	heredity	علم توارث
infection	عفونت	biology	علم حیات
infestation	عفونت انگلی	taxonomy	علم رده بندی
bacterial infection	عفونت باکتریائی	chemical genetics	علم ژنتیک شیمیائی
flesh-eating infection	عفونت خورنده	technology	علم صنعت
infective	عفونت زا	meteorology	علم کاینات هوا
infectivity	عفونت زایی	overt	علنی
opportunistic infection	عفونت فرصت طلب	feedstock	علوفه
cross-infection	عفونت متقاطع	genomic sciences	علوم ژنومی
bacterial infection	عفونت میکروبی	feedstock	علیق
contaminant, infectious, infective	عفونی	bulk	عمده
		lifetime	عمر
fragile X mental retardation	عقب افتادگی ذهنی اکس. شکننده	immune function	عملکرد ایمنی
		good clinical practice	عملکرد بالینی مفید
retrograde	عقب رونده	biological operation	عملکرد بیولوژیکی
mental retardation	عقب ماندگی ذهنی	production function	عملکرد تولید
X-linked recessive mental retardation	عقب ماندگی ذهنی مغلوب مرتبط به کروموزوم اکس.	utility function	عملکرد مطلوب
		contained work	عمل کنترل شده
node	عقده	operator	عملگر
ganglion	عقده عصبی	generic, systeomics	عمومی
hand	عقربه	prevalence	عمومیت
sterile	عقیم	retroelements	عناصر پسرو
sterilization	عقیم سازی	upstream promoter elements	عناصر پیشبرنده بالا دست
sterility	عقیمی	ARS element	عنصر ای.آر.اس.
immune response	عکس العمل ایمنی	long interspersed nuclear element	عنصر پراکنده هسته ای طولانی
antidote	علاج		
affinity	علاقه	non-viral retroelement	عنصر پسگرای غیر ویروسی
device, label, marker, symptom	علامت	transposable element	عنصر ترانهادی
		enhancer element	عنصر تقویت گر
prodrome	علامت اولیه	regulatory element	عنصر تنظیمی
biological marker	علامت بیولوژیکی	autonomous replicating sequence (ARS) element	عنصر توالی همانند ساز اتونوموس
gene expression marker	علامت تظاهر ژن (ی)		
signaling	علامت دادن	autonomous replicating sequence (ARS) element	عنصر توالی همانند ساز خودمختار
cell signalling	علامت دهی سلولی		
biomarker	علامت زیستی	transposable element, transposon	عنصر جا به جا شدنی
magnetic labeling	علامت گذاری مغناطیسی		
labling	علامت گذاشتن	autonomous replicating segment (ARS) element	عنصر جزء همانند ساز خودمختار
symptomatic	علامتی		
etiology	علت شناسی	short interspersed nuclear	عنصر دارای فواصل هسته ای

element	کوتاه	medifoods	غذاهای طبی
transposable genetic element	عنصر ژنتیکی جا به جا شدنی	optimum food	غذای بهینه
extrachromosomal genetic element	عنصر ژنتیکی خارج کروموزومی	genetically modified food (GMF)	غذای تغییر داده شده بطریق ژنتیک(ی)
retroviral-like element	عنصر شبه رتروویروس	optimum food	غذای مطلوب
sieve element	عنصر غربالی	trophic	غذایی
contact zone element	عنصر منطقه تماس	immunological screening	غربال ایمونولوژیکی
nonspecific symptom	عوارض غیر مشخص	target-ligand interaction screening	غربال برهم کنش هدف- لیگاند
hepato-renal syndrome	عوارض کبدی-کلیوی	genetic screening	غربال ژنتیکی
vulgaris	عوامانه	screening	غربال کردن
clotting factors	عوامل انعقاد	mass screening	غربال کردن گسترده
relapse	عود	high-content screening	غربالگری با حجم بالا
reversion	عودت	in-silico screening	غربالگری کامپیوتری
grade, titer	عیار	plus and minus screening	غربالگری مثبت و منفی
assay	عیار سنجی	molecular sieve	غربال ملکولی
assay	عیارگیری	flooding	غرق کردن
multiplex assay	عیارگیری مرکب	membrane	غشاء
overt	عیان	plasmalemma	غشاء خارجی پروتوپلاسم
deficiency	عیب	intracellular membrane	غشاء درون سلولی
		plasma membrane	غشاء دشتینه ای
	غ	nanotube membrane	غشاء ریز لوله
		antibody-laced nanotube membrane	غشاء ریز لوله ای با آستر آنتی بادی
insidious	غافلگیر کننده	biofilm	غشاء زیستی
dominant	غالب	cell membrane (cell wall)	غشاء سلولی
autosomal dominant	غالب اتوزومی	cytoplasmic membrane, plasma membrane	غشاء سیتوپلاسمی
immunodominant	غالب ایمن		
X-linked dominant	غالب متصل به اکس.	peritoneal membrane	غشاء صفاقی
sex influenced dominance	غالبیت تحت تاثیر جنس	mucous membrane	غشاء مخاطی
incomplete dominance	غالبیت ناقص	unit, unit membrane	غشای واحد
dust	غبار	plasmic membrane	غشای یاخته
sub-mandibular glands	غدد زیر فکی/آرواره ای	oversight	غفلت
gland, growth, tumor	غده	case, legume	غلاف
endocrine gland	غده اندوکرین	invagination	غلاف شدگی
prostate	غده پروستات	prevalence	غلبه
sebaceous gland	غده چربی	incomplete dominance	غلبه جزئی
endocrine gland	غده درون ریز	concentration, consistency, density, viscosity	غلظت
anterior pituitary gland	غده زیر وحشه جلویی		
oncology	غده شناسی	densitometer	غلظت سنج
pituitary gland	غده هیپوفیز	lethal concentration	غلظت کشنده
anterior pituitary gland	غده هیپوفیز قدامی		

reference concentration	غلظت مرجع	phage	فاژ
benchmark concentration	غلظت معیار	virulent phage	فاژ بیماری زا
mean corpuscular hemaglobin concentration	غلظت میانگین هموگلوبین کورپوسکولار	transducing phage	فاژ ترانسدیوسر
		Charon phage	فاژ شارون
		helper phage	فاژ کمک کننده
bioenrichment	غنی سازی زیستی	lambda phage, λ (lambda) phage	فاژ لاندا
gigantism	غول پیکری		
abiotic, inroganic	غیر آلی	temperate phage	فاژ ملایم
nontraumatic	غیر تروماتیک	decay	فاسد شدن
noninvasive	غیر تعارضی	genetic distance	فاصله ژنتیکی
asexual, somatic	غیر جنسی	synaptic gap	فاصله سیناپسی
nonenteric	غیر روده ای	map distance	فاصله نقشه
abiotic	غیر زنده	synaptic gap	فاصله هموری
acellular	غیر سلولی	wastewater	فاضلاب
nontraumatic	غیر ضربه ای	sewerage	فاضلاب داری
adventitious	غیرطبیعی	sewer	فاضلاب رو
nonvolatile	غیر فرار	asexual	فاقد اندام جنسی
insertion inactivation	غیر فعال سازی درون جای دهی	asexual	فاقد میل جنسی
		spermatogenesis factor 3	فاکتور اسپرم زائی 3
laser inactivation	غیرفعال سازی لیزری	fogger	فاگر
antisense	غیر قابل ترجمه	Fontan fogger	فاگر فونتان
non-polar	غیر قطبی	London Fog foggers	فاگر لندن فاگ
antisense	غیر کد کننده	phagosome	فاگوزوم
nonsecretor	غیر مترشحه	phagocyte	فاگوسیت
antiparallel	غیر موازی	phagocytize	فاگوسیتایز
non-target	غیر هدف	phagocytosis	فاگوسیتوز
antiparallel	غیر هم جهت	chromatophore	فام بر
anaerobe, anaerobic	غیر هوازی	chromosome	فام تن
		x chromosome	فام تن اکس.
	ف	metacentric chromosome	فام تن پس میانپار
		accessory chromosome	فام تن فرعی
pharming	فارمینگ	chromosomal	فام تنی
farnesyl transferase	فارنزیل ترانزفراز	chromonema	فام راک
phase	فاز	chromatography	فام نگاری
S phase (DNA replication phase), s-phase	فاز اس.	sib	فامیل
		chromatin	فامینه
gap phase	فاز بینابینی	phytase	فایتاز
G1 phase of the cell cycle	فاز جی.1 از چرخه سلولی	use	فایده
G2 phase of the cell cycle	فاز جی.2 از چرخه سلولی	efficacy	فایده بخشی
liquid-crystalline phase	فاز کریستالین مایع	photoperiod	فتو پریود
phasmis	فازمیز	photosynthesis	فتو سنتز

English	Persian	English	Persian
photophosphorylation	فتو فسفریزه شدن	phragmoplast	فراگموپلاست
estrogen	زا حلف	anabolism	فراگوهرشز
product	فرآورده	metafemale	فرا ماده
metabolic product	فرآورده دگر گوهرشی	ultracentrifuge	فرامیان گریز
metabolic product	فرآورده سوخت وساز	occurrence	فراوانی
metabolite	فرآورده سوخت وساز	gene frequency	فراوانی ژن (ی)
petrochemicals	فرآورده های پتروشیمی	species richness	فراوانی گونه
primary productivity	فرآوری اولیه	co-factor recycle	فرایند بازتولید کوفاکتور
pathologic process	فرآیند آسیب شناختی	adipose	فربه
amylo process	فرآیند آمیلو	callipyge	فربهی
activated sludge process	فرآیند اسلاج فعال شده	person	فرد
pathologic process	فرآیند بیماری زایی	previously-vaccinated individual	فرد پیش واکسینه شده
post-transcriptional processing	فرآیند پسا نسخه برداری	unimmunized individual	فرد مصونیت نیافته
diagnostic procedure	فرآیند تشخیصی (طبی)	personal	فردی
aeration process	فرآیند تهویه	milling	فرز کاری
biological rotating disc process	فرآیند دیسک چرخان زیستی	milled	فرزکاری شده
		sibling	فرزندان یک خانواده
bioprocess	فرآیند زیستی	ablation	فرساب
Symba process	فرآیند سیمبا	genetic erosion	فرسایش ژنتیکی
aeration process	فرآیند هوادهی	fatigue	فرسودگی
overatropinization	فرا آتروپین سازی	equilibrium theory	فرضیه تعادل
parataxonomist	فرا آرایه شناس	sequence hypothesis	فرضیه توالی
ultraviolet	فرا بنفش	shifting balance theory	فرضیه جابجائی توازن
biological monitoring	فرابینی زیستی	selfish DNA hypothesis	فرضیه دی.ان.ای. خودخواه
environmental monitoring	فرابینی محیط زیست	null hypothesis	فرضیه صفر
ultrafiltration	فرا پالایش	Gaia hypothesis	فرضیه گایا
signal transduction	فرا پیام انتقالی	wobble hypothesis	فرضیه وابل
overwinding	فراپیچی	yin-yang hypothesis (of biological control)	فرضیه یین و یانگ
uptake	فراجذب		
immune escape	فرار مصون	branch	فرعی
anabolism	فراساخت	discriminator	فرق گذار
ultrafilter	فرا صافی	evolution	فرگشت
ultrafiltration	فرا صافی کردن	phyletic evolution	فرگشت تباری
ultrasonic	فرا صوتی	in-vitro evolution	فرگشت در محیط کشت
ultrasonication	فرا صوتی کردن	dendritic evolution	فرگشت دندریتی
ultrafilter	فرا فیلتر	molecular evolution	فرگشت مولکولی
metastasis	فراگستری	directed evolution	فرگشت هدایت شده
anabolism	فراگشت	convergent evolution	فرگشت همگرا
anabolite	فراگشته	NAD, reduced form	فرم کاهش یافته ان.ای.دی.
anabolic	فراگشتی	NADP, reduced form	فرم کاهش یافته ان.ای.دی.پی.

113

English	Persian
fermenter	فرمنتر
formulation	فرمول بندی
formulation	فرموله کردن
ferrobacteria	فروباکتری
aspirate	فروبردن
ingestion	فروبری
breakdown, fragmentation	فروپاشی
aspirate	فرو دادن
downstream	فرودست
ferrodoxin	فرودوکسین
depression	فرورفتگی
primary constriction	فرورفتگی ابتدائی
major groove	فرورفتگی بزرگ
minor groove	فرورفتگی کوچک
catabolism	فروساخت
bioleaching	فرو شست زیستی
photosynthesis	فروغ آمایی
catalysis	فروکافت
catalysis	فروکافتن
catalyst	فروکافنده
fructan	فروکتان
fructooligosaccharide (FOS)	فروکتوالیگوساکارید
fructose	فروکتوز
fructose-1,6-bis-phosphate	فروکتوز-6،1-بیس فسفات
fructose oligosaccharide	فروکتوز الیگوساکارید
fructification	فروکتیفیکاسیون
embedding	فرو کردن
settle	فروکش کردن
ferrochelatase	فروکلاتاز
catabolism	فروگشت
catabolite	فروگشته
catabolic	فروگشتی
catabolic	فروگوهرشی
pheromone	فرومون
quelling, repression	فرونشانی
post-transcriptional gene silencing	فرونشانی پسا نسخه برداری ژنی
ferritin	فریتین
extremozyme	فرین زیمایه
amplification	فزون سازی

English	Persian
morbidity	فساد
biodegradation	فساد بیولوژیکی
cystic fibrosis	فساد کیستی الیاف
necrosis	فساد نسج
phosgene	فسژن
phosphate	فسفات
fructose-1-phosphate	فسفات-1-فروکتوز
glucose-1-phosphate	فسفات-1-گلوکز
glyceraldehyde-3-phosphate	فسفات-3-گلیسرآلدهید
fructose-6-phosphate	فسفات-6-فروکتوز
glucose-6-phosphate	فسفات-6-گلوکز
phosphatase	فسفاتاز
acid phosphatase	فسفاتاز اسیدی
alkaline phosphatase	فسفاتاز قلیایی
phosphorylation	فسفات افزایی
pyridoxal-5-phosphate	فسفات پیریدوکسال-5
sodium phosphate	فسفات سدیم
inorganic phosphate	فسفات غیر آلی
phosphatidylethanolamine	فسفاتیدیلاتانولآمین
phosphatidyl-ethanolamine	فسفاتیدیل-اتانولآمین
phosphatidyl-serine	فسفاتیدیل سرین
phosphatidyl choline, phosphatidyl-choline	فسفاتیدیل کولین
phosphatidyl serine	فسفاتیدین سرین
phosphorus	فسفر
bioluminescence	فسفر افکنی
highly available phosphorous (HAP)	فسفر شدیدا در دسترس
phosphorus	فسفری
photosynthetic phosphorylation	فسفریزه شدن فتو سنتزی
photosynthetic phosphorylation	فسفریزه شدن نور ساختی
cyclic phosphorylation	فسفری شدن چرخه ای
photophosphorylation	فسفری شدن نوری
phosphorylation	فسفریلاسیون
phosphorylation	فسفریله کردن
phospho-enol-pyruvate	فسفو-انول-پیروویت
phospho enol-pyruvate carboxylase	فسفو-انول-پیروویت کاربوکسیلاز
phosphotransferase	فسفوترانسفراز

114

phosphotyrosine	فسفوتیروزین
phosphodiester	فسفو دی استر
phosphodiesterase	فسفو دی استراز
phosphoramidite	فسفورآمیدایت
phosphramidon	فسفورآمیدون
thymidine phosphorylase	فسفوریلاز تیمیدین
purine nucleoside phosphorylase	فسفوریلاز نوکلئوزید پیورین
oxidative phosphorylation	فسفوریلاسیون اکسایشی
phosphoserine	فسفوسرین
phosphofructokinase	فسفو فروکتوکیناز
phosphoglucomutase	فسفوگلوکوموتاز
phospholamban	فسفولامبان
phospholipase	فسفو لیپاز
phospholipid	فسفو لیپید
phosphonomycin	فسفونومایسین
phosphovitin	فسفو ویتین
phosphinothricin, phosphinotricine	فسفینو تریسین
phosphinothricin acetyltransferase	فسفینو تریسین استیل ترانسفراز
pressure	فشار
strain	فشار
osmotic pressure	فشار اسمزی
selection pressure	فشار انتخاب(ی)
turgor pressure	فشار ترگور
high blood pressure	فشار خون بالا
injector	فشاننده
premature chromosome condensation	فشردگی کروموزوم نابهنگام/زود هنگام
bioastronautics	فضانوردی زیستی
perinuclear space	فضای پری نوکلئار
subarachnoid space	فضای ساب آراکنوئیدی
solid	فضایی
labile	فعال
activator	فعال ساز
enzyme activation	فعال سازی آنزیم
transactivation	فعال سازی ترانس
macrophage activiation	فعال سازی درشت خوار
photoreactivation	فعال سازی نوری
complement acvtivation	فعال شدن/کردن/سازی مکمل

activator, catalyst, promoter	فعال کننده
tissue-type plasminogen activator	فعال کننده بافت ماند پلاسمینوژن
tissue plasminogen activator	فعال کننده پلاسمینوژن بافتی
transcriptional activator	فعال کننده رونویسی
surfactant	فعال کننده سطح
transcription activator	فعال کننده نسخه برداری
constitutive promoter	فعال کننده نهادی
enzyme activity	فعالیت آنزیم
specific activity	فعالیت اختصاصی
biological activity, biological operation	فعالیت زیست شناختی
bioactivity	فعالیت زیستی
cellulolytic activity	فعالیت سلولایتیک
optical activity	فعالیت نوری
aleukia	فقدان پلاکتهای خون
fluorophore	فلئوروفور
flap endo nuclease	فلاپ اندونوکلئاز
flagella	فلاژلا
flagellin	فلاژلین
microbial floc	فلاک میکروبی
flavonoid	فلاونوئید
flavoprotein	فلاووپروتئین
flavonols	فلاوونول (ها)
flavin	فلاوین
flavinoid	فلاوینوید
flavivirus	فلاوی ویروس
flaviviridae	فلاوی ویروس (ها)
phlebovirus	فلبو ویروس
paresis	فلج خفیف
scrapie	فلج گوسفندی
paresis	فلج ناقص
alkaline earth metal	فلز قلیلیی خاکی
phloem	فلوئم
fluorescence	فلو ئورسانس
fluorophore	فلوئورسانس ساز
fluorescence in situ hybridization (FISH)	فلئوروسنس هم تیرگی در محل
fluoroquinolone	فلئوروکینولون
phenylmethylsulfonyl fluoride (PMSF)	فلوئورید فنیلمتیلسولفونیل

English	Persian
Fluidizer	فلوئیدایزر
flora	فلورا
phloretin	فلورتین
phloroglucinol	فلوروگلوسینول
phlorizin	فلوریزین
flow cytometry	فلو سایتومتری
flocculation	فلوکولاسیون
flocculant	فلوکولانت
bulk	فله ای
technology	فن
antisense	فنآوری آنتی سنس/ضد
technology)anti-sense technology(	سنس/غیر کد کننده/غیر قابل ترجمه
traditional breeding technique	فن اصلاح نژاد سنتی
technology	فناوری
fermentation technology	فناوری تخمیر
chip DNA technology	فناوری تراشه دی.ان.ای.
fermentation technology	فناوری رشد موجودات ذره بینی
microsystems technology	فناوری ریز سیستمها
microbial technology	فناوری میکروبی
phentolamine	فنتولامین
technology	فن شناسی
phenotype	فنوتیپ
quantitative phenotype	فنوتیپ کمی
Bombay phenotype	فنوتیپ نوع بمبئی
phenocopy	فنوکپی
phenomics	فنومیکز
phenylalanine	فنیل آلانین
phenyl-thio-carbomide	فنیل تیوکاربومید
phenyl-thio-hydration	فنیل تیوهیدراسیون
phenylketonuria	فنیل کتونوری
phenylmethylsulfonylfluoride	فنیل متیل سولفونیل فلوئورید
photoautotroph	فوتواتوتروف
photoorganotroph	فوتوارگانوتروف
phototroph	فوتوتروف
photorhabdus luminescens	فوتورابدوس لومینسانس
photoluminescence	فوتولومینسنس
photolysis	فوتولیز
photon	فوتون

English	Persian
furanose	فورانوز
furanocoumarin	فورانوکومارین
formite	فورمایت
formite	فورمیت
furocoumarin	فوروکومارین
emergency, express	فوری
biological emergency	فوریت زیست شناختی
fusion	فوزیون
centric fusion	فوزیون مرکزی
fusarium	فوساریوم
fusarium graminearum	فوساریوم گرامینئاروم
fusarium moniliforme	فوساریوم مونیلیفورم
phosgene	فوسژن
hyperimmune	فوق ایمن
ultracentrifuge	فوق سانتریفوژ
preparative ultracentrifuge	فوق سانتریفوژ آماده سازی
superactivated	فوق فعال
metafemale	فوق ماده
positive supercoiling	فوق مارپیچ مضاعف
focus group	فوکاس گروپ
phocomelia	فوکوملی
fulminant	فولمینان
fumarase	فوماراز
fumonisin	فومونیسین
fauna	فون
fauna	فونا
inventorying	فهرست برداری
sense	فهم
soluble fiber	فیبر قابل حل
water soluble fiber	فیبر محلول در آب
fibroblast	فیبروبلاست
cystic fibrosis	فیبروز سیستی
fibronectin	فیبرونکتین
fibrin	فیبرین
fibrinogen	فیبرینوژن
phytate	فیتات
phytoalexin	فیتو آلکسین
phytoene	فیتوئین
phyto-sterol	فیتو استرول
phytophthora	فیتوفتورا
phytochrome	فیتو کروم

feedstock	فید استاک
physostigmine	فیزوستیگمین
physiology	فیزیو لوژی
bacterial physiology	فیزیولوژی باکتریایی
pathophysiology	فیزیولوژی بیماری شناسی
microbial physiology	فیزیولوژی میکروبی
fission	فیسیون
centric fission	فیسیون مرکزی
ficoll	فیکول
phylasalaemin	فیلاسالائمین
drum filter	فیلتر بشکه ای/طبلی
high-efficiency particulate air filter	فیلتر تصفیه ذرات هوا با کارآیی بالا
trickle filter	فیلتر چکه ای
HEPA filter	فیلتر هپا/اچ.ئی.پی.ای.
microbial filaments	فیلمانهای میکروبی
film yeast	فیلم ییست
filovirus	فیلوویروس
filoviridae	فیلوویروس ها
fimbria	فیمبریا
fusion	فیوژن

ق

aerosolization	قابل افشاندن کردن
communicable	قابل انتقال
explosive	قابل انفجار
refillable	قابل پر شدن مجدد
refillable	قابل پر کردن مجدد
biodegradable	قابل تجزیه بیولوژیکی
respirable	قابل تنفس
soluble	قابل حل
sustainable	قابل دوام
capacity, competency	قابلیت
applicability	قابلیت اجرا
applicability	قابلیت اعمال
vagility	قابلیت انتشار
transmissibility	قابلیت انتقال
mutagenicity	قابلیت ایجاد جهش
transformation efficiency	قابلیت تبدیل
transformation efficiency	قابلیت تغییر شکل

heritability	قابلیت توارث
absorbance	قابلیت جذب
viability	قابلیت زیست
infectibility	قابلیت سرایت
infectibility	قابلیت عفونت
natural killer	قاتل طبیعی
bactericidal	قادر به باکتری کشی
fungus	قارچ
parasite	قارچ انگلی
mycorrhizae	قارچ-ریشه
mycotoxin, vomitoxin	قارچ زهر
fungicide	قارچ کش
slime fungi	قارچ گل و لای
cellular slime mold	قارچ مخاطی سلولی
myxomycetes	قارچهای مخاطی
fungus	قارچی
stringency	قاطعیت
mold	قالب
template	قالب
mold	قالب گرفتن
dominance	قانون بارزیت/برتری/غلبه
law of independent assortment	قانون تفکیک مستقل (ژنها)
law of segregation	قانون جداشدن یا انفصال
chakrabarty	قانون چاکرابارتی
law of dominance	قانون غالبیت
kefauver rule	قانون کفاور
Hardy-Weinberg law	قانون هاردی-واینبرگ
valuation	قدر
efficacy	قدرت تاثیر
invasiveness	قدرت تهاجم
hybrid vigor (=heterosis)	قدرت دو رگه
resolving power	قدرت رفع
carcinogenicity	قدرت سرطان زایی
virulence	قدرت عفونت زایی
powered	قدرت گرفته
aging	قدمت
convention	قرارداد
encapsidation	قرار دادن در رویوش/روکش/پوسته
exposure	قرار دادن در معرض چیزی

encapsidation	قرار گرفتن در روپوش/روکش/پوسته	basophilic	قلیا خواه
acute exposure	قرار گرفتن شدید/حاد در معرض(چیزی)	alkaline	قلیایی
inhalation exposure	قرارگیری در معرض استنشاق	glucose	قند
lupus	قرحه آکله	nucleoside diphosphate sugar	قند دفسفات نوکلئوزید
congo red	قرمز کنگوئی	glycobiology	قند-زیست شناسی
redness	قرمزی	glycine	قند سریشم
quarantine	قرنطینه	fructose	قند میوه
quarantine	قرنطینه کردن	sucrose	قند نیشکر
genus	قسم	saccharification	قندی شدن
polarimeter	قطب سنج	template	قواره
polarization	قطبش	ingestion	قورت دادن
polar	قطبی	ethnobiology	قوم زیست شناسی
polarity	قطبیت	aptitude	قوه
polarimeter	قطبیت سنج	retrograde	قهقرایی
fluorescence polarization	قطبیت فلوئورسانس	analogue	قیاسی
polarize	قطبیدن		
polarized	قطبیده		
polarity	قطبی شدن/بودن		**ک**
droplet	قطرک	capsid	کاپسید
contact zone thickness	قطر منطقه تماس	capillary isotachophoresis	کاپیلاری ایزوتاکوفورسیس
pinocytosis	قطره خواری	capillary isotechophoresis	کاپیلاری ایزوتکوفورسیس
excision	قطع	catabolism, katabolism	کاتابولیسم
Kappa particles	قطعات کاپا	catabolic	کاتابولیک
joining (J) segment	قطعه اتصال	katal (Kat)	کاتال
stuffer fragment	قطعه استافر	catalysis	کاتالیز
N-segment	قطعه ان.	catalyst	کاتالیزور
DNA fragmentation	قطعه قطعه کردن دی.ان.ای.	biocatalyst	کاتالیزور حیاتی
seizure	قفل کردن	catalyst	کاتالیست
denaturation	قلب ماهیت	katanin	کاتانین
denature	قلب ماهیت کردن	catechin	کاتچین
hypo plastic right heart	قلب هیپوپلاستیک سمت راست	cathespin	کاتسپین
airbrush	قلم رنگ پاش	catechin	کاتشین
domain	قلمرو	blue vitriol	کات کبود
B-domain	قلمرو بی.	catalase	کاتلاز
domain of protein	قلمرو پروتئین	catenin	کاتنین
graf	قلمه	catecholamine	کاته کولامین
transplantation	قلمه زنی	cation	کاتیون
alkali	قلیا	cationic	کاتیونی
		cadherin	کادهرین
		field trial	کارآزمایی صحرایی

English	Persian	English	Persian
physiology	کار اندام	caryotype, karyotype	کاریوتیپ
physiologic, physiological	کار اندام شناختی	incomplete karyotype	کاریوتیپ ناقص
physiologist	کار اندام شناس	karyosome	کاریوزوم
physiology	کار اندام شناسی	caryokinesis, karyokinesis	کاریوکینزیس
carbetimer	کاربتیمر	karyogamy	کاریوگامی
user	کاربر	caryogram, karyogram	کاریوگرام
application, use	کاربرد	caryon, karyon	کاریون
sustainable use	کاربرد پایدار	casein	کازئین
new drug application	کاربرد داروی جدید	caseinogen	کازئینوژن
new animal drug application	کاربرد داروی جدید حیوانی	Cosmid	کازمید
sustainable use	کاربرد قابل دوام	kasugamycin	کازوگامیسین
structure-functionalism	کاربرد گرایی ساختاری	cassava	کاساوا (نشاسته)
computer assisted new drug application	کاربری داروی جدید به کمک کامپیوتر	caspase	کاسپاز
		cassette	کاست
contained use	کاربری کنترل شده	terminator cassette	کاست پایان دهنده
application	کاربست	meiosis	کاستمان
carboxyl proteinase	کاربوکسیل پروتئیناز	Cosmid	کاسمید
plant	کارخانه	plant	کاشتن
molecular knife	کارد مولکولی	cachectin	کاشکتین
cardenolide	کاردنولید	caffeine	کافئین
hypertrophic cardiomyopathy	کاردیومیوپاتی هایپرتروفیک	glycolysis	کافت گلیکوزی
		lyse	کافتن
genotoxic carcinogen	کارسینوژن ژنوتوکسیک	lysis	کافتندگی
carcinoma	کارسینوما	lyse	کافته شدن
carcinomatosis	کارسینوماتوز	lytic	کافتی
immune function	کارکرد ایمنی	lysogenic	کافتی زا
production function	کارکرد تولید	lysogeny	کافتی زایی
Good Manufacturing Practice	کارکرد تولیدی مفید	lysozyme	کافتی زیما
		lysozyme	کاف زیما
utility function	کارکرد مطلوبیت	lytic	کافنده
mounted	کار گذاشته شده	lysosome	کافنده تن
hand	کارگر	camphor	کافور
risk worker	کارگر خطر پذیر	lysosome	کافینه تن
operator	کارگردان	non-exclusive goods	کالاهای غیر انحصاری
activation energy	کارمایه کنش ور سازی	scaffold, template	کالبد
karnal bunt	کارنال بونت	dissection	کالبد شکافی
carnosine	کارنوزین	autopsy	کالبد گشایی
carnitine	کارنیتین	calpain	کالپین
carotene	کاروتن	calpain-10	کالپین 10
beta carotene	کاروتن بتا	caltractin	کالتراکتین
		caltrin	کالترین

119

English	Persian	English	Persian
calorie	کالری		زا/شب
calcitonin	کالسیتونین		نمایی/پرتوافشانی/شارندگ
calciphorin	کالسیفورین		ی زا
chalcone isomerase	کالکون ایزومراز	hybridization probing	کاوشگری با هیبریدسازی
calmodulins	کالمودولین ها	heterologous probing	کاوشگری ناهمگن
chalmydospore	کالمیدواسپور	nucleic acid probe	کاووشگر اسید نوکلئیک
callus	کالوس	dissolved oxygen probe	کاووشگر اکسیژن محلول
calomys colosus	کالومیس کولوسوس	oligonucleotide probe	کاووشگر اولیگونوکلئوتید
chalone	کالون	DNA probe	کاووشگر دی.ان.ای
callipyge	کالیپیجی	bioprobe	کاووشگر زیستی
kallidin	کالیدین	gene probe	کاووشگر ژن (ی)
colicin	کالیسین	RNA probes	کاووشگر های آر.ان.ای.
kallikerin	کالیکرین	hybridization probe	کاووشگر هیبریداسیون
calicol	کالیکول	caveolae	کاوئولا
Campbell Hausfeld	کامپبل هاوسفلد	caveolin	کاوئولین
camptothecin	کامپتوتسین	reduction	کاهش
truck	کامیون	ozone depletion	کاهش ازن
van	کامیون کوچک	hypoxia	کاهش اکسیژن بدن
cannabinoid	کانابینوئید	inbreeding depression	کاهش درون آمیزی
membrane channel	کانال غشاء	risk reduction	کاهش ریسک
kanamycin	کانامیسین	nitrate reduction	کاهش نیترات
canavanine	کاناوانین	surfactant	کاهنده کشش سطحی
con-till (= conservation tillage)	کان-تیل	reduction	کاهیدگی
		kirromycin	کایرومایسین
CANDA (= computer assisted new drug application)	کاندا	kairomone	کایرومون
		liver	کبد
		callus	کبره
candida	کاندیدا	cyanosis	کبودی پوست
candicidin	کاندیسیدین	capsid	کپسول
gene conversion	کانورژن ژن (ی)	capsule	کپسول
canola	کانولا	encapsulated	کپسولی
focal point	کانون	mold	کپک
nucleus	کانون هسته سلول	mold	کپک زدن
inroganic	کانی	slime molds	کپک گل و لای
cannitracin	کانیتراسین	gene duplication	کپی برداری ژن (ی)
lumen	کاواک	copy number	کپی نامبر
probe (= DNA probe)	کاوشگر	capping	کپینگ
multi-locus probe	کاوشگر چند نقطه ای	coat protein	کت پروتئین
genetic probe	کاوشگر ژنتیکی	contamination	کثافت
fluorogenic probe	کاوشگر	dirty	کثیف
	فلوئوروژنیک/فلوئورسانس	dirty	کثیف شدن

dirty, fouling	کثیف کردن		(helicoverpazea = H.zea)
dirty	کثیف کننده	rootworm	کرم ریشه
aberration	کج راهی	corn rootworm	کرم ریشه ذرت
code	کد	northern corn rootworm	کرم ریشه ذرت شمالی
turbidity	کدری	pink bollworm	کرم غوزه
codon	کد ژنتیکی سه گانه	bollworms	کرم های پنبه
code	کد گزاری کردن	nematodes	کرم های حلقوی
hereditary code	کد وراثتی	nematodes	کرم های رشته ای
turbidity	کدورت	bollworms	کرم های غوزه
codon	کدون	nematodes	کرم های گرد
ocher codon	کدون آشر	nematodes	کرم های نخی شکل
initiation codon	کدون آغاز	strain	کرنش
start codon	کدون آغازین	chromatophore	کروماتوفور
in-frame start codon	کدون آغازین در چارچوب	chromatography	کروماتوگرافی
amber codon	کدون امبر	preparative chromatography	کروماتوگرافی آماده سازی
opal codon	کدون اوپال	affinity chromatography	کروماتوگرافی بر اساس میل
chain termination codon	کدون پایان بخش زنجیره		ترکیبی
termination codon	کدون پایانی	column development	کروماتوگرافی پیشرفت/توسعه
in-frame stop codon	کدون پایانی در چارچوب	chromatography	ستون
degenerate codon	کدون دژنره	exclusion chromatography	کروماتوگرافی تخریبی
initiator codon	کدون شروع کننده	affinity chromatography	کروماتوگرافی تمایلی
codegenerate codon	کدون کودژنره	antibody affinity	کروماتوگرافی خویشاوندی آنتی
sense codon	کدون معنی دار	chromatography	بادی ها
kratein	کراتئین	exclusion chromatography	کروماتوگرافی طردی
keratosulfate	کراتوسولفات	reverse phase	کروماتوگرافی فاز معکوس
keratocyte	کراتوسیت	chromatography	
keratin	کراتین	cell affinity	کروماتوگرافی کشش سلولی
keratinocyte	کراتینوسیت	chromatography	
keratinization	کراتینیزه سازی	affinity chromatography	کروماتو گرافی کششی
creatinine	کراتینین	gas chromatography	کروماتوگرافی گازی
crossing-over	کراسینگ اور	chromatid	کروماتید
somatic crossing over	کراسینگ اور سوماتیکی/بدنی	sister chromatid	کروماتید خواهر
crown gall	کراون گال	chromatin	کروماتین
activated carbon	کربن آکتیو شده	sex chromatin	کروماتین جنسی
activated carbon	کربن فعال شده	euchromatin	کروماتین حقیقی
carbohydrate	کربو هیدرات	heterochromatin	کروماتین ناجور
castor bean	کرچک	human artificial episomal	کروموزوم اپیزومی انسانی
mouse-ear cress	کرس گوش موشی	chromosome	مصنوعی
pilus	کرک	human artificial episomal	کروموزوم اپیزومی مصنوعی
earthworm	کرم خاکی	chromosome	انسان
corn earworm	کرم ذرت	chromosome	کروموزوم

English	Persian
derivative chromosome	کروموزوم اشتقاقی
accessory chromosome	کروموزوم اضافی
satellite chromosome	کروموزوم اقماری
x chromosome	کروموزوم اکس.
attached X chromosome	کروموزوم اکس. پیوسته/متصل
x-chromosome	کروموزوم ایکس.
polytene chromosome	کروموزوم پلی تن
sex chromosome	کروموزوم جنسی
homologous chromosome	کروموزوم جور
structural chromosome	کروموزوم ساختاری
autosome	کروموزوم غیر جنسی
philadelphia chromosome	کروموزوم فیلادلفیا
lampbrush chromosome	کروموزوم لمپ براش
supernumerary chromosome	کروموزوم مازاد
metacentric chromosome	کروموزوم متاسانتریکی
metaphase chromosome	کروموزوم متافازی
human artificial chromosome	کروموزوم مصنوعی انسان
bacterial artificial chromosome	کروموزوم مصنوعی باکتریایی
mammalian artificial chromosome	کروموزوم مصنوعی پستانداران
mega-yeast artificial chromosome	کروموزوم مصنوعی کلان-مخمر
yeast artificial chromosome	کروموزوم مصنوعی مخمر
matrix-associated region marker chromosome	کروموزوم نشانگر منطقه وابسته به ماتریکس
recombinant chromosome	کروموزوم نوترکیب
y chromosome	کروموزوم وای.
cell growth Y chromosome	کروموزوم وای. رشد سلولی
isoderivative chromosome	کروموزوم هم اشتقاقی
homologous chromosome	کروموزوم همساخت
chromosomal	کروموزومی
chromogenic substrate	کروموژنیک سوب استرات
chromomere	کرومومر
chromonema	کرومونما
chronobiochemistry	کرونوبیوشیمی
chronic effect	کرونیک افکت
deafness	کری
creatine	کریاتین

English	Persian
cryptobiosis	کریپتوبایوسیز
cryptobiosis	کریپتوبیوزی
vacuoles	کریچه ها
liquid crystal	کریستال مایع
chimera	کژزاد
DNA chimera	کژزاد دی.ان.ای.
accession	کسب
agricultural	کشاورزی
con-till (= conservation tillage)	کشاورزی حفاظتی
culture	کشت
aquaculture	کشت آبی
axenic culture	کشت آزنیک
soft agar culture	کشت آگار نرم
agnotobiotic culture	کشت آگنوتوبیوتیک
anoxic culture	کشت آنوکسیک
aquaculture	کشتاب ورزی
shake culture	کشت ارتعاشی/شیک
incremental feeding culture	کشت ازدیاد تغذیه ای
spinner culture	کشت اسپینر
stock culture	کشت استاک/پایه
stab culture	کشت استبی/سوزنی
slide culture	کشت اسلاید
slide cell culture	کشت اسلاید سل
organ culture	کشت اندام
tissue culture	کشت بافت
aquaculture	کشت بدون خاک
single cell culture	کشت تک سلولی
slide culture	کشت تیغه ای
slide cell culture	کشت تیغه ای سلولی
stationary culture	کشت ثابت
multiple cropping	کشت چندگانه
quadruple cropping	کشت چهار گانه
pure culture	کشت خالص
dormant seeding	کشت خفته
habituated culture	کشت خوگرفته
culture	کشت دادن
silviculture	کشت درختان
in-vitro culture	کشت در شیشه
mixed cropping	کشت درهم
intercropping	کشت درهم ردیفی

batch culture	کشت دسته جمعی/گروهی	cellular slime mold	کفک مخاطی سلولی
double cropping	کشت دوگانه	clubbing	کلابینگ
aftercrop	کشت دوم	clathrin	کلاترین
dialysis culture	کشت دیالیزی	chelation	کلات سازی
sewage farm	کشتزار آبیاری شده با فاضلاب	glomerulus	کلافک
surface culture	کشت سطحی	cluster	کلاله
cell culture	کشت سلول	megabase	کلان باز
mammalian cell culture	کشت سلول پستاندار	macrophage	کلان خوار
animall cell culture	کشت سلول جانوری/حیوانی	submunition	کلاهک خوشه ای/فرعی
insect cell culture	کشت سلول حشره	autopsy	کلبد شکافی
cell culture	کشت سلولی	chlorine	کلر
suspension culture	کشت سوسپانسیونی	chlorination	کلر دار کردن/شدن
tripple cropping	کشت سه گانه	chlorination	کلر زدن
submerged culture	کشت غرقی	chlorination	کلر زنی
fed-batch culture	کشت فد-بچ	chloroplast	کلروپلاست
shifting agriculture	کشت کوچان	chlorosome	کلروزوم
starvation culture	کشت گرسنگی	chlorosis	کلروسیز
slant culture	کشت مایل	chloroform	کلروفورم
restoration	کشت مجدد	chlorophyll	کلروفیل
mixed culture	کشت مخلوط	caesium chloride	کلرید سزیم (نمک)
lyse	کشتن	brassica napus	کلزا
con-till (= conservation tillage)	کشت و زرع استحفاظی	cholesterol	کلسترول
		cholesterol oxidase	کلسترول اکسیداز
conservation tillage	کشت و زرع حفاظتی	familial hypercholesterolaemia	کلسترول بالای ارثی
cultivar	کشته		
synchronous culture	کشت همگاه/همزمان	calcyclin	کلسیکلین
hanging-drop culture	کشت هنگینگ-دراپ	calcium	کلسیم
affinity, strain, tropism	کشش	calcium oxalate	کلسیم اکسالات
chemotaxis	کشش شیمیائی	colchicine	کلشیسین
whey	کشک	raft	کلک
lethal, malignant	کشنده	total solids (TS)	کل مواد جامد (کمج)
zygotic lethal	کشنده زیگوتی	total dissolved solids (TDS)	کل مواد جامد محلول (کم جم)
genetic lethal	کشنده ژنتیکی	colloid	کلوئید (ی)
country providing genetic resource	کشور تامین کننده منابع ژنتیکی	colloidal	کلوئیدی
		agglomeration	کلوخه شدن
country of origin of genetic resource	کشور مبداء منابع ژنتیکی	clostridium	کلوستریدیوم
		clone	کلون
extension	کشیدگی	gene cloning	کلون سازی ژن (ی)
bed, benthos	کف	colony	کلونی
kephalin	کفالین	knockout	کله پا کردن
benthos	کف زی	codon	کله رمز

English	Persian	English	Persian
generic	کلی	gene complex	کمپلکس ژن (ی)
clindamycin	کلیندامایسین	synaptinemal complex	کمپلکس سیناپتینمال
cleave	کلیواژ	origin recognition complex	کمپلکس شناسائی منشاء
autosomal dominant polycystic kidney	کلیه پلی سیستیک غالب اتوزومی	Golgi complex	کمپلکس گلژی
hypoxia	کم اکسیژنی	open promoter complex	کمپلکس مبلغ باز
deficiency	کمبود	immune-stimulating complexes	کمپلکس های محرک ایمنی
adenylate kinase deficiency	کمبود آدنیلات کیناز	RC Replication Complex	کمپلکس همانند سازی آر.سی.
acid maltase deficiency	کمبود اسید مالتاز	major histocompatibilty complex	کمپلکس هیستوکمپتیبیلیته اساسی
oxygen deficiency	کمبود اکسیژن	complexation	کمپلکسیشن
ornithine transcarbamylase deficiency	کمبود اورنیتین ترانسکارباآمیلاز	complement	کمپلمان
multiple sulfatase deficiency	کمبود سولفاتاز چندگانه	complementation	کمپلیمانتاسیون
glutathione synthetase deficiency	کمبود گلوتاتیون سینتتاز	complementation	کمپلیمنتیشن
glucose 6 phosphate dehydrogenase deficiency	کمبود گلوکز 6 فسفات دهیدروژناز	composting	کمپوست درست کردن
		attenuated	کم توان شده
		TS (= total solids)	کمج
lactase deficiency	کمبود لاکتاز	TDS (total dissolved solids)	کم جم
myelo-peroxidose deficiency	کمبود مایلو-پروکسی دوز	anemia	کم خونی
isolated human growth hormone deficiency	کمبودمجزای هورمون رشد انسانی	congenital dyserythropoietic anaemia	کم خونی دیسرتروپوئیتیک مادر زادی
immunodeficiency	کمبود مصونیت	Fanconi anaemia	کم خونی فانکونی
congenital myeloperoxidase deficiency	کمبود میلو پراکسیداز مادر زادی	iron deficiency anemia	کم خونی ناشی از کمبود آهن
oligomer	کم پار	aplastic anemia	کم خونی ناشی از ناسازی
compressor	کمپرسور	hypoglycemia	کم قند خونی
Air America compressors	کمپرسورهای ایر آمریکا	co-repressor	کمک بازدارنده
Stanley Bostitch oil-free air compressor	کمپرسور هوای غیر روغنی استانلی بوستیچ	co-chaperonin	کمک چاپرونین
		helper T cell	کمک کننده تی. سل/سلول تی
oligotrophic	کم پرور	co-enzyme	کمک مخمر
campesterol, campestrol, campsterol	کمپسترول	lariat	کمند
major histocompatibility complex	کمپلکس بافت سازگاری اصلی	chemokine	کموکین
light-harvesting complex	کمپلکس برداشت سبک	latency	کمون
pre-replication complex	کمپلکس پیش از نسخه برداری	communism	کمونیزم
pyruvate de-hydrogenase complex	کمپلکس دهیدروژناز پایروویت	codex alimentarius commission	کمیسیون قوانین غذائی
		kanR	کن آر
		niche	کنام
		biological control	کنترل بیولوژیکی

medical control	کنترل پزشکی	composting	کود آلی درست کردن
autogenous control	کنترل خودزا	Crusader Garden Powder	کود باغچه کروسیدر
biological control	کنترل زیست شناختی	fertilization	کود دهی
acceptor control	کنترل گیرنده	nitrogenous fertilizer	کود نیتروژنی
positive control	کنترل مثبت	nonsense codon	کودون بی معنی
negative control	کنترل منفی	core enzyme	کور آنزیم
meter	کنتور	cortisol	کورتیزول
metered	کنتور دار	corticotropin	کورتیکوتروپین
probe (= DNA probe)	کندو کاو	(corticotrophin)	
construct	کنستراکت	corticotrophin	کورتیکوتروفین
consortium	کنسرسیوم	(=corticotropin)	
diversity biotechnology	کنسرسیوم تنوع بیوتکنولوژی	core DNA	کور دی.ان.ای.
consortium		cordyceps simensis	کوردیسپ سایمنزیس/زیمنزی
diversity biotechnology	کنسرسیوم تنوع زیست فناوری	curcumin	کورکومین
consortium		blast furnace	کوره بلند
consortia	کنسرسیوم ها	choriogonadotropin	کوریوگنادوتروپین
weak interaction	کنش متقابل ضعیف	choriomamotropin	کوریوماموتروپین
gauche conformation	کنفورمیشن گاش	koseisho	کوسیشو
absolute configuration	کنفیگوراسیون مطلق	co-factor	کوفاکتور
milled	کنگره دار	co-factor recycle	کوفاکتور ریسایکل
milling	کنگره دار کردن	fatigue	کوفتگی
convention	کنوانسیون	coformycin	کوفورمایسین
acarophobia	کنه ترسی	collagen	کولاژن
acariasis, acarinosis	کنه زدگی	collagenase	کولاژناز
acarodermatitis	کنه زدگی پوست	cold-shock protein	کولد شوک پروتئین
acarine	کنه سان	colon	کولون
acarina	کنه سانان	colitis	کولون آماس
coenzyme A	کوآنزیم آ.	colectomy	کولون برداری
coenzyme A	کوآنزیم ای.	colonoscopy	کولون بینی
coenzyme Q	کوآنزیم کیو.	colony	کولونی
reduced coenzyme Q	کوآنزیم کیو. کاهش یافته	colony lift	کولونی لیفت
quercetin	کو ارستین	cholecystokinin	کوله سیستوکینین
co-enzyme	کوآنزیم	cholecalcin	کوله کالسین
co-evolution	کو اوولوشن	colistin	کولیستین
cobalamin(e)	کوبالامین	colicin	کولیسین
copolymer	کوپلیمر	cholecalciferol	کولیکالسیفرول
co-translational	کوترانسلیشنال مودیفیکیشن	choline	کولین
modification		cholinesterase	کولین استراز
nanism	کوتولگی	cholinergic	کولینرژیک
short limbed dwarfism	کوتولگی کوتاهی عضو	combinatorics	کومبیناتوریکز
mini	کوچک	(combinatorial	

English	Persian	English	Persian
chemistry)		chymotrypsin	کیموتریپسین
comutagen	کوموتاژن	chymosin	کیموزین
Konzo (= lathrism)	کونزو	chemometrics	کیمومتری
b-conglycinin	کونگلیسینین بی.	kinase	کیناز
covariation	کو واریانس	cyclin dependent protein kinase	کیناز پروتئین متکی به سیکلین
covarion	کو واریون		
coccus	کوییزه	janus kinase	کیناز جانوس
aging	کهنگی	muscle creatine kinase	کیناز کریاتین عضلانی
chronic	کهنه	cyclin-dependent kinase	کیناز وابسته به سایکلین
aging	کهنه شدن	kinetosome	کینتوزوم
senescence	کهولت	kinetochore	کینتوکور
rash	کهیر	enzyme kinetics	کینتیک آنزیمی
chiasma	کیاسما	kinetin	کینتین
kpnl	کی.پی.ان.ال.	kinesin	کینزین
decontamination kit	کیت ضد عفونی	quinolone	کینولون
ketoprofen	کیتوپروفن	kinome	کینوم
ketose	کیتوز	kinomere	کینومر
ketosis	کیتوزیس	kinin	کینین
ketogenesis	کیتوژنز	kininase	کینیناز
ketone	کیتون	kininogen	کینینوژن
ketonemia	کیتونمی	high molecular weight kininogen	کینینوژن با وزن ملکولی بالا
ketonuria	کیتونوری		
chitin	کیتین	Q-RT-PCR (= quantitative reverse transcriptase PCR)	کیو.-آر.تی.-پی.سی.آر.
chitinase	کیتیناز		
cyst	کیست		
kistrin	کیسترین	QH2 (= reduced coenzyme Q)	کیو.اچ.2.
cell	کیسه		
cyst	کیسه	Q-band	کیو. باند
gall	کیسه صفرا	QB	کیو.بی.
grade	کیفیت		
protein quality	کیفیت پروتئین	**گ**	
kb (= kilobase)	کیلو باز		
kilobase (Kb)	کیلوباز	wagon	گاری چهار چرخ
kilodalton	کیلودالتون	gas	گاز
chimeraplasty	کیمرا پلاستی	biogas	گاز بیولوژیکی
chimeric antibody	کیمریک آنتی بادی	landfill gas	گاز زباله دان
chemoorganotroph (heterotroph)	کیموارگانوتروف	carbon dioxide	گاز کربنیک
		greenhouse gas	گاز گلخانه ای
chemoprophylactic	کیموپروفیلاکتیک	gas	گازی
chemoprophylaxis	کیموپروفیلاکسی	gastric	گاستریکی
chymotrypsin	کیموتریپزین	gastrin	گاسترین

126

circularization	گردشی کردن
dust	گرد گیری
platelet	گرده
open pollination	گرده افشانی باز
cross-pollination	گرده افشانی چلیپائی/متقاطع
blood platelet, platelet	گرده خون
nephron	گردیزه
plague	گرفتار کردن
derived	گرفته شده
thermal	گرم
thermophile	گرما دوست/خواه
calorimeter	گرماسنج
thermal	گرمایی
gram-positive	گرم مثبت
gram-negative	گرم منفی
cluster, colony, company	گروه
prosthetic group	گروه افزایشی
company	گروهان
prosthetic group	گروه پروستتیک
linkage group	گروه پیوسته
French-American-British Cooperative Group	گروه تعاون فرانسوی-آمریکائی-انگلیس ی
thiol group	گروه تیول
ABO blood group	گروه خونی ای.بی.او/آ.ب.او
X chromosome linked blood group	گروه خونی متصل به کروموزوم اکس.
age group	گروه سنی
functional group	گروه عامل/عمل کننده
functional group	گروه عامل(ی)
nonpolar group	گروه غیر قطبی
phosphate group	گروه فسفات
activating group	گروه فعال ساز
polar group	گروه قطبی
carboxyl group	گروه کاربوکسیل
focus group	گروه کانونی
affected party	گروه مبتلا/متاثر
inverted micelle	گروه ملکولی وارونه
incompatibility group	گروه ناهمخوان/ناخوانا
nucleophilic group	گروه هسته دوست/خواه
multiple cloning group	گروه همسانه ساز چندگانه

galactose	گالاکتوز
juvenile galactosia lidosis	گالاکتوزیا لیدوز نابالغ
alpha galactoside	گالاکتوزید آلفا
galactosemia	گالاکتوسمی
galactomannan	گالاکتومانان
gamma globulin	گاما گلوبولین
gamete	گامت
gametophyte	گامتوفیت
gamone	گامون
ganglion	گانگلیون
chiselplow	گاو آهن قلمی
gynandromorph	گایاندرومورف
flux	گداز آور
transition	گذار
pathway	گذرگاه
metabolic pathway	گذرگاه دگر گوهرشی
metabolic pathway	گذرگاه سوخت وسازی
environmental pathway	گذرگاه محیط زیست
serial passage	گذر متوالی
osmosis	گذرندگی
labile	گذرنده
Gramacidin	گراماسیدین
viscosity	گران روی
metachromatic granule	گرانول متاکروماتیکی
granulocyte	گرانولوسیت
granulocidin	گرانولوسیدین
volutin granule	گرانول ولوتین
granuloma	گرانولوما
tropism	گرایش
chemotaxis	گرایش شیمیائی
ionotropic	گرایش یونی
dust, powder	گرد
data collection	گرد آوری داده ها
consequence management	گردانندگی پی آمد(ی)
common property resource management	گردانندگی منابع املاک مشترک/عمومی
operator	گرداننده
spontaneous assembly	گردایش فوری/خودبخودی
Crusader Garden Powder	گرد باغچه کروسیدر
nitrogen cycle	گردش ازت
powdered	گرد شده

English	Persian	English	Persian
node	گره		تیروکسین
nodule	گره ریز	immune globulin,	گلوبولین ایمنی
ganglion	گره عصبی	immunoglobulin	
nodule	گرهک	accelerator globulin	گلوبولین شتابدهنده/تسریع
avoidance	گریز		کننده
centrifuge	گریز از مرکز	antihemophilic globulin	گلوبولین ضد هموفیلی
centrifuge	گریزانه	globin	گلوبین
centrifuge	گریز دادن	glutathione	گلوتاتیون
greenleaf technologies	گرین لیف تکنولوجیز	glutathione (reduced form)	گلوتاتیون (تضعیف شده)
green leafy volatile	گرین لیفی ولتایل	glutamate	گلوتامات
notification	گزارش	glutamate dehydrogenase	گلوتامات دهیدروژناز
reporter	گزارشگر	glutamine	گلوتامین
xanthophyll	گزانتوفیل	glutamine synthetase	گلوتامین سنتتاز
idiochromatin	گزین تنده	glutelin	گلوتلین
selection	گزینش	gluten	گلوتن
stabilizing selection	گزینش استوار کننده/پایدار	glutenin	گلوتنین
normalizing selection	گزینش بهنجار کننده	glufosinate, gluphosinate	گلوفوزینات
in-vitro selection	گزینش در محیط کشت	glucagon	گلوکاگون
disruptive selection	گزینش گسسته	glucan	گلوکان
lac selection	گزینش لاک	beta-glucan	گلوکان بتا
artificial selection	گزینش مصنوعی	glucose	گلوکز
idiochromosome	گزین فام تن	glucose oxidase	گلوکز اکسیداز
idiotype	گزین مونه	glucose isomerase	گلوکز ایزومراز
idiogram	گزین نگاره	glucose phosphate	گلوکز فسفات ایزومراز
idiovariation	گزین وردایی	isomerase	
guessmer	گسمر	UDP-glucose	گلوکز- یو.دی.پی.
vasodilator	گشاد کننده عروق	glucosidase	گلوکسیداز
fecundity	گشنیدگی	beta-glucuronidase	گلوکورونیداز بتا
slime	گل	beta-d-glucuronidase	گلوکورونیداز بتا-د
turbidity	گل آلودگی	glucosinolates	گلوکوزینولات (ها)
leukocyte, white corpuscle	گلبول سفید	glucocerebrosidase	گلوکوسربروسیداز
white blood cell	گلبول سفید خون	gluconeogenesis	گلوکونئوژنز
erythrocyte	گلبول قرمز	sewage sludge	گل ولای فاضلابی
red blood cell	گلبول قرمز خون	magic bullet	گلوله جادویی
GoldenRiceTM	گلدن رایس.	glomalin	گلومالین
garden	گلستان	nasopharynx	گلو و بینی
cruciferae	گلمیان	glia	گلیا
daffodil	گل نرگس (زینتی)	gliadin	گلیادین
rosetting	گل وبوته دادن	glycetein	گلیستئین
globulin	گلوبولین	glycitin	گلیستین
thyroxine-binding-globuline	گلوبولین اتصال دهنده	glyceraldehyde	گلیسرالدهید

128

English	Persian	English	Persian
glycitein	گلیسیتئین	guanosin, guanosine	گوانوزین
glycine	گلیسین	guanosine tri-phosphate	گوانوزین تری فسفات
glycine max	گلیسین ماکس	guanosine tri-phosphatase	گوانوزین تری فسفاتاز
glycinin	گلیسینین	guanosine mono-phosphate	گوانوزین مونو فسفات
glyphosate	گلیفسات	guanylate	گوانیلات
glyphosate oxidase	گلیفسات اکسیداز	guanylate cyclase	گوانیلات سیکلاز
glyphosate oxidoreductase	گلیفسات اکسیدو ردوکتاز	guanine	گوانین
glycoalkaloid	گلیکوالکالوئید	depression	گودی
glycobiology	گلیکوبیولوژی	ghost	گوست
glycoprotein	گلیکو پروتئین	granulation tissue	گوشت نو
variable surface glycoprotein	گلیکو پروتئین دارای سطح تغییر پذیر	otocephaly	گوش سری
glycoside	گلیکوزید	biodesulfurization	گوگرد زدایی بیولوژیکی
cardiac glycoside	گلیکوزید کاردیاکی	placebo	گول زنک
glycosidic	گلیکوزیدی	gonado blastoma Y	گونادوبلاستومای نوع وای.
glycosylation	گلیکوزیلاسیون	human chronic gonadotropin	گونادوتروپین انسانی کرونیک
glycosyltransferase	گلیکوزیل ترانسفراز	human chronic gonadotropin	گونادوتروپین تدریجی انسانی
glycosinolate	گلیکوزینولات		
glycogen	گلیکوژن	diversity, interspecific variation, variation, variegation	گوناگونی
variable surface glycoprotein	گلیکوژن سطح متغیر		
glycosidases	گلیکوسیداز ها	genetic diversity	گوناگونی زاد شناختی
DNA glycosylase	گلیکوسیل دی.ان.ای.آنزیم	heteromorphism	گوناگونی طبیعی یک کروموزوم
glycoform	گلیکوفورم	molecular diversity	گوناگونی ملکولی
glycocalyx	گلیکوکالیکس	beta-conglycinin	گونکلیسینین بتا
glycolipid	گلیکولیپید	species	گونه
glycolysis	گلیکولیز	breed not at risk	گونه امن
determinant	گمارنده	reactive oxygen species	گونه باز فعال اکسیژن
antigenic determinant	گمارنده پادگنی	native species	گونه بومی
capacity, volume	گنجایش	peromyscus species	گونه پرومایسکوس
genome	گنجینه توارثی	threatened species	گونه تهدیدشده
sepsis	گند	compensating variation	گونه جبرانی
infective	گند زا	umbrella species	گونه چتری
infectivity	گند زایی	alien species	گونه خارجی
antiseptic, disinfectant	گند زدا	breed at risk	گونه در خطر
disinfection	گندزدایی	breed at risk	گونه ریسکی
decay	گندیدن	speciation	گونه زایی
gnotobiosis	گنوتوبیوزیس	allopatric speciation	گونه زایی دگر بوم
digestion	گوارش	sympatric speciation	گونه زایی همبوم
complete digestion	گوارش کامل	sympatric speciation	گونه زایی همجا
digest	گواریدن	cultivated species	گونه زراعی

English	Persian
indicator species	گونه شاخص
indicator species	گونه شناساگر
cognate tRNA	گونه صحیح تی.آر.ان.ای.
alien species	گونه غیر بومی
breed not at risk	گونه غیر ریسکی
cultivated species	گونه کشت شده
genetic diversity	گونه گونی ژنتیکی
keystone species	گونه مبنا
flagship species	گونه مرغوب
invasive species	گونه مهاجم
indicator species	گونه نشانه
introduced species	گونه وارد شده
domesticated species	گونه های اهلی شده
exotic species	گونه های بیگانه
new species	گونه های جدید
incipient species	گونه های در حال تظاهر
exotic species,	گونه های غیر بومی
non-indigenous species	
sibling species	گونه هم نیا
gem	گوهر
substrate	گوهر مایه
granulocyte	گویچه سفید دانه ای
blood cell	گویچه
monocyte	گویچه تک هسته
red blood cell	گویچه سرخ خون
white corpuscle	گویچه سفید
polymorphonuclear	گویچه سفید چند هسته ای
leukocyte	
polymorphonuclear	گویچه سفید دانه ای چند
granulocyte	هسته ای
erythropoiesis	گویچه قرمز سازی
polar body	گویچه قطبی
spheroplast	گوییزه دش
phytochrome	گیارنگ
plant	گیاه
halophile	گیاه آب شور
flora	گیاهان
phytoplankton	گیاهان شناور
phytoremediation	گیاه پالائی
phytoplankton	گیاه پلانکتونی
hybrid	گیاه پیوندی

English	Persian
legume	گیاه خوردنی
phytopharmaceutical	گیاه-داروئی
phytopharmaceutical	گیاه دارویی
halophile	گیاه شوره زی
phytochemical	گیاه شیمیائی
parasite	گیاه طفیلی
herbicide	گیاه کش
transgenic plant	گیاه واریخته/تراژنی
plant	گیاهی
biogeography	گیتا شناسی زیستی
affinity	گیرائی
cellular affinity	گیرایی سلولی
acceptor, receptor	گیرنده
adrenergic receptor	گیرنده آدرنرژیک
x receptor	گیرنده اکس .
transferrin receptor	گیرنده ترانسفرین
protein-coupled receptor	گیرنده جفتی پروتئین
homing receptor	گیرنده جهت یاب (ی)
cellular adhesion receptor	گیرنده چسبندگی سلولی
homing receptor	گیرنده حسی بازگشت به خانه
di-hydro testosterone	گیرنده دی.-هایدروتستوسترون
receptor	
retinoid X receptor	گیرنده رتینوئید ایکس
T-cell receptor	گیرنده سلول تی.
T cell receptor alpha	گیرنده سلول تی. آلفا
farnesoid X receptor (FXR)	گیرنده فارنسوئیدی اکس.
leptin receptor	گیرنده لپتین
complement receptor	گیرنده مکمل
photoreceptor	گیرنده نوری
orphan receptor	گیرنده های جفت نشده
toll-like receptor	گیرنده های شبه باج گیر
nuclear receptor	گیرنده هسته ای
nuclear hormone receptor	گیرنده هورمون هسته ای
hormone	گیزن

ل

English	Persian
lathrism	لاتریسم
lathyrism	لاتیریسم
lazaroid	لازاروئید ها
laccase	لاکاز

lactalbumin	لاکتآلبومین	Feulgen stain	لکه فئولگن
Lactobacillus	لاکتوباسیل	macule	لکه کوچک پوستی/جلدی
lactoperoxidase	لاکتوپروکسیداز	leghemoglobin	لگوگلوبین
lactose	لاکتوز	legumin	لگومین
lactogen	لاکتوژن	leghemoglobin	لگ هموگلوبین
lactoferritin	لاکتوفریتین	lentinan	لنتینان
lactoferricin	لاکتوفریسین	lymph	لنف
lactoferrin	لاکتوفرین	axillary lymphadenopathy	لنفا دنوپاتی آکزیلاری
lactoglobulin	لاکتوگلوبولین	cervical lymphadenitis	لنفادنیت سرویکال
lactonase	لاکتوناز	lymphocyte	لنفوسیت
lactic dehydrogenase	لاکتیک دهیدروژناز	B lymphocyte	لنفوسیت بی.
laminin	لامینین	peripheral blood	لنفوسیت خون جانبی
lantibiotic	لانتیبیوتیک	lpymhocyte	
lanolin	لانولین	cytotoxic killer lymphocyte	لنفوسیت کشنده سم سلولی
lavendustin	لاوندوستین		
bed	لایه	tumor-infiltrating	لنفوسیت نفوذ گر در تومور
ozone layer	لایه ازن	lymphocyte	
lawn	لایه باکتری (روی محیط کشت)	lymphocytosis	لنفوسیتوز
black-layered (corn)	لایه سیاه (ذرت)	B-lymphocytes	لنفوسیت های بی.
slime layer	لایه گل و لای	t lymphocytes,	لنفوسیت های تی.
respiratory mucosa	لایه مخاطی دستگاه تنفسی	t-lymphocytes	
lobe	لپ	cytotoxic T lymphocytes	لنفوسیت های سیتوتوگزیک نوع
leptotene	لپتوتین		تی.
leptonema (= leptotene)	لپتونما	lymphokine	لنفوکاین
leptin	لپتین	lymphogranuloma	لنفوگرانولوما
slime	لجن	Burkhit lymphoma	لنفوم بورکیت
slough	لجنزار	lymphatic	لنفی
thrombus	لخته	lymphocyte	لنف یاخته
coagulation	لخته شدگی/شدن/کردن	t lymphocytes	لنف یاخته های تی.
gelation	لخته شدن	port	لنگر گاه
blood clotting	لخته شدن خون	wobble	لنگ زدن
chills	لرز	equipment	لوازم
wobble	لرزش	Croplands equipment	لوازم کراپلند
chill	لرزیدن	legume	لوبیا
lecithin	لسیتین	leupeptin	لوپپتین
slime	لعاب	lupus	لوپوس
labile	لغزنده	lupus erythematosus	لوپوس اریتماتوز
fertilization	لقاح	systemic lupus	لوپوس اریتماتوز سیستمی
wobble	لقوه	erythematosus	
lectin	لکتین	luteolysis	لوتئولایز
smut	لکه	luteolysin	لوتئولایزین

English	Persian
luteolin	لوتئولین
lutein	لوتئین
laurate	لوریت (یک اسید چرب)
loricrin	لوریکرین
sodium lauryl sulfate	لوریل سولفات سدیم
diamond	لوزی
luciferase	لوسیفراز
luciferin	لوسیفرین
leucine	لوسین
loci	لوکای
acute promyelocytic leukemia	لوکمی پرومیلوسیتیک حاد
acute nonlymphocytic leukemia	لوکمی غیر لنفوسیتی حاد
acute lymphocytic leukemia	لوکمی لنفوسیتی حاد
chronic lymphocytic leukemia	لوکمی لنفوسیتی مزمن
acute myeloblastic leukemia	لوکمی میلوبلاستیک حاد
chronic myelocytic leukemia	لوکمی میلوسیتی مزمن
acute myelomonacytic leukemia	لوکمی میلوموناسیتیک حاد
leukopenia	لوکوپنی
leukotriene	لوکوترین
locus	لوکوس
leukocyte	لوکوسیت
leukocytosis	لوکوسیتوز
levulose	لو لز
hose	لوله (خرطومی)
hose	لوله (لاستیکی)
airway	لوله (های) هوا
luliberin	لولیبرین
lumen	لومن
lumiflavin	لومیفلاوین
lumichrome	لومیکروم
luminophore	لومینوفور
levulose	لولولوز
mash	له
lyase	لیاز
luminescence	لیانندگی
luminescence	لیانی
lipase	لیپاز
pancreatic lipase	لیپازپانکراتیکی
lipemia	لیمی
lipoamide	لیپوآمید
lipoprotein	لیپو پروتئین
very low-density lipoprotein (very-low-density lipoprotein)	لیپوپروتئین با غلظت خیلی کم
low-density lipoprotein	لیپوپروتئین با غلظت پائین
intermediate-density lipoprotein	لیپوپروتئین با غلظت متوسط
high-density lipoprotein	لیپو پروتئین پر غلظت/غلیظ
low-density lipoprotein	لیپو پروتئین کم غلظت
lipo-protein lipase	لیپو پروتئین لیپاز
lipo-poly saccharide	لیپو پلی ساکارید
lipopolysaccharide (LPS)	لیپوپلی ساکارید
lipotropin	لیپو تروپین
liposome	لیپوزوم
lipogenesis	لیپوژنز
lipofection	لیپوفکشن
lipofuscin	لیپوفوسین
lipocaic	لیپوکائیک
lipoxidase	لیپوکسیداز
lipoxygenase	لیپوکسی ژناز
lipoxygenase null	لیپوکسی ژناز صفر
lipolysis	لیپولایز
lipovitellin	لیپوویتلین
lipid (= fat)	لیپید
lipophilic	لیپید دوست
lipidomics	لیپیدومیکز
lithotroph	لیتوتروف
protein-based lithography	لیتوگرافی مبتنی بر پروتئین
molecular lithography	لیتوگرافی ملکولی
lear	لیر
lyse, lysis	لیز
lysate	لیزات
alkaline lysis	لیز قلیائی/آلکالینی
lysostaphin	لیزواستافین
lysosome	لیزوزوم
lysozyme	لیزوزیم
lysogen	لیزوژن

thermoinducible lysogen	لیزوژن القا شونده با حرارت
lysogenesis	لیزوژنز
lysogeny	لیزوژنی
lysogenic	لیزوژنیک
lysophosphatidylethanolami ne	لیزوفسفاتیدیل اتانولامین
lysine	لیزین
listeria monocytogene	لیستریا مونوسیتوژن
leishmaniasis	لیشمانیاز
lycopene	لیکوپین
ligase (DNA), synthetase	لیگاز
T4 RNA ligase	لیگاز آر.ان.ای. تی.4
fatty acid synthetase	لیگاز اسید چرب
T4 DNA ligase	لیگاز دی.ان.ای. تی.4
ligation	لیگاسیون
ligand	لیگاند
bidentate ligand	لیگاند دو دندانه ای
ligandin	لیگاندین
lignan	لیگنان
lignocellulose	لیگنوسلولز
lignin	لیگنین
ligate	لیگیت
limonene	لیمونن
lincomycin	لینکومایسین
alpha linolenic	لینولنیک آلفا
lyochrome	لیو کروم
lyonization	لیونیزاسیون

م

matrix metalloproteinase	ماتریکس متالو پروتئیناز
nuclear matrix	ماتریکس هسته ای
maturase	ماتوراز
derived	ماخوذ
infrared	مادون قرمز
contaminant	ماده آلوده کننده
bulk	ماده اضافی
primer, substrate	ماده اولیه
placebo	ماده بی اثر
bulk	ماده بی مصرف
regulated article	ماده تنظیمی

lignin	ماده چوب(ی)
insecticide	ماده حشره کش
protoplast	ماده حیاتی سلولی
hazardous substance	ماده خطرناک
Delaney clause	ماده دیلانی
substrate	ماده زمینه
genetic material	ماده ژنتیکی
genome	ماده ژنتیکی مجموعه ژنها
carcinogen	ماده سرطان زا
keratin	ماده شاخی
hirudin	ماده ضد انعقاد خون
antiseptic, disinfectant	ماده ضد عفونی کننده
infectious material	ماده عفونی
vitafood	ماده غذائی طبی
essential nutrient	ماده غذائی اساسی/ضروری
phytonutrient	ماده غذایی گیاهی
surfactant	ماده فعال سطحی
saponin	ماده کف آور
adjuvant, co-factor	ماده کمکی
slime	ماده لزج
adjuvant	ماده محرک ایمنی
intercalating agent	ماده موتاسیون زا
active ingredient	ماده موثر
analyte	ماده مورد تجزیه
gynandromorph	ماده-نر ریخت
cross reacting material	ماده واکنشگر متقاطع
hereditary material	ماده وراثتی
helix	مارپیچ
alpha helix	مارپیچ آلفا
helix	مارپیچ استوانه ای
a-helix	مارپیچ ای./آلفا/آ.
double helix	مارپیچ دوگانه/مضاعف
helix breaker	مارپیچ شکن
alpha-helice	مارپیچهای آلفا
redundancy	مازاد بر احتیاج بودن
redundancy	مازادی
gall, tannin	مازو
mask	ماسک
high-efficiency particulate air filter mask	ماسک تصفیه ذرات هوا با کارآیی بالا
multi-layered	ماسک چند لایه کاربرد بالای

133

English	Persian	English	Persian
high-efficiency particulate air mask	ذرات معلق در هوا	smooth muscle	ماهیچه صاف
mask	ماسک زدن	myocardium	ماهیچه قلب
machine	ماشین	myograph	ماهیچه نگار
gene machine	ماشین ژن (ی)	myography	ماهیچه نگاری
washer	ماشین شوینده	myrothecium verrucaria	مایروتسیوم وروکاریا
molecular machine	ماشین ملکولی	myristoylation	مایریستولاسیون
machine	ماشینی	myxomycetes	مایزومایست
ultrasonic	مافوق صوت	myxovirus	مایزوویروس
macrophage	ماکروفاژ	mycelium	مایسلیوم
macrolide	ماکرولاید	maysin	مایسین
macromutation	ماکروموتاسیون	liquid	مایع
macromolecule	ماکرومولکول	ascites fluid	مایع آب شکم
macronutrient	ماکرونوترینت	amniotic fluid	مایع آمنیون
maxicell	ماکسی سل	lachrymal fluid	مایع اشکی/لاکریمال
macule	ماکول	nozzle	مایع پاش
maculopapular	ماکولاپاپولار	supernatant	مایع رویی
magainin	ماگائینین	shock fluid	مایع شوکی
multiple sclerosis	مالتیپل اسکلروز	supercritical fluid	مایع فوق بحرانی
fluorescence multiplexing	مالتی پلکسینگ	MicronAir	مایکرون ایر
	باپرتوافشانی/شب	Micronair AU4000	مایکرون ایر ای.یو. 4000
	نمایی/شارندگی/تشعشع	mycoagglutinin	مایکوآگلوتینین
	ماهتابی	mycobacterium tuberculosis	مایکوباکتری سل
proprietary	مالکانه	mycobiont	مایکوبیونت
mannan	مانان	cassava	ماینوک (گیاه)
mannan oligosaccharide	مانان الیگوساکارید	double minute	ماینیوت دوبل
mannanoligosacchariddes	مانانولیگو ساکارید ها	double minute	ماینیوت مضاعف
hereditary persistence of fetal hemoglobin	ماندگاری ارثی هموگلوبین جنینی	vaccine	مایه
genetic block	مانع ژنتیکی	vaccine	مایه آبله
repressor	مانع شونده	inoculum	مایه آبله کوبی/تلقیح
steric hindrance	مانع فضایی	anion exchanger	مبادله کننده آنیون
mannose	مانوز	ion exchanger	مبادله کننده یونی
mannosidostreptomycin	مانوسیدواستریتومایسین	gas exchange	مبادله گازی
mannogalactan	مانوگالاکتان	plague	مبتلا کردن
manometry	مانومتری	pharmacognosy	مبحث داروشناسی
mannitol	مانیتول	origin	مبداء
ultraviolet, ultra-violet	ماوراء بنفش	origin of replication	مبداء همانند سازی
ultrasonic	ماوراء صوت	Met (= methionine)	مت
uv1	ماورای بنفش نوع اول	metabolome	متابولوم
myocard)ium(	ماهیچه دل	metabolomics	متابولومیک
		metabolon	متابولون
		metabolite	متابولیت

134

English	Persian
anabolism	متابولیزم سازنده
metabolism	متابولیسم
nitrogen metabolism	متابولیسم ازت
intermediary metabolism	متابولیسم واسطه ای
metabonomics	متابونومیک
metaplasia	متاپلازی
metastasis	متاستاز
phenazine methosulphate	متاسولفات فنازین
metaphase	متافاز
metallothionein	متالوتیونئین
metalloflavoprotein	متالوفلاووپروتئین
metafemale	متا ماده
metamodel	متا مدل
methanol	متانول
metanomics	متانومیکز
crystallization	متبلور سازی
analog, homogeneous	متجانس
vagile	متحرک
BESS t-scan method	متد تی.اسکن بس
meter	متر
antagonist	متضاد
standard	متعارف
fluctuant	متغیر
light-chain variable	متغیر زنجیره سبک
heavy-chain variable	متغیر زنجیره سنگین
solid	متفق
antiparallel	متقابل
precursor	متقدم
methotrexate	متوترکسات
median lethal dose	متوسط دوز کشنده
geometric mean titer	متوسط غلظت هندسی
medium intake rate	متوسط نرخ ورودی
enzyme repression	متوقف شدن فعالیت آنزیم.
enzyme immobilization	متوقف کردن آنزیم
fatty acid methyl ester	متیل استر اسید چرب
methylation	متیلاسیون
DNA methylation	متیلاسیون دی.ان.ای.
methyl-guanine-DNA methyl-transferase	متیل ترانسفراز متیل-گوانین-دی.ان.ای.
methylated	متیل دار شده
methylation	متیل دار کردن

English	Persian
DNA methylation	متیل دار کردن/شدن دی.ان.ای.
methyl jasmonate	متیل ژاسمونات
methyl salicylate	متیل سالیسیلات
methylophilus methylotrophus	متیلوتوفوس متیلوفیلوس
DNA methylation	متیله شدن دی.ان.ای.
methionine	متیوناین
methionine	متیونین
FALSE positive	مثبت کاذب
polyclinic	مجتمع درمانی
lumen, pathway	مجرا
membrane channel	مجرا غشاء
respiratory tract	مجرای تنفسی
sodium channel	مجرای سدیمی
ion channel	مجرای یون
operator	مجری
fluorescence activated cell sorter (FACS)	مجزا کننده سلولی فعال شده با پرتوافشانی/شب نمایی/شارندگی/تشعشع ماهتابی
SS (= suspended solids)	مجم
FSS (= fixed suspended solids)	مج مث
VSS (= volatile suspended solids)	مج مف
abortive complex	مجموعه بی نتیجه
antigen-antibody complex	مجموعه پادگن-پادتن
open promoter complex	مجموعه پروموتر باز
pre-initiation complex	مجموعه پیش آغاز
open promoter complex	مجموعه پیش محرک باز
American Type Culture Collection (ATCC)	مجموعه کشت تیپ آمریکا
protectant	محافظ
conservation	محافظت
moribund	محتضر
proteome	محتوای پروتئینی سلول
sex limited	محدود به جنس
domain	محدوده
minimized domain	محدوده مینیم شده/مینیمایز شده
miniprotein domain	محدوده یک مینی پروتئین

restriction	محدودیت	internal ribosome entry site	محل ورود ریبوزوم داخلی
phylogenetic constraint	محدودیت دودمانی	stock solution	محلول استاک/پایه
apoptosis	محدودیت ژنتیکی طول عمر سلولها	buffer solution	محلول بافر
		aqueous	محلول در آب
host-controlled restriction	محدودیت مهار شده توسط میزبان	Ringer's solution	محلول رینگر
		gel	محلول ژلاتینی
nutrient limitation	محدودیت های مواد مغذی	Hank's balanced salt solution	محلول متعادل نمکی هنک
host restriction	محدودیت های میزبان		
long-acting thyroid stimulator	محرک تیروئیدی دراز مدت	phosphate-buffered saline	محلول نمکی بافری فسفاتی
		restriction sites	محلهای محدود کننده
crop, product	محصول	endemic, localized	محلی
blood derivative	محصولات مشتق از خون	DNA backbone	محور دی.ان.ای.
primary productivity	محصول دهی اولیه	medium	محیط
herbicide-tolerant crop	محصول زراعی تحمل کننده علف کش	open environment	محیط باز
		medium	محیط بدن
herbicide-resistant crop	محصول زراعی مقاوم نسبت به علف کش	biological environment	محیط بیولوژیکی
		potential receiving environment	محیط تحویل گیرنده بالقوه
standing crop	محصول سرپا/قد کشیده		
enhanced nutrition crop	محصول غذائی تقویت شده	production environment	محیط تولید
metabolite	محصول متابولیسم	in-vivo	محیط زنده
end product	محصول نهائی	environment	محیط زیست
lumen	محفظه مرکزی غده	accessible ecosystem	محیط زیست سریعا در دسترس
subarachnoid space	محفظه نخاعی	biological environment	محیط زیست شناختی
characterization assay	محک خصلت	medium	محیط عمل
solid	محکم	culture, culture medium, medium	محیط کشت
binding site	محل اتصال		
RNA binding site	محل اتصال آر.ان.ای.	basic medium	محیط کشت بازی
double-stranded DNA binding site	محل اتصال دی.ان.ای. دو رشته ای	biological medium	محیط کشت بیولوژیکی
		defined medium	محیط کشت تعریف شده
transcription factor binding site	محل اتصال عامل رونویسی	undefined medium	محیط کشت تعریف نشده
		phyto-hem-agglutinin-stimulated lymphocyte	محیط کشت تنظیم شده با لنفوسیت تحریک شده توسط فیتو-هم-آگلوتینین
acceptor junction site	محل اتصال گیرنده		
heteroduplex binding site	محل اتصال هترودوپلکس/مضاعف ناهمگن	conditioned medium	
		minimal medium	محیط کشت حداقل/مینیمم
		differential medium	محیط کشت دیفرانسیلی
ribosome-binding site	محل پیوند ریبوزوم	enriched medium	محیط کشت غنی شده
combining site	محل ترکیب شدن	maximal medium	محیط کشت ماکسیمال
biotope	محل زیست	environmental media	محیط کشت محیطی
habitat	محل سکونت	transmission	مخابره
catalytic site	محل کاتالیزوری	protein signaling	مخابره پروتئینی
acceptor site	محل گیرنده/اکسپتور	signaling	مخابره کردن

fluorescence multiplexing	مخابره هم زمان چند پیام با پرتوافشانی/شب نمایی/شارندگی/تشعشع ماهتابی	animal model	مدل جانوری
		animal model	مدل حیوانی
		modelling, simulation	مدل سازی
		homology modeling	مدل سازی همسانی
hazard	مخاطره	environmental fate model	مدل سرنوشت محیط زیست
mucoid	مخاط مانند	null model	مدل صفر
mucoid	مخاطی	structure-activity model	مدل فعالیت ساختاری
antagonist	مخالف	lock-and-key model	مدل قفل و کلیدی
antagonism	مخالفت	sliding filament model	مدل میلک لغزان
species specific	مختص به گونه	immune modulator	مدولاتور ایمنی
anal	مخرجی	selective estrogen receptor modulator (SERM)	مدولاتور گیرنده استروژن انتخابی
acrocephalic	مخروط سر		
acrocephalia	مخروط سری	modulation	مدولاسیون
library	مخزن	consequence management	مدیریت پی آمد(ی)
animal genetic resources databank	مخزن اطلاعاتی منابع ژنتیکی حیوانی	integrated pest management (IPM)	مدیریت تلفیقی آفات
seed bank	مخزن بذر	integrated disease management	مدیریت تلفیقی بیماری
chromosome jumping library	مخزن پرشی کروموزومی	integrated crop management	مدیریت تلفیقی گیاهان زراعی
animal genetic resources databank	مخزن داده های منابع ژنتیکی حیوانی	risk management	مدیریت ریسک
DNA bank	مخزن دی.ان.ای.	common property resource management	مدیریت منابع املاک مشترک/عمومی
cDNA library	مخزن زنجیره واحد دی.ان.ای		
		waste management	مدیریت مواد پسماند
animal genome (gene) bank	مخزن ژن حیوانی	environmental media	مدیوم محیطی
in-situ gene bank	مخزن ژن در محل/در جای خود/درجای طبیعی/در محل طبیعی/در جا/به موضع	male	مذکر
		chaperone	مراقب
		ecosystem service	مراقبت از بوم سازگان
animal genome (gene) bank	مخزن ژنوم جانوری	supportive care	مراقبت پشتیبانی کننده
genomic library	مخزن ژنومی	molecular chaperone	مراقب ملکولی
gene bank/library/pool, gene-bank	مخزن ژن (ی)	centers of origin	مراکز اصالت
		centers of origin and diversity	مراکز اصالت و تنوع/گوناگونی
animal gene bank	مخزن ژنی حیوانی		
expression library	مخزن نمایانی/تظاهر/هویدائی	centers of genetic diversity	مراکز گوناگونی/تنوع ژنتیکی
mixing	مخلوط کردن	atopic	مربوط به آلرژی
yeast, zymogen	مخمر	bronchial	مربوط به برونشی/نایژه
baker's yeast	مخمر نانوائی	avian	مربوط به پرندگان
environmentalist	مدافع محیط زیست	lytic	مربوط به تجزیه
lifetime	مدت عمر/زندگی	diagnostis	مربوط به تشخیص طبی
intake	مدخل	atmospheric	مربوط به جو
template	مدل	murine	مربوط به جوندگان

adipose	مربوط به چربی	programmed cell death	مرگ سلولی برنامه ای/برنامه ریزی شده
biological	مربوط به زیست		
cannabinoid	مربوط به شاهدانه/کانابیس	apoptosis	مرگ سلولی برنامه ریزی شده
cyanotic	مربوط به کبودی پوست		
portal	مربوط به مدخل باب کبدی	necrosis	مرگ نسج
educator	مربی	cellular necrosis	مرگ نسج سلولی
competent authority	مرجع ذیصلاح	tissue-necrosis	مرگ نسجی
M phase	مرحله ام.	restoration	مرمت
nucleosome phasing	مرحله ای کردن نوکلئوزوم	merozygote	مروزیگوت
lag phase	مرحله تاخیر/تاخر	meromixis	مرومیکزی
stationary phase	مرحله ثابت	conjugate	مزدوج شدن
growth phase	مرحله رشد	chronic	مزمن
exponential growth phase	مرحله رشد تصاعدی	mesoscale	مزواسکیل
log growth phase	مرحله رشد لگاریتمی	mesotocin	مزوتوسین
zygotene stage	مرحله زیگوتین	mesosome	مزوزوم
logarithmic phase	مرحله لگاریتمی	cilia	مژک ها
logarithmic phase	مرحله نمایی	cilia	مژه ها
slough	مرداب	free-rider problem of public goods	مسئله استفاده رایگان کالاهای عمومی
eutrophication	مردابی شدن		
stillbirth	مرده زائی	liability	مسئولیت
necrobiosis	مرده زیستی	palindrome	مساوی الطرفین
port	مرز	induration	مستحکم سازی
infection	مرض	capable of being transmitted from one person to another	مستعد انتقال از شخصی به شخص دیگر
Kennedy disease	مرض کندی		
mercaptoethanol	مرکاپتواتانول		
multiplex	مرکب	intoxication	مستی
complexation	مرکب شدن	channel-blockers (= calcium channel-blockers)	مسدود کننده مجرا
nucleus	مرکز		
center of diversity	مرکز تنوع		
active center	مرکز فعال/کنش ور	calcium channel-blockers	مسدود کننده های مجرای کلسیم
density gradient centrifugation	مرکز گریزش به روش تغییرات چگالی		
		communicable, contagious, infectious	مسری
containment facility	مرکز محدود کننده/باز دارنده	level	مسطح
mercury	مرکور	analgesic, palliative	مسکن
expiration	مرگ	habitat	مسکن طبیعی
necrosis	مرگ بافت	cocking	مسلح کردن
apoptosis	مرگ برنامه ریزی شده سلولها	toxicity	مسموم کنندگی
		cytotoxic	مسموم کننده سلول
genetic death	مرگ ژنتیکی (جهش)	poisoning	مسموم کننده(گی)
necrosis	مرگ سلولها	intoxication, poisoning	مسمومیت
cell death	مرگ سلولی	toxemia	مسمومیت خونی

MessengerTM	مسنجر	adaptive immunity	مصونیت مزاجی
lane, pathway	مسیر	plague	مصیبت
anabolic pathway	مسیر آنابولیک	duplication	مضاعف سازی
ubiquitin-proteasome pathway	مسیر اوبیکیتین -پروتئازوم	prospective study	مطالعه آتی
sweepstake route	مسیر بخت آزمایی/سوئیپ استیک	etiology	مطالعه علل بیماری
		observational study	مطالعه مشاهده ای
airway	مسیر تنفس	case study	مطالعه موردی
amphibolic pathway	مسیر دو حرکتی	biologic indicator of exposure study	مطالعه نشانگر زیستی تماس
lytic pathway	مسیر کافتی		
central metabolic pathway	مسیر متابولیک/سوخت و ساز مرکزی	cohort study	مطالعه همزادگان
		Hardy-Weinberg equation	معادله هاردی-واینبرگ
anaplerotic metabolic pathway	مسیر متابولیکی آناپلروتیک	adoptive cellular therapy	معالجه سلولی انتخابی
		convention	معاهده
salvage pathways	مسیرهای بازیابی	Convention on Biological Diversity (CBD)	معاهده بین المللی حفظ و بهره برداری از منابع بیولوژیکی جهان
airway	مسیر هوا		
airway	مسیر هوایی		
analog, analogue, homologous	مشابه	Biological and Toxin Weapons Convention (BTWC)	معاهده جنگ افزار های بیو لوژیکی و سمی
clone, simulation	مشابه سازی	probe (= DNA probe)	معاینه
genetic counseling	مشاوره ژنتیکی	pathway	معبر
blood derivative	مشتقات خونی	validation	معتبر سازی
derived	مشتق شده	inroganic	معدنی
endothelium-derived	مشتق شده از بافت تروپوشی/اندوتلیوم	gastric	معدی
		agent, reagent	معرف
inherited characteristic	مشخصه اکتسابی	nuclease-free reagent	معرف بدون نوکلئاز
consultation	مشورت	bioindicator	معرف زیستی
consumption	مصرف	Millon's reagent	معرف میلون
oxygen consumption	مصرف اکسیژن	introduction	معرفی
acceptable daily intake	مصرف قابل قبول روزانه	aromatic	معطر
adequate intake	مصرف کافی	inverted	معکوس
use	مصرف کردن	inversion	معکوس شدگی
user	مصرف کننده	educator	معلم
public good	مصلحت همگانی	inbreeding coefficient	معیار هم خونی
immunofluorescence	مصون پرتوافشانی	nutrient	مغذی
adoptive immunization	مصون سازی انتخابی	nucleus	مغز
immunodominant	مصون غالب	recessive	مغلوب
immunization	مصون کردن	autosomal recessive	مغلوب اتوزومی
immunity	مصونیت	X- linked recessive	مغلوب مرتبط به اکس.
humoral-mediated immunity	مصونیت متعادل مزاجی	dormancy	مغلوبیت
		pharmacognosy	مفردات پزشکی

synovial	مفصلی	interallelic complementation	مکمل سازی بین آللی
import, sense	مفهوم	genetic complementation	مکمل سازی ژنتیکی
ideal protein concept	مفهوم پروتئین ایده آل	in vitro marker complementation	مکمل سازی نشانگر این ویترو
resistant	مقاوم		
thermoduric	مقاوم به گرما	complementary	مکمل (همدیگر)
persistence, resistance	مقاومت	aspirate	مکیدن
systemic acquired resistance	مقاومت اکتسابی سازگانی/سیستمیک	megabase	مگاباز
		megaplasmid	مگاپلاسمید
aluminum resistance	مقاومت به آلومینیم	mega base pair	مگا جفت باز
streptomycin resistance	مقاومت به استرپتو مایسین	megaDalton	مگا دالتون
drug resistance, tolerance	مقاومت به دارو	drosophila	مگس سرکه/میوه
multi-drug resistance	مقاومت داروئی چند گانه	Mediterranean fruit fly	مگس میوه/سرکه مدیترانه ای
antibiotic resistance	مقاومت در مقابل آنتی بیوتیک		
		mole	مل
cold tolerance	مقاومت درمقابل سرما	melatonin	ملاتونین
systematic activated resistance	مقاومت فعال شده روش مند	melanocyte	ملانوسیت
		melanoma	ملانوما
herbicide resistance	مقاومت نسبت به علف کش	melanoidin	ملانویدین
hardening	مقاوم سازی	melanin	ملانین
cold hardening	مقاوم سازی در مقابل سرما	concomitant	ملزوم
antibiotic resistance	مقاوم شدن نسبت به آنتی بیوتیک	capture molecule	ملکول اتصال
		biological molecule	ملکول بیولوژیکی
bioassay	مقایسه زیستی	signaling molecule	ملکول پیام دهنده
dose, volume	مقدار	intercellular adhesion molecule	ملکول چسبندگی بین سلولی
chlorine residual	مقدار کلر ته نشین شده		
chlorine demand	مقدار کلر مطلوب/مورد نیاز	cellular adhesion molecule	ملکول چسبندگی سلولی
dose	مقدار مصرف	informational molecule	ملکول خبر رسان
chlorine dose	مقدار مصرف کلر	macromolecule	ملکول درشت
vent	مقعد	chimeric DNA	ملکول دی.ان.ای نوترکیب/کیمریک
anal	مقعدی		
mesoscale	مقیاس متوسط	toxic molecule	ملکول سمی
locus	مکان	chaperone molecule	ملکول شاپرون
locus	مکان هندسی	chemical molecule	ملکول شیمیائی
pathway feedback mechanisms	مکانیزم فید بک مسیر	polar molecule	ملکول قطبی
		sugar molecule	ملکول قند
environmental media and transport mechanism	مکانیسم انتقال و واسطه گری محیطی	gram molecular weight	ملکول گرم
		amphipathic molecules	ملکول های آمفی پاتیک
transport mechanism	مکانیسم نقل وانتقال	amphipathic molecules	ملکول های دو بخشی
intake	مکش	knottins	ملکولهای گره ای
complement	مکمل	melibiose	ملیبیوز
oxygen supplementation	مکمل سازی اکسیژنی		

140

English	Persian
melitriose	ملیتریوز
insulin-dependent diabetes mellitus	ملیتوس دیابتی وابسته به انسولین
prevention	ممانعت
co-repressor	ممانعت کننده
resource	منابع
biotic resource	منابع زیستی
homogeneously staining regions	مناطق رنگ آمیزی همگن
dominant control regions	مناطق کنترل/مهار غالب
renewable resource	منبع تجدید پذیر
biological resource	منبع زیستی
genetic resource	منبع ژنتیکی
natural source	منبع طبیعی
indirect source	منبع غیر مستقیم
non-point source	منبع نان-پوینت/عمومی
point-source	منبع نقطه ای
disseminating	منتشر کردن
disseminator	منتشر کننده
vector	منتقل کننده
helix	منحنی پیچ
growth curve	منحنی رشد
canola	منداب روغنی کانادایی
integrated	منسجم
origin	منشاء
replication origin	منشاء همانند سازی
origin of DNA replication	منشاء همانند سازی دی.ان.ای.
bifurcate, branch	منشعب شدن
area of release	منطقه آزاد کردن
double- stranded RNA binding domain	منطقه اتصال آر.ان.ای. دو رشته ای
scaffold attachment region	منطقه اتصال داربست
precautionary zone	منطقه احتیاطی/اخطاری
localized	منطقه ای
inhibitory zone	منطقه بازداری
buffer zone	منطقه بافر
pest free area	منطقه بدون آفت
intergenic region	منطقه بین نژادی
terminator region	منطقه پایان دهنده
untranslated region (UTR)	منطقه ترجمه نشده
contact zone	منطقه تماس
constant region	منطقه ثبات
flanking region	منطقه جانبی/کناری
buffer zone	منطقه حائل
hot spot	منطقه حساس
protected area	منطقه حفاظت شده
hybrid zone	منطقه دورگه
coding region	منطقه رمز ساز(ی)
adaptive zone	منطقه سازشی
pseudo autosomal region	منطقه شبه اتوزومی
buffer zone	منطقه ضربه گیر
protective action zone	منطقه عمل حفاظتی
catalytic domain	منطقه کاتالیزوری
crossover region	منطقه کراس اوور
locus control region	منطقه کنترل لوکوس
variable region	منطقه متغیر
thymus-dependent area	منطقه متکی به تیموس
systematics	منظم
inhibition	منع
explosive	منفجره
vent	منفذ
voltage-gated ion channel	منفذ ترا شامه ای یون
stomatal pore	منفذ روزنه ای
ion channel	منفذ یون
extinct	منقرض
meningitis	مننژیت
plague meningitis	مننژیت همه گیر
strand, µ (= micro)	مو
biomimetic materials	مواد بیومیمتیک
fixed suspended solids (FSS)	مواد جامد معلق ثابت (مج مث)
volatile suspended solids (VSS)	مواد جامد معلق فرار (مج مف)
suspended solids (SS)	مواد جامد معلق (مجم)
feedstock	مواد خام
push package	مواد دفع شدنی
auxins	مواد رشد
paint	مواد رنگ کاری
nutraceuticals	مواد غذائی طبی
functional food	مواد غذائی کارکردی/عملی/کاربردی

141

English	Persian	English	Persian
balance, compensation	موازنه	habitat	موطن
prior informed consent	موافقت از قبل اعلام شده	loci	موقعیت
mutein	موتئین	niche	موقعیت مناسب
mutase	موتاز	acid mucopolysaccharide	موکو پلی ساکارید اسیدی
mutagen	موتاژن	muco-poly-saccharidosis	موکو-پلی-ساکاریدوز
cassette mutagenesis	موتاژنز کاستی	Mucor	موکور
mutation	موتاسیون	mole	مول
down promoter mutation	موتاسیون داون پروموتر	generating, generator	مولد
mutant	موتان	zymogen	مولد آنزیم
muton	موتون	Marple Aerosol Generator	مولد افشانه ماریل
effector	موثر	pathogenic	مولد بیماری
mole	موج شکن	molecule	مولکول
microorganism	موجود ذره بینی	amphiphilic molecules	مولکول آمفی فیل
organism	موجود زنده	effector molecule	مولکول اثر کننده
living modified organism	موجود زنده تغییر یافته	Molecular Breeding TM	مولکولار بریدینگ
model organism	موجود زنده مدل	Molecular Pharming TM	مولکولار فارمینگ
standing crop	موجودی گیاهی	monomer	مولکول تک واحدی
modon	مودون	adhesion molecule	مولکول چسبندگی
murein	مورئین	recombinant DNA molecule	مولکول دی.ان.ای. نوترکیب
murine	موراین	mole	مولکول گرم
case, incidence	مورد	nanocrystal molecules	مولکولهای ریز کریستال
case-by-case	مورد به مورد	monensin	موننزین
case	مورد بیماری	monoiodo-tyrosine	مونوئیدو تیروزین
contraindication	مورد عدم استعمال	monoploid	مونوپلوئید
case-finding	مورد یابی	mono-thio-glycerol	مونو-تیو-گلیسرول
morphology	مورفولوژی	monosome	مونوزوم
safety-pin morphology	مورفولوژی سنجاق قفلی ای	monosomic	مونوزومی
mosaic	موزائیک (ی)	mono zygote	مونوزیگوت
severe, childhood,	موسکولار دیستروفی مغلوب	monosaccharide	مونوساکارید
autosomal, recessive	اتوزومی بچگانه حاد	monocistron	مونوسیسترون
muscular dystrophy		xanthosine mono phosphate	مونوفسفات زانتوزین
Becker muscular dystrophy	موسکولار دیستروفی نوع بکر	hexose mono-phosphate	مونوفسفات هگزوز
mucin	موسین	uridine mono-phosphate	مونوفسفات یوریدین
rocket	موشک	monocotyledon	مونوکوتیلدون
mole	موش کور	monolepsis	مونولپسی
knockout mouse	موش ناک اوت	monomer	مونومر
allosteric site	موضع آلوستریک	mononucleotide	مونونوکلئوتید
niche	موضع بوم شناختی	flavin mononucleotide	مونو نوکلئوتید فلاوین
allosteric site	موضع دگر ریختار	flavin mono-nucleotide	مونونوکلئوتید فلاوین
localized	موضعی	wild type	مونه وحشی
case	موضوع	asset	موهبت

fog, mist	مه	mean corpuscular	میانگین هموگلوبین
turbidity	مه آلودگی	hemaglobin	کورپوسکولار
cumulative feedback	مهار بازخورد تراکمی	cytoplasm	میان مایه
inhibition		interband	میان نوار
medical control	مهار پزشکی	cytoplasm	میان یاخته
biological control	مهار زیستی	cytoplasmic	میان یاخته ای
stringent control	مهار قاطع	mithramycin	میترا مایسین
inhibition	مهار کنندگی	mithridatism	میتریداتیزم
repressor	مهار کننده	mitosis	میتوز
Cro repressor	مهار کننده کرو	mitogen	میتوز زا
acceptor control	مهار گیرنده	mitogen	میتوژن
lethal, malignant	مهلک	mitochondrion	میتوکندری
munition	مهمات	mitochondria	میتوکندری ها
multi-agent munition	مهمات چند عاملی	mitomycin	میتومایسین
	مهمات خوشه ای/فرعی	contour-clamped	میدانهای الکتریکی همگن
protein engineering	مهندسی پروتئین	homogeneous electric	کنتور-کلامپ
biomolecular engineering	مهندسی زیست ملکولی	fields	
bioengineering	مهندسی زیستی	grade, unit	میزان
genetically engineered	مهندسی (ژنتیک) شده	incidence rate	میزان اتفاق/بروز
genetic engineering	مهندسی ژنتیک(ی)	contact rate	میزان تماس
gene engineering	مهندسی ژن (ی)	uptake	میزان جذب
metabolic engineering	مهندسی سوخت وساز	intake rate	میزان درون جذب
carbohydrate engineering	مهندسی کربوهیدرات	prevalence rate	میزان شیوع
engineering	مهندسی کردن	LD 50	میزان کشندگی 50 درصدی
genetic predisposition	مهیا سازی ژنتیکی	case-fatality rate	میزان کشندگی موردی
centromere	میانپار	medium intake rate	میزان متوسط جذب درونی
mesoderm	میان پوست	intake rate	میزان ورودی
mesodermal, mesodermic	میان پوستی	host	میزبان
centrosome	میان تن	immunocompromised host	میزبان به خطر افتاده از لحاظ
mesophile	میان دما		ایمنی
centriole	میانک	Mistblower	میست بلوئر
centrifugation	میان گریزش	mischarging	میسشارژ کردن
differential centrifugation	میان گریزش دیفرانسیلی	micelle	میسل
density gradient	میان گریزش شیب چگالی	lipid micelle	میسل لیپید
centrifugation		reversed micelle	میسل معکوس
centrifuge	میان گریز کردن	micro-RNAs	میکرو آر.ان ای. ها
tris-acetate buffer (TAB)	میانگیر تریس-استات	microorganism,	میکروارگانیسم
TE buffer	میانگیر تی.ئی	micro-organism	
mean corpuscular volume	میانگین حجم کورپوسکولار	marine microorganism	میکروارگانیسم دریائی
mean lifetime	میانگین طول عمر	marsh gas microorganisms	میکروارگانیسم های گاز مرداب
mean residue weight	میانگین وزن رسوبی	recombinant DNA	میکروارگانیسم های نوترکیب

143

microorganism		meiosis	میوز
microbe	میکروب	myosin	میوزین
microbalance	میکرو بالانس	myofibril	میوفیبریل
microbology	میکروب شناسی	myocard(ium)	میوکارد
microbicide	میکروب کش	myoglobin	میوگلوبین
aerobe	میکروب هوازی	machine	میونگ
aerobic microbe	میکروب هوازی	seedless fruit	میوه بدون بذر
microbiology	میکروبیولوژی	fructification	میوه دهی
food microbiology	میکروبیولوژی تغذیه		
microperoxisome	میکروپراکسی زوم	ن	
microplasts	میکروپلاست ها		
microtome	میکروتوم	neoendemics	نئواندمیکز
ultraviolet microscope	میکروسکوپ فرابنفش	neoplasia	نئوپلازی
microscopy	میکروسکوپی	multiple endocrine	نئوپلازی اندوکرین چندگانه
electron microscopy	میکروسکوپی الکترونی	neoplasia	
atomic force microscopy	میکروسکوپی نیروی هسته ای	neoplasm	نئوپلاسم
microphtalmic	میکروفتالمیک	neocarzinostatin	نئوکارزینواستاتین
microbodymicrocarrier	میکروکارير میکروبادی	neolignan	نئولیگنان
microgram	میکروگرم	neomycin	نئومایسین
micro-meter, micrometer	میکرومتر	apomixis	نا آمیزی
micro-meter	میکرون	solid	ناب
micron	میکرون	adventitious	نابجا
mycotoxin	میکوتوکسین	extinction	نابودی
mycorrhizae	میکوریز	contamination	نا پاکی
ectotrophic mycorrhiza	میکوریزای اکتوتروفیک	labile	ناپایدار
willingness to accept	میل به پذیرفتن	variance	نا پایداری
willingness to pay	میل به پرداختن	incapacitation	ناتوان سازی
affinity	میل ترکیبی	incapacitate	ناتوان کردن
myeloblastin	میلوبلاستین	heterogeneous	ناجور
myeloma	میلوما	hetero-	ناجور-
multiple myeloma	میلومای چند گانه	heterozygote	ناجور تخم
bacillus	میلیزه	heterozygosity	ناجور تخمی
myelin	میلین	heterocyclic	ناجور حلقه ای
mimetics	میمتیک	heteroallels	ناجور دگره ها
artillery mine	مین توپی (توپخانه ای)	heterodimer	ناجور دوپار
mineralocorticoid	مینرالوکورتیکوئید	heteroduplex	ناجور دوتایی
mini	مینی	heterochromatin	ناجور رنگینه
miniprotein	مینی پروتئین	constitutive	ناجور رنگینه نهادی
mitochondrial myopathy,	میوپاتی، انسفالوپاتی، لاکتیک	heterochromatin	
encephalopathy, lactic	اسیدوز و شبه سکته	heterogametic	ناجور زامه ای
acidosis and stroke-like	میتوکندریائی	heteromorphism	ناجورشکلی

heterokaryon	ناجور هسته	carrier, contagious, transfer, vector, vehicle	ناقل
heterokaryosis	ناجور هسته بودن	SIN vector (= self-inactivating vector)	ناقل اس.ای.ان.
heterogeneity, variance, xenogenesis	ناجوری	electron carrier	ناقل الکترون
domain	ناحیه	vector borne	ناقل برد
donor splicing site	ناحیه اسپلایس دونر/دهنده	intercellular transport	ناقل بین سلولی
intercistronic region	ناحیه بین سیسترونی	biological vector	ناقل بیولوژیکی
intergenic region	ناحیه بین گونه ای	Pre Con vector (= promoter conversion vector)	ناقل پری کان
garden	ناحیه حاصلخیز		
breakpoint cluster region	ناحیه خوشه شکننده	promoter conversion vector	ناقل تبدیل پروموتور
active site	ناحیه فعال	replacement vector	ناقل جایگزینی
locus control region	ناحیه کنترل لوکوس	self-inactivating vector	ناقل خود خنثی گر
heterozygote, heterozygous	نا خالص	insertion vector	ناقل درون جای دهی
heterozygosity	ناخالصی	bifunctional vector	ناقل دو کاربردی/بایفانکشنال
heterotroph	ناخود پرور	retroviral vector	ناقل رترو ویروسی
heterotrophic	ناخود پروری	shuttle vector	ناقل شاتلی
self-incompatibility	ناخود سازگاری	neurotransmitter	ناقل عصبی
morbidity	ناخوشی	molecular vehicle	ناقل ملکولی
epizootic	ناخوشی همه گیر حیوانی	host-vector	ناقل-میزبان
respiratory distress	ناراحتی تنفسی	transporter associated with antigen presentation	ناقل وابسته به تظاهر آنتی ژن
deficiency	نارسایی		
immunodeficiency	نارسایی ایمنی	cloning vector	ناقل همسانه سازی
acridine orange	نارنجی آکریدینی	RNA vectors	ناقلین آر.ان.ای.
naringen	نارینژن	knockin	ناک این
sterile	نازا	knockdown	ناک داون
abiogenesis, spontaneous generation	نازیست زائی	nondisjunction	نا گسستگی
abiotic	نازیوا	naloxone	نالوکسون
bronchitis	ناژه آماس	heterogeneous	نامتجانس
antagonism	ناسازگاری	heterogeneity	نامتجانسی
hereditary fructose intolerance	ناسازگاری فروکتوزی ارثی	generic	نام جنسی
		arrhythmia	نا موزونی
contaminant	نا سالم کننده	apomixis	نامیختگی
contamination	ناسالمی	apogamous, apomict, apomictic	نامیخته
dissimilation	ناشبیه ساختن		
deafness	ناشنوایی	immortal	نامیرا
nonvolatile	نا فرار	immortalization	نامیرا سازی
nafcillin	نافسیلین	apogamy	نامیزش
inactivation	نافعالی	apospory	نامیزی
malformation	ناقص الخلقگی	carbon nanotube	نانو تیوب کربن(ی)
teratogenesis	ناقص الخلقه سازی	nanocochleate	نانوکوکلیات

145

English	Persian
nanogram	نانوگرم
nanometer	نانومتر
dip-pen nanolithography	نانو نقش زدن نوک قلمی
dip-pen nanolithography	نانو نگارش نوک قلمی
nozzle	ناودانک
dissimilation	ناهماند سازی
allopatric	نا هم بوم
heterologous	نا همساخت
antagonism	ناهمسازی
heterogeneous	ناهمگن
heterogeneity, heterology	ناهمگنی
genetic heterogeneity	ناهمگنی ژنتیکی/وراثتی
heterogeneous	ناهمگون
dissimilation	ناهمگون سازی
teratogenesis	ناهنجار سازی
teratology	ناهنجار شناسی
malformation	ناهنجاری
anaerobe	ناهوازی
bronchitis	نایژاماس
bronchiole	نایژک
bronchus	نایژه
bronchial	نایژه ای
bronchoscope	نایژه بین
bronchoscopy	نایژه بینی
bronchi	نایژه ها
nystatin	نایستاتین
plant	نبات
nebulizer	نبولایزر
nebulin	نبولین
knapsack	نپساک
netilmicin	نتیلمایسین
delivery	نجات
embryo rescue	نجات جنین
maker rescue	نجات سازنده
strand	نخ
waste	نخاله
spinning	نخ تابی
nematodes	نخسانه ها
hyphae	نخینه ها
nodule	ندول
nodulation	ندولاسیون

English	Persian
male	نر
contact rate	نرخ تماس
specific production rate	نرخ تولید اختصاصی
per capita intake rate	نرخ جذب سطحی سرانه
specific growth rate	نرخ رشد اختصاصی
incidence rate	نرخ شیوع
intake rate	نرخ مکش
intake rate	نرخ ورودی
per capita intake rate	نرخ ورودی سرانه
spermatid	نر زامچه
spermatium	نر زامه بی تاژک
male-sterile	نرعقیم
flaccid	نرم
hermaphrodite, monoecious	نر-ماده
maleness	نرینگی
paravertebral	نزدیک مهره ها
retrograde	نزولی
breed, pedigree, strain	نژاد
landrace	نژاد بومی
endangered breed	نژاد به خطر افتاده
critical-maintained breed and endangered-maintained breed	نژاد در خطر بحرانی نگه داشته شده و نژاد در معرض انقراض نگه داشته شده
critical breed	نژاد در خطر وخیم/بحرانی
endangered-maintained breed	نژاد در معرض انقراض نگه داشته شده
genotype	نژادگان
molecular breeding	نژادگیری مولکولی
genotype	نژادمانه
extinct breed	نژاد منقرض
isogenic stocks	نژادهای ایزوژنیک
attributable proportion	نسبت آتریبیوتبل (به چیزی)
risk ratio	نسبت خطر
attributable proportion	نسبت ناشی از (چیزی)
granulation tissue	نسج التیامی
transcript	نسخه
transcription	نسخه برداری
primary transcript	نسخه برداری اولیه
latency-associated transcript	نسخه برداری مربوط به نهفتگی
reverse transcription	نسخه برداری معکوس

quantitative reverse transcriptase PCR	نسخه برداری معکوس کمی پی.سی.آر.	tumor marker	نشانگر تومور
hybrid-release translation (HRT)	نسخه برداری هیبرید رها شده	multi-locus probe	نشانگر چند نقطه ای
		DNA probe, DNA marker	نشانگر دی.ان.ای
hybrid-arrested translation	نسخه برداری هیبرید متوقف شده	anonymous DNA marker	نشانگر دی.ان.ای. ناشناس
		radioactive marker	نشانگر رادیواکتیو
copy DNA	نسخه دی.ان.ای.	bioprobe	نشانگر زیستی
gene transcript	نسخه ژن (ی)	biological marker of exposure	نشانگر زیستی سطح تماس
generation, strain	نسل		
extinct breed	نسل منقرض	genetic marker, genetic probe	نشانگر ژنتیکی
amyloplast	نشادیسه		
starch	نشاسته	bcr-abl genetic marker (= genetic marker)	نشانگر ژنتیکی بی.سی.آر.-ای.بی.ال.
glycogen	نشاسته حیوانی		
transplantation	نشاکاری	c-kit genetic marker	نشانگر ژنتیکی کیت سی.
marker	نشان	gene probe	نشانگر ژن (ی)
affinity tag	نشان دار	fluorogenic probe	نشانگر
labling	نشاندار کردن		فلوئوروژنیک/فلوئورسانس زا/شب
embedding, settle	نشاندن		نمایی/پرتوافشانی/شارندگ
syndrome	نشانگان		ی زا
Beckwith-Weideman syndrome	نشانگان بکویت-ویدمان	RNA probes	نشانگر های آر.ان.ای.
		hybridization probe	نشانگر هیبریداسیون
hepato-renal syndrome	نشانگان کبدی-کلیوی	trait	نشانویژه
Klinefelter syndrome	نشانگان کلاین فلتر	production trait	نشانویژه تولید
Lesch Nyhan syndrome	نشانگان لش نایهان	indicator, marker, symptom, tag	نشانه
systemic inflammatory response syndrome	نشانگان واکنش التهابی سازگانی		
		prodrome	نشانه اولیه
target	نشانگاه	symptomatic	نشانه ای
hybridization probing	نشانگذاری با دورگه سازی	symptom, symptomatic	نشانه بیماری
marker, probe (= DNA probe)	نشانگر	sequela	نشانه زخم
		biomarker	نشانه زیستی
nucleic acid probe	نشانگر اسید نوکلئیک	pattern biomarker	نشانه زیستی الگو
dissolved oxygen probe	نشانگر اکسیژن محلول	selectable marker	نشانه قابل انتخاب
oligonucleotide probe	نشانگر اولیگونوکلئوتید	tracer	نشانه گذار
in vitro marker	نشانگر این ویترو	tag	نشانه گذاری
biological marker	نشانگر بیولوژیکی	diffusion	نشت
biological marker of exposure	نشانگر بیولوژیکی سطح تماس	vegetative propagation	نشر گیاهی
		mounted	نصب شده
linked gene/marker	نشانگر پیوسته	zygote	نطفه
genetic marker	نشانگر توارثی	oversight	نظارت
bcr-abl genetic marker (= genetic marker)	نشانگر توارثی بی.سی.آر.-ای.بی.ال.	DNA extraction and quality control	نظارت بر کیفیت و استخراج دی.ان.ای.

147

structure	نظام	gene map	نقشه ژن (ی)
innate immune system	نظام ایمنی مادر زادی	physical map (of genome)	نقشه فیزیکی (ژنوم)
technology protection system	نظام حمایت تکنولوژی	chromosome map	نقشه کروموزومی
		mapping	نقشه کشی
immune system	نظام دفاعی بدن	fluorescence mapping	نقشه کشی/برداری فلوئورسانس
firefly luciferase-luciferin system	نظام لوسیفراز-لوسیفرین حشره(کرم) شب تاب	gene mapping	نقشه کشی ژن (ی)
systematics	نظام مند	cellular pathway mapping	نقشه کشی مسیر سلولی
non-equilibrium theory	نظریه غیر تعادلی	contig map (= contiguous map)	نقشه کن-تیگ
theory of local existence	نظریه وجود محلی		
structure	نظم	macrorestriction map	نقشه ماکرو ریستریکشن
enzyme regulation	نظم آنزیم	restriction map	نقشه محدود کننده
homologous	نظیر	circular restriction map	نقشه محدودیت حلقوی/دایره ای
asset	نعمت		
gas, wind	نخخ	haplotype map	نقشه نیمگان جور
person	نفر	linkage map	نقشه وابستگی/پیوستگی
nephrocalcin	نفروکالسین	haplotype map	نقشه هاپلوتیپ
wind	نفس	ordered clone map	نقشه همانندسازی منظم
tachypnea	نفس تندی	deficiency	نقص
public good	نفع عمومی	deficiency	نقصان
nephelometer	نفلومتر	genetic defect	نقص ژنتیکی
diffusion, infiltration, osmosis	نفوذ	malformation	نقص عضو
		neural tube defects	نقص لوله عصبی
dermal penetration	نفوذ جلدی/از راه پوست	R point (= restriction point in the cell cycle)	نقطه آر.
infiltrate	نفوذ دادن/کردن		
population	نفوس	achromic point	نقطه بدون رنگ/آکرومیک
mask	نقاب	compensation point	نقطه جبران
convalescence	نقاهت	startpoint	نقطه شروع
imprinting	نقش بندی/پذیری	focal point	نقطه کانونی
gene imprinting	نقش بندی ژن (ی)	quantum dot	نقطه کوانتومی
genomic imprinting	نقش زدن ژنومیکی	restriction point in the cell cycle	نقطه منع در چرخه سلولی
dip-pen lithography	نقش زدن نوک قلمی		
gene mapping	نقشه برداری ژن (ی)	transfer	نقل
contiguous map	نقش(ه) پیوسته/هم جوار/پیاپی	gated transport	نقل و انتقال مهار شده
		necrobiosis	نکروبیوز
genetic linkage map	نقشه پیوند/اتصال ژنتیکی	necrosis	نکروز
haplotype map	نقشه تک لاد مونه	tissue-necrosis	نکروز بافتی
sequence map	نقشه توالی	negative supercoiling	نگاتیو سوپر کویلینگ
dentogram	نقشه دندانها/دهان	dip-pen lithography	نگارش نوک قلمی
genetic map	نقشه ژنتیکی	mapping	نگاشت
genome map	نقشه ژنوم	gene mapping	نگاشت ژن (ی)

precautionary approach	نگرش احتیاطی	environmental sample	نمونه محیطی
revealed preference approach	نگرش ترجیحی آشکار شده	voucher specimen	نمونه واچر
protectant	نگهدارنده	circadian rythm	نواخت سیرکادیان/شبانه روزی
cryogenic storage	نگه داری سرد خانه ای/کرایوژنیک	r-banding	نوار بندی آر
		preprophase band	نوار پریپروفاز
culture preservation	نگهداری کشت	high resolution chromosome banding	نوارگذاری کروموزومی با دقت بالا
nematodes	نماتود ها	hybrid zone	نواره آمیخته
karyotype	نمادکروموزومی	periodicity	نوبه ای
expression	نمایانی	de novo	نو پدید
gene expression profiling	نمای تظاهر ژن (ی)	regeneration	نوپدیدی
transcriptional profiling	نمای رونویسی	naturaceuticals (=nutraceuticals)	نوتراسوتیکال
genetic profiling	نمای ژنتیکی		
phage display	نمایش باکتری خوار	neutraceuticals	نوتراسوتیکال(ها)
differential display	نمایش دیفرانسیلی	recombinant	نوترکیب
metabolite profiling	نمای متابولیتی	recombination	نوترکیبی
molecular profiling	نمای ملکولی	additive recombination	نو ترکیبی افزایشی
oleate	نمک اسید اولئیک	reciprocal recombination	نوترکیبی دوطرفه
nitrite	نمک اسید نیترو	genetic recombination	نوترکیبی ژنتیکی
nitrate	نمک اسید نیتریک	non-homologous recombination	نوترکیبی غیر همتا/ناهمتا
halophile	نمک دوست		
glyphosate isopropylamine salt	نمک گلیفسات ایزوپروپیلامین	homologous recombination	نوترکیبی همگون
		neutropenic	نوتروپنیک
growth	نمو	neutrophil	نوتروفیل
expressivity	نمود	nutrigenomics	نوتریژنومیکس
amino acid profile	نمودار اسید آمینه	nutriceuticals (=nutraceuticals)	نوتریسوتیکال
accession, case	نمونه		
type specimen	نمونه استاندارد	neutriceuticals	نوتریسوتیکال(ها)
accession	نمونه ای در بانک ژن	nutricine	نوتریسین
biopsy	نمونه بافت	nojirimycin	نوجیریمایسین
biopsy	نمونه برداری	neoplasm	نودشتینه
chorionic villus sampling	نمونه برداری از بافت جنینی (پرده کوریون)	luminescence	نور
		photoorganotroph	نورآلی پرور
chorionic villus sampling	نمونه برداری از پرزهای جفتی	luminesce	نور افشان
percutaneous umbilical blood sampling	نمونه برداری زیر جلدی از خون بند ناف	luminescence	نورافشانی
		neural	نورال
		neuraminidase	نورامینیداز
acute sample	نمونه پیشرفته	bioluminescence	نور تابی زیستی
continuous sample	نمونه پیوسته	chemiluminescence	نورتابی شیمیائی
acute sample	نمونه حاد	northern blot/transfer	نورترن بلات/ترانسفر
modelling	نمونه سازی	northern blotting	نورترن بلاتینگ
type specimen	نمونه شاخص		

149

photoautotroph	نورخودپرور	slough	نهر باتلاقی
photoperiod	نور دوره	dormancy, latency	نهفتگی
exposure	نور دهی	covert, recessive	نهفته
luminesce	نورزا	thalamus	نهنج
luminescence	نورزایی	demand	نیاز
optrode	نورکاو	biochemical oxygen demand (BOD)	نیاز بیو شیمیائی به اکسیژن
phototropic	نورگرا		
phototropism	نورگرایی	biological oxygen demand	نیاز زیستی به اکسیژن
plus ultra-violet light of the A wavelength	نور مثبت ماورای بنفش با طول موج ای.	case, legume	نیام
		legume	نیامک
optical activity	نورورزی	nitrate	نیترات
neuropsychiatric	نوروسایکیاتریک	nitrate reductase	نیترات ردآکتاز
neuro-fibromatosis	نورو-فیبروماتوز	nitrocellulose	نیترات سلولز
neuron	نورون	nitrification	نیتراته شدن
regeneration	نوزائی	nitrogenase	نیتروژناز
leader	نوساقه	nitrocellulose	نیتروسلولز
antidote	نوشدارو	nitrite	نیتریت
breed, genus, grade, species	نوع	nitric oxide	نیتریک اکسید
mating type	نوع جفتگیری	nitric oxide synthase (NOS)	نیتریک اکسید سینتاز
biotype	نوع زیستی	nitrilase	نیتریلاز
serotype	نوع سرمی	synergy	نیرو زایی
wild type	نوع وحشی	adhesive force	نیروی چسبندگی
nuclease	نوکلئاز	nisin	نیسین
s1 nuclease	نوکلئاز اس.1	moiety	نیم
nucleoid	نوکلئوئید	haptene	نیم پادگن
nucleoprotein	نوکلئو پروتئین	hemizygous	نیم جور تخم
nucleoplasm	نوکلئوپلاسم	monoploid	نیم دانه
nucleotid, nucleotide	نوکلئوتید	biological half-time	نیم عمر بیولوژیکی
flavin nucleotide	نوکلئوتید فلاوین	monoploid	نیمگان
complementary nucleotide	نوکلئوتید مکمل	haplotype	نیمگان جور
nucleosome	نوکلئوزوم	moiety	نیمه
nucleoside	نوکلئوزید	subclinical	نیمه تشخیصی
nucleosidase	نوکلئوزیداز	biological half-time	نیمه عمر زیستی
nuclein	نوکلئین	serum half life	نیمه عمر سرم
nogalamycin	نوگالامایسین	subculture	نیمه کشت
nullisomy	نولیزومی	sub-clone	نیمه کلون
novobiocin	نووبیوسین	subclinical	نیمه کلینیکی
latency	نهان بودگی	monoploid	نیمه گان
covert	نهان (ی)	subcloning	نیمه همسانه سازی
insidious, latency	نهانی		
nadir	نهایت افت		

و

English	Persian
affinity, linkage, tag	وابستگی
streptomycin dependence	وابستگی به استرپتو مایسین
sex linkage	وابستگی جنسیتی
paternal	وابسته به پدر
sex linked	وابسته به جنسیت
wobble	وابل
dissociation	واپاشی
repression	واپس رانی
reannealing	واتاباندن
soft laser desorption	واجذب لیزری نرم
voucher specimen	واچر اسپسیمن
unit	واحد
taxon	واحد آرایه شناختی
chromosomal packing unit	واحد بسته بندی کروموزومی
taxon	واحد سیستماتیک
transcription unit	واحد نسخه برداری
map unit	واحد نقشه
colony forming units	واحد های تشکیل دهنده کلونی
introduction	وارد سازی
import, introgression	وارد کردن
site-directed mutagenesis	واردکردن یا حذف جهت یافته
importer	وارد کننده
Warfarin	وارفارین
inversion	وارونش
conversion, inversion	وارونگی
gene conversion	وارونگی ژن (ی)
somatic variant	واریانت یاخته غیر جنسی
variance	واریانس
covariation	واریانس مشترک
cultivar	واریته
disabled strain	واریته خنثی شده
inbred strain	واریته درون آمیخته
Abelson strain of murine	واریته (مربوط به جوندگان) آبلسونی
live vaccine strain	واریته واکسن زنده
transgenic	واریخته
vasopressin	وازوپرسین
vasotocin	وازوتوسین
inversion	واژگونی

English	Persian
paracentric inversion	واژگونی پاراسنتریکی
pericentric inversion	واژگونی پریسنتریک
vaginosis	واژنوز
denature	واسرشتن
denaturation	واسرشتی
enzyme denaturation	واسرشتی آنزیم
agent, mediator, medium	واسطه
basic medium	واسطه بازی
undefined medium	واسطه تعریف نشده
differential medium	واسطه دیفرانسیلی
defined medium	واسطه کشت تعریف شده
transition-state intermediate	واسطه وضعیت انتقال
washer	واشر
patent	واضح
depolarization	واقطبش
depolarizer	واقطبنده
depolarize	واقطبیدن
depolarized	واقطبیده
upstream	واقع در قسمت علیای رودخانه
event	واقعه
vaccine	واکسن
polyvalent vaccine	واکسن چند بنیانی
edible vaccine	واکسن خوراکی
DNA vaccine	واکسن دی.ان.ای.
Sabin vaccine	واکسن سابین
killed vaccine	واکسن کشته
response	و اکنش
SOS response	واکنش اس.او.اس.
addition reaction	واکنش افزایشی
oxidation-reduction reaction	واکنش اکسایش-کاهش
group-transfer reaction	واکنش انتقال گروهی
endergonic reaction	واکنش اندرگونیک
endergonic reaction	واکنش اندوترمیک
endergonic reaction	واکنش انرژی خواه
exergonic reaction	واکنش انرژی ده
immune reaction, immune response	واکنش ایمنی
antibody-mediated immune response	واکنش ایمنی بواسطه آنتی بادی ها
innate immune response	واکنش ایمنی درون ذاتی
serum immune response	واکنش ایمنی سرم

English	Persian	English	Persian
cellular immune response	واکنش ایمنی سلولی	contagion	واگیری
humoral immune response	واکنش ایمنی مزاجی	pheromone	واگیزن
polyclonal antibody response	واکنش پادتن چند کلونی	parents	والدین
adaptive response	واکنش تطبیقی	valine	والین
periodic acid-Schiff reaction	واکنش تناوبی اسید-شیف	continuous perfusion	وامیختگی پیوسته
side effect	واکنش ثانوی	van	وانت
cross reaction	واکنش چلیپائی	vancomycin	وانکومایسین
polyclonal response	واکنش چند کلونی	Yq (= Y long arm)	وای اندیس کیو.
hypersensitive response	واکنش خیلی حساس	YAC (= yeast artificial chromosome)	وای.ای.سی.
idling reaction	واکنش درجا		
polymerase chain reaction (PCR)	واکنش زنجیره ای پلیمراز	genetics, heredity	وراثت
		horizontal inheritance	وراثت اتوزومی مغلوب
adaptive response	واکنش سازشی	heritability	وراثت پذیری
cellular response	واکنش سلولی	polygenic inheritance	وراثت چند ژنی/فاکتوری
delayed-type hypersensitivity reaction	واکنش فوق حساسیتی تاخیر دار	genetic inheritance	وراثت ژنتیکی
		cytoplasmic inheritance	وراثت سیتوپلاسمی
reagent	واکنشگر	non-Mendelian inheritance	وراثت غیر مندلی
cross reactivity	واکنش گری/پذیری متقاطع	cytoplasmic inheritance	وراثت غیر مندلی/غیر هسته ای
cross reaction	واکنش متقابل	cytoplasmic inheritance	وراثت مادری
protein-protein interaction	واکنش متقابل پروتئین-پروتئین	heredity	وراثتی
cross reaction	واکنش متقاطع	variance, variation	وردایی
humoral response	واکنش مزاجی	sport	ورزش
adverse reaction	واکنش معکوس	β-sheet	ورقه بتا
Maillard reaction	واکنش میلارد	node	ورم
Millon reaction	واکنش میلون	gall	ورم بافت گیاه
adverse reaction	واکنش نا مطلوب	arthritis	ورم مفاصل
photochemical reaction	واکنش نورشیمیائی	introgression	ورود
ninhydrin reaction	واکنش نینهیدرین	income, intake	ورودی
Wasserman reaction	واکنش واسرمن	atomic weight	وزن اتمی
hydroxylation reaction	واکنش هیدروکسیله کردن	blower	وزنده
catalyst	واکنش یار	molecular mass, molecular weight	وزن ملکولی
vacuoles	واکوئل ها		
quantum yield	واگذاری کوانتمی	gram molecular weight	وزن مولکولی بر حسب گرم
conversion	واگردانی	equivalent weight	وزن هم ارز
truck, van, wagon	واگن باری	vesicle, vesicule	وزیکول
contagious	واگیر	coated vesicle	وزیکول پوشش دار
infectious	واگیر دار	coated vesicle	وزیکول پوشیده
communicable	واگیردار	coated vesicle	وزیکول روپوش دار
contaminate	واگیر داشتن	equipment	وسایل
		decontamination kit	وسایل رفع آلودگی
		western blot	وسترن بلات

152

western blotting	وسترن بلاتینگ
broad-specturm	وسیع الطیف
applicator, device, medium	وسیله
injector	وسیله تزریق
biotinylatio	وصل بیوتینی
emergency	وضع اضطراری
parturition	وضع حمل
case	وضعیت
contained casualty setting	وضعیت تلفات مهار شده/تحت کنترل
in-situ condition	وضعیت در جا
dose	وعده
ribosomal adaptor	وفق دهنده ریبوزومی
recombination frequency	وفور نوترکیبی
gap period, inhibition	وقفه
event, incidence, occurrence	وقوع
vector borne	وکتور برن
parturition	ولادت
avidity	ولع
Wolff-Parkinson-White	ولف-پارکینسون-وایت
vomitoxin	ومیتوکسین
volicitin	وولیسیتین
pica	ویار
VLA (= very late antigen)	وی.ال.ای.
VLDL (= very low-density lipoprotein)	وی.ال.دی.ال.
VNTR (= variable number of tandem repeats)	وی.ان.تی.آر.
v-oncogene	وی.-انکوژن
vibrio cholerae	ویبریو کولرا
vitamer	ویتامر
vitamin	ویتامین
ascorbic acid	ویتامین ث
VDD (= vitamin d-dependency rickets)	وی.دی.دی.
proofreading	ویرایش
kinetic proofreading	ویرایش کینتیک
virus	ویروس
poxvirus	ویروس آبله
cowpox virus	ویروس آبله گاوی
cowpea mosaic virus	ویروس آفت لوبیای چشم

(CpMV)	بلبلی
alphavirus	ویروس آلفا
amphitropic virus	ویروس آمفیتروپیک
Epstein Barr virus	ویروس اپشتاین بار
FBJ osteosarcoma virus (=oncogene FOS)	ویروس استئوسارکومای اف.بی.جی.
Human Immunodeficiency Virus	ویروس افت ایمنی انسان
satellite virus	ویروس اقماری
sin nombre virus (SNV)	ویروس بی شماره
xenotropic virus	ویروس بیگانه گرا
newcastle disease virus	ویروس بیماری نیوکاسل
papovavirus	ویروس پاپوا
papillomavirus	ویروس پاپیلوما
bovine papilloma virus (BPV)	ویروس پاپیلومای گاوی
poxvirus	ویروس پوکس
dengue fever virus	ویروس تب دانگ
respiratory syncytial virus	ویروس تنفسی سینسیتیال
tumor virus	ویروس تومور
mammary tumor virus	ویروس تومور پستان
mouse mammary tumor virus	ویروس تومور پستان موش
latent virus	ویروس خفته/نهفته
virion	ویروس رسیده
simian virus 40 (SV40)	ویروس سیمیان/بوزینه ای 40
sindbis virus	ویروس سیندبیس
pseudo-rabies virus	ویروس شبه هاری
helper virus	ویروس کمک کننده
lymphocytic choriomeningitis virus	ویروس کوریومننژیت لنفاوی
avian leukemia virus	ویروس لوکمی طیور
bovine leukemia virus	ویروس لوکمی گاوی
murine leukemia virus	ویروس لوکمی موشی
Moloney murine leukemia virus	ویروس لوکمی موشی مولونی
myxovirus	ویروس مخاطی
adeno-associated virus	ویروس مرتبط به آدنو
replication-competent virus	ویروس مستعد همانند سازی
defective virus	ویروس معیوب
tobacco mosaic virus	ویروس موزائیک توتون

English	Persian
mucosa rota virus	ویروس موکوسا روتا
defective virus	ویروس ناقص
simian immunodeficiency virus	ویروس نقص ایمنی سیمیان/بوزینه ای
varicella zoster virus	ویروس واریسلا زوستر
hepatitis B virus	ویروس هپاتیت بی.
herpes virus saimiri	ویروس هرپس سایمیری
herpes simplex virus	ویروس هرپس سیمپلکس
viral	ویروسی
virion	ویریون
icosahedral virion	ویریون آیکوزاهدرال
VZV (= varicella zoster virus)	وی.زد.وی.
visfatin	ویزفاتین
specificity, trait	ویژگی
drought tolerance trait	ویژگی تحمل خشکسالی
production trait	ویژگی تولید
toxicity characteristic leaching procedure	ویژگی سمیت فرآیند آبشویی
adaptation traits	ویژگی های سازگاری
v-gene	وی.-ژن
WhiskersTM	ویسکرز
viscosity	ویسکوزیته
apparent viscosity	ویسکوزیته ظاهری
vicilin	ویسیلین
virus	ویش
virus disease	ویش زدگی
virion	ویشه
vinblastine	وینبلاستین
vincristine	وینکریستین
vivaparous	ویواپاروز

ه

English	Persian
haptoglobin	هاپتوگلوبین
misdivision haploid	هاپلوئید میسدیویژن
haplotype	هاپلوتیپ
haplophase	هاپلوفاز
haploid, monoploid	هاپلوئید
hot spot	هات اسپات
harpin	هارپین

English	Persian
H2O2 (= hydrogen peroxide)	هاش دو او. دو
spore	هاگ
sporulation	هاگ آوری
sporangium	هاگدان بر
sporulation	هاگ زایی
sporozoite	هاگ زیا
sporulation	هاگ سازی
sporulation	هاگ گذاری
spore	هاگه
corona	هاله ماه
holoenzyme	هام زیمایه
hantavirus	هانتاویروس
hyperoxaluria	هایپراوکسالوری
hyperprolinemia	هایپرپرولینمی
hyperplasia	هایپرپلازی
congenital adernal hyperplasia	هایپرپلازی آدرنال مادرزادی
hyperpolarization	هایپر پولاریزاسیون
hypertrophy	هایپرتروفی
delayed hypersensitivity	هایپر سنسیتیویته با تاخیر
hyperfiltration	هایپر فیلتراسیون
hyperchromicity	هایپرکرومیسیته
hypercholesterolemia	هایپرکلسترولمی
familial hypercholesterolaemia	هایپرکلسترولمی خانوادگی
hyphae	هایفا
hymenium	هایمنیوم
hepatatrophic	هپاتاتروفیک
heparin	هپارین
hapten	هپتن
hapmap	هپ مپ
heteroacryon	هترواکریون
heteroallels	هترواآلل ها
heteroantigen	هتروآنتی ژن
heteroecious	هترو اسیوس
heteroploid (= aneuploid)	هتروپلوئید
heterotroph	هتروتروف
heterodimer	هترو دیمر
heterosis	هتروز
heterozygote	هتروزیگوت

154

compound heterozygote	هتروزیگوت مرکب	airborne precaution	هشدار ذرات معلق در هوا
heterogeneous	هتروژن	airborne precaution	هشدار (ذرات) هوائی/هوابرد
heterocyclic	هتروسیکلیک	digestion	هضم
heterokaryosis	هتروکاریوزیس	anaerobic digestion	هضم آن-ایروبیک/غیر هوازی
heterokaryon	هتروکاریون	membrane digestion	هضم غشائی
heterochromatin	هتروکروماتین	complete digestion	هضم کامل
constitutive heterochromatin	هتروکروماتین ساختمانی	digest	هضم کردن/شدن
constitutive heterochromatin	هترو کروماتین نهادی	partial digestion	هضم ناقص
		hexadecyltrimethylammonium bromide	هگزادسیل تری متیل آمونیم برومید
heterogametic	هتروگامتی	hex A (= hexosaminidase A)	هگز ای.
heterogamic	هتروگامی		
heteromorphism	هترومورفیزم	hex B (= hexosaminidase B)	هگز بی.
onset	هجوم	hexose	هگزوز
target	هدف	hexosaminidase A	هگزوزآمینیداز ای.
panmixia	هر آمیزی	hexosaminidase B	هگزوزآمینیداز بی.
hermaphrodite	هرمافرودیت	helicase	هلیکاز
true hermaphrodism	هرمافرودیسم واقعی	DNA b helicase	هلیکاز بی. دی.ان.ای.
external cost	هزینه خارجی	helix	هلیکس
opportunity cost	هزینه فرصت	helicoverpa zea, helicoverpazea (=corn earworm)	هلیکوورپازیا
social opportunity cost	هزینه فرصت اجتماعی		
private opportunity cost	هزینه فرصت اختصاصی		
nucleolus, nucleus	هستک	heme	هم
nucleus	هسته	guild	هم آشیانه
imidazole	هسته اصلی هیستیدین	hemagglutinin	هم آگلوتینین
nucleocapsid	هسته پوشه	conjugation	هم آوری
nucleosome	هسته تن	consistency	هماهنگی
karyokinesis	هسته جنبی	hematopoietin	هماتوپوئیتین
mitosis	هسته جنبی	hematochezia	هماتوکزی
eukaryote	هسته دار	substantial equivalence	هم ارزی اساسی
karyogamy	هسته زامی	cooxidation	هم اکسایش
nucleoid	هسته سا	analogous, concomitant, homogeneous, homologous	همانند
synergid	هسته قرینه		
synergid	هسته کمکی		
karyotype	هسته مونه	assimilation, replication	همانند سازی
karyotyper	هسته مونه گر	theta replication (θ)	همانند سازی تتا
incomplete karyotype	هسته مونه ناقص	genetic assimilation	همانند سازی توارثی
karyogram	هسته نگاره	rolling circle replication	همانند سازی حلقه رولینگ
nucleoid	هسته واره	DNA replication	همانند سازی دی.ان.ای.
core octamer	هسته هشت واحدی	gene cloning, gene duplication	همانند سازی ژن (ی)
cell nucleus, karyon	هسته یاخته		

155

English	Persian
semiconservative replication	همانند سازی نیمه محافظه کارانه
virus replication	همانند سازی ویروسی
homology	همانندی
generic	همانی
coagulation	هم بستی
sympatric	همبوم
isomer	همپار
stereoisomer	همپار فضایی
covariation	هم پراشی
isotope	هم پروتون
syntrophism	هم پروری
composting	همپوده درست کردن
conformation	هم پیکری
syngraft	هم پیوند
homologous	همتا
replication	همتا سازی
conformation	همتاشی
Golgi complex	همتافت گلژی
abortive complex	همتافت ناموفق
synaptinemal complex	همتافته نوار همموردی
homology	همتایی
co-transduction, co-transformation	هم تبدیلی
analogous	هم تراز
hybridization	هم تیرگی
nucleic acid hybridization	هم تیرگی اسید نوکلئیک
allele- specific oligonucleotide hybridization	هم تیرگی اولیگونوکلئوتید مختص به آلل
southern hybridization	هم تیرگی به روش ساترن
dot hybridization	هم تیرگی داتی
in situ hybridization	هم تیرگی در محل
introgressive	هم تیرگی درون رونده
dynamic allele-specific hybridization	هم تیرگی دینامیکی مختص آلل
northern hybridization	هم تیرگی شمالی
colony hybridization	هم تیرگی کولونی
cross-hybridization	هم تیرگی متقاطع
sympatric	همجا
homogeneous	همجنس
agglomeration, fusion	همجوشی
gene fusion	همجوشی ژن (ی)
centric fusion	همجوشی مرکزی
cell fusion	همجوشی یاخته ای
dextrorotary	هم جهت عقربه ساعت
comutagen	هم جهش زا
agglutination	هم چسبی
consistency	همخوانی
consanguineous, sib	هم خون
company, concomitant	همراه
allele	هم ردیف ژنی
homology	هم ردیفی
cohort	همزادان
syngamy	هم زامی
symbiotic	هم زیست
symbiotic	هم زیستانه
symbiosis	همزیستی
biocoenosis	هم زیستی جانوران
homologous	همساخت
co-adaptation	همسازی
isotope	هم سان
analog, analogue, homologous	همسان
radioactive isotope	همسان پرتوزا
homogenization	همسان سازی
clone	همسانه
cDNA clone	همسانه زنجیره واحد دی.ان.ای
cloning	همسانه سازی
cocloning	همسانه سازی اضافی
directional cloning	همسانه سازی جهت دار
gene cloning	همسانه سازی ژن (ی)
positional cloning	همسانه سازی موقعیتی
homology	همسانی
TATA homology	همسانی تی.ای.تی.ای.
cosuppression	هم سرکوبی
commensalism	همسفرگی
commensal	همسفره
substantial equivalence	هم ظرفیتی بنیادی
commensalism	هم غذایی
co-evolution	هم فرگشت

156

isomer	همفرمول	hemicellulose	همی سلولز
commensalism	همکاری	conjugate	هم یوغ شدن
co-management	هم گردانی	conjugation	هم یوغی
cohort	هم گروه	sexual conjugation	هم یوغی جنسی
homogeneous	همگن	air	هوا
homologous	همگون	airplane	هواپیما
assimilation	همگون سازی	air	هوا دادن
genetic assimilation	همگون سازی ژنتیکی	aeration	هوا دهی
sequence homology	همگونی در توالی	aerobe, aerobic	هوازی
co-management	هم مدیری	obligate aerobic	هوازی اجباری
synkaryon	هم نقشه	aerobic	هوازیست
clade, sibling	هم نیا	aerobic	هوازیستی
cladistics, sibship	هم نیائی	meteorology	هواشناسی
sympatric	همنیاک	aerosol	هواویز
hemostasis	همو استاز	aerosol	هواویزه
hommopolymer	هموپلیمر	aerobic	هوایی
hemopoietin	هموپوئیتین	eukaryotic	هوبر
hemopoiesis	هموپوئیزیس	eutrophication	هو پرور شدگی
homothalism	هموتالیسم	hormone	هورمون
synapse, syndesis	همور	adrenocorticotropic hormone	هورمون آدرنوکورتیکوتروپیک
hemorrhagic pleural effusion	هموراژیک پلورال افیوژن	antidiuretic hormone (ADH)	هورمون آنتی دیورتیک
covariation	هم وروائی	interstitial cell-simulating hormone (ICSH)	هورمون اینترستیشیال محرک سلول
homogenization	هموژنیزه کردن	antidiuretic hormone (ADH)	هورمون پادپیشابی
haemophilia, hemophilia	هموفیلی	parathormone	هورمون پاراتیروئیدی
vascular hemophilia	هموفیلی عروقی	pancreatic hormone	هورمون پانکراتیکی
homograft	هموگرافت	follicle stimulating hormone	هورمون تحریک کننده فولیکول
hemoglobin	هموگلوبین	endocrine hormone	هورمون درون ریز
adult hemoglobin	هموگلوبین بالغ	growth hormone, somatotropin	هورمون رشد
fetal hemoglobin	هموگلوبین جنینی		
hemolymph	همولنف	human growth hormone	هورمون رشد انسان
hemolysis	همولیز	bovine somatotropin	هورمون رشد گاوی
hemolysin	همولیزین	releasing hormone	هورمون رها سازی
homokaryon, isomer	هم هسته	endocrinology	هورمون شناسی
pandemic	همه گیر	anti-mullerian hormone	هورمون ضدمولریان
epizootic	همه گیری دامی	activation hormone	هورمون فعال کننده
behavioral epidemic	همه گیری رفتاری	phenolic hormone	هورمون فنولی
pandemic	همه گیری گسترده	adrenocorticotropic	هورمون قطعه قدامی غده
mutualism	همیاری		
hemizygous	همیزایگوس		
hemicellulase	همی سلولاز		

English	Persian
hormone	صنوبری
heparin	هورمون کبدی
adrenal cortical hormone	هورمون کورتیکال آدرنال
phytohormone, plant hormone	هورمون گیاهی
thyroid stimulating hormone	هورمون محرک تیروئید
human thyroid-stimulating hormone	هورمون محرک تیروئید انسان
luteinizing hormone	هورمون محرک جسم زرد
luteinizing hormone	هورمون محرک یاخته بینابینی پرولان بی
gastrointestinal hormone	هورمون معده-روده ای
juvenile hormone	هورمون نابالغ
auxins	هورمون های گیاهی
anesthesia, narcosis	هوش بری
euploid	هولاد
holoenzyme	هولو آنزیم
holotype	هولوتیپ
holin	هولین
homeostasis	هومئواستازی
immunological homeostasis	هومئواستازی ایمونولوژیکی
homeobox	هومئوباکس
homeo box region 1	هومئوباکس منطقه 1
homeo box region 2	هومئوباکس منطقه 2
homeo box region 3	هومئوباکس منطقه 3
homocysteine	هوموسیستئین
homocystinuria	هوموسیستینوری
homokaryon	هوموکاریون
homologous	هومولوگ
homing receptor	هومینگ ریسپتور
eucaryote	هوهسته ای
expression	هویدایی
configuration	هیئت
examination body	هیئت بررسی
hybrid	هیبرید
hybridization	هیبریداسیون
northern hybridization	هیبریداسیون نورترن
hybridization	هیبریدشدگی
annealing	هیبرید شدن زنجیره دی.ان.ای. یا آر.ان.ای
hybridized	هیبرید شده
hybridoma	هیبریدوما
hypothalamus	هیپوتالاموس
hypotrichosis	هیپوتریکوزی
hypoxanthine	هیپوزانتین
hypoxanthine–aminopterin-thymidine	هیپوزانتین-آمینوپترین-تیمیدین
hypoxantine-guanine phospho-ribosyl transferase	هیپوزانتین-گوانین فسفو-ریبوسیل ترانسفراز
hypostasis	هیپوستازی
hypoxanthine phospho ribosyl transferase	هیپوگزانتین فسفوریبوزیل ترانسفراز
carbohydrate	هیدرات کربن
hydrazine	هیدرازین
hydrazinolysis	هیدرازینولیز
hydrogen	هیدروژن
hydrogenation	هیدروژن دار سازی
dehydrogenation	هیدروژن زدایی
dehydrogenation, hydrogenation	هیدروژن گیری
hydrogenation	هیدروژنه کردن
hydrophobic	هیدروفوب
hydrophilic	هیدروفیلیک
hydroxy praline	هیدروکسی پرالین
aluminum hydroxide	هیدروکسید آلومینیم
hydroxylysine	هیدروکسیلایزین
S-adeno homocysteine hydrolase	اس.-آدنوهوموسیستئین هیدرولاز
hydrolysis	هیدرولیز
hydrolyzate	هیدرولیزات
alkaline hydrolysis	هیدرولیز قلیایی
hydrolyze	هیدرولیز کردن
hairpin loop	هیرپین لوپ
hirudin	هیرودین
histamine	هیستامین
hysteresis	هیسترزیس
histone	هیستون
histidine	هیستیدین
hyphae	هیف ها

English	Persian
cytochrome	یاخته رنگ
epithelial cell	یاخته روپوشه ای
filler epithelial cell	یاخته روپوشه ای پرکننده
germ cell	یاخته زایشی
cell cytometry	یاخته شماری
cytology	یاخته شناسی
neuron	یاخته عصبی
flow cell	یاخته فلو/جریان
natural killer cell (NK cell)	یاخته قاتل طبیعی
cytolysis	یاخته کافت
cytolysis	یاخته کافتی
cytomegalovirus	یاخته کلان ویروس
platelet	یاخته لخته شده
blood platelet	یاخته لخته کننده
spermatogonium cell	یاخته مادر اسپرم
mast cell	یاخته ماستوسیت
histoblast	یاخته متشکل بافت
ES cells (= embryonic stem cells)	یاخته(های) ئی.اس.
beta cells	یاخته های بتا
entrapped cells	یاخته های در تله/گیر افتاده
eukaryotic cell	یاخته هوبر
adjuvant	یاری کننده
adjuvant	یاور
lyophilization	یخ خشکانی
potassium iodide	یدید پتاسیم
unit	یکان
solid	یکپارچه
homogeneous	یک دست
homogeneous	یکسان
isotope	یکسانگرد
monosaccharide	یک قندی
mononuclear	یک هسته ای
cointegrate	یکی کردن
U (= uracil)	یو.
U-RNA	یو.-آر.ان.ای.
UMP (= uridine mono-phosphate)	یو.ام.پی.
UAG (= termination codons)	یو.ای.جی.
euploid	یوپلوئید

ی

English	Persian
menopause	یائسگی
cell	یاخته
cytotoxic	یاخته آزار
cytotoxicity	یاخته آزاری
stem cell	یاخته اصلی
mast cell	یاخته بافت هم بند
fibroblast	یاخته بافت همبند
somatic cell	یاخته بدنی
stem cell	یاخته بنیاد
mesodermal adult stem cell	یاخته بنیادی بالغ میان پوستی
embryonic stem cell	یاخته بنیادی جنینی
stem cell one	یاخته بنیاد یک
mesenchymal stem cell	یاخته بنیادی میان آگنه ای
B cell (= Bursa dependent cell)	یاخته بی.
phagocytic cell	یاخته بیگانه خوار
stem cell	یاخته پایه
plasma cell	یاخته پلاسما
germ cell	یاخته تناسلی
suppressor t cell	یاخته تی بازدارنده
gamete, germ cell	یاخته جنسی
phagocytosis	یاخته خواری
granulocyte	یاخته دانیزه ای
endothelial cell	یاخته درون پوشه ای/اندوتلیال
plasma cell	یاخته دشتینه
stem cell	یاخته دودمانی
adult stem cell	یاخته دودمانی بالغ
multipotent adult stem cell	یاخته دودمانی بالغ چند خاصیتی (قابل تمایز)
mesenchymal adult stem cell	یاخته دودمانی بالغ میان آگنه ای
mesodermal adult stem cell	یاخته دودمانی بالغ میان پوستی
totipotent stem cell	یاخته دودمانی پر توان
embryonic stem cell	یاخته دودمانی جنینی
human embryonic stem cell	یاخته دودمانی جنینی انسان
hematopoietic stem cell	یاخته دودمانی خون ساز
mesenchymal stem cell	یاخته دودمانی میان آگنه ای
stem cell one	یاخته دودمانی یک

uteroglobin	یوترو گلوبین
UTP (= uridine tri-phosphate)	یو.تی.پی.
UDP (= uridine di-phosphate)	یو.دی.پی.
UDPG (= UDP-glucose)	یو.دی.پی.جی.
urobilin	یوروبیلین
urobilinogen	یوروبیلینوژن
urokinase	یوروکیناز
uricotelic	یوریکوتلیک
eucaryote, eukaryote	یوکاریوت
eukaryotic	یوکاریوتیک
euchromatin	یوکروماتین
ion	یون
pyro-phosphate ion	یون پیروفسفات

ionization	یونش
lyate ion	یون لیات
cation	یون مثبت
anion	یون منفی
ionizing	یوننده
ionization	یونیزاسیون
ionization	یونیزه شدن
ionization	یونیزه کردن
ionizing	یونیزه کننده
unicellular	یونی سلولار
univalent	یونی والان
uv (= ultraviolet), UV (= ultraviolet)	یو.وی.
uv1	یو.وی. وان/1
yohimbine	یوهیمبین

English-Persian

A

A (= adenine)	آ.
A (= adenovirus)	آ.
A chain	زنجیره آ.
A form DNA	دی.ان.ای نوع آ.
A I (= first meiotic anaphase)	آ.یک، ای.وان.
A II (= second meiotic anaphase)	آ.دو، ای.تو.
A,P site	جایگاه ای.،پی.
aa (amino acid)	آا
AAG (= acid alpha-glucosidase)	ای.ای.جی.
AAT (= alpha-antitrypsin)	ای.ای.تی.
AAV (= adeno-associated virus)	ای.ای.وی.
ab (= antibody)	ای.بی.
ab initio gene prediction	پیش‌بینی ژن اب اینیتیو
ABC transporter	انتقال دهنده ای.بی.سی.
Abelson strain of murine	جوره/واریته/رقم/سویه (مربوط به جوندگان) آبلسونی
aberration	کج راهی
abiogenesis	تولید خود بخود، تولید مثل خود به خودی، نازیست زایی
abiotic	غیر آلی، غیر زنده، نازیوا
abiotic stress	تنش عوامل فیزیکی محیط، تنش ابیوتیک
abl (=Abelson strain of murine)	ای.بی.ال
ABL (=Abelson strain of murine)	ای.بی.ال.
ablation	فرساب، سایش، ذوب یخ
ABO (= ABO blood group)	ای.بی.او.
ABO blood group	گروه خونی ای.بی.او/آ.ب.او
abortive complex	مجموعه بی نتیجه، همتافت ناموفق
abortive transduction	انتقال بی نتیجه، ترارسانی ناموفق

abrachia (= abrachiatism)	
abrachiatism	بی دستی مادرزاد
abrachiocephalia	بی سرو دستی مادر زاد
abrachiocephalus	بی سر و دست مادر زاد
abrachius	بی دست مادر زاد
abrin	ابرین
abscisic acid	ابسیزیک اسید
absolute configuration	پیکر بندی/کنفیگوراسیون/آرای ش مطلق
absorbance	جذب کنندگی، درآشامندگی، درکشیدگی، قابلیت جذب
absorption	جذب، جذب سطحی، درآشامی، درکشی، درون جذبی
abzyme	آبزیم، ابزایم
AC (= adenyl cyclase)	آ.ث.، ای.سی.
acanthella	خار سرک
acanthocephala	خار سران
acanthocephalan (= acanthocephalus)	
acanthocephalus	خار سر
acanthor	خار سرچه
acardia	بی قلبی مادر زاد
acardiac	بی قلب مادر زاد
acardiacus (=acardiac)	
acardius (=acardiac)	
acariasis	کنه زدگی
acarina	کنه سانان
acarine	کنه سان
acarinosis (= acariasis)	کنه زدگی
acarodermatitis	کنه زدگی پوست
acarophobia	کنه ترسی
accelerated maturation	بلوغ شتابدار/تسریع شده
accelerator globulin	گلوبولین شتابدهنده/تسریع کننده
accelerin	آکسلرین
acceptable daily intake	مصرف/درون جذب قابل قبول روزانه
acceptable level of risk	سطح ریسک قابل قبول

English	Persian
acceptor	پذیرا، گیرنده، پذیرنده
acceptor control	کنترل/مهار گیرنده
acceptor junction site	محل اتصال گیرنده، سایت اتصال دریافت کننده
acceptor site	سایت/محل گیرنده/اکسپتور
accessible ecosystem	محیط زیست سریعا در دسترس
accession	بروز، حمله، شیوع، کسب، نمونه (در بانک ژن)
accessory cell	سلول جانبی/فرعی
accessory chromosome	کروموزوم اضافی، فام تن فرعی
accident	اتفاق، تصادف
accidental release	آزادسازی/رهایی تصادفی/اتفاقی
acclimatization	اقلیم/بوم/خو پذیری/گیری، اقلیمی شدن، سازش با محیط
Ac-CoA (= acetyl-coenzyme A)	
accumulation	انباشت، انباشتگی، انبوهی، تجمع، تراکم
acellular	بی یاخته، غیر سلولی
acentric	بدون سانترومر
acetic acid bacteria	باکتری جوهر سرکه
acetolactate synthase	سینتاز استو لاکتیت
acetolactate synthase gene	ژن سینتاز استو لاکتیت
acetone	استون
acetyl carnitine	استیل کارنیتین
acetyl co-enzyme A	استیل کوآنزیم آ.
acetyl spiramycin	استیل اسپیرامایسین
acetylcholine	استیل کولین
acetylcholinesterase	استیل کولین استراز
acetylcholinesterase acid	اسید استیل کولین استراز
acetyl-CoA	استیل کو آ.
acetyl-CoA carboxylase	استیل کو آ. کربوکسیلاز
acetyl-coenzyme A	استیل کوآنزیم آ
ACH (= acetylcholinesterase)	ای.سی.اچ.
Achondroplasia	آکوندروپلازی، بیماری کوتاهی دست و پا
achromic point	نقطه بدون رنگ/آکرومیک
achromotrichia	سفید مویی مادرزاد
acid	اسید
acid alpha-glucosidase	اسید آلفا گلوکوزیداز
acid maltase deficiency	کمبود اسید مالتاز
acid mucopolysaccharide	موکو پلی ساکارید اسیدی
acid oxide	اکسید اسیدی
acid phosphatase	فسفاتاز اسیدی
acid protease	پروتئاز اسیدی
acid rain	باران اسیدی
acid saccharification process	پروسه ساکاریفیکاسیون اسیدی
acid-fast stain	رنگ مقاوم به اسید، رنگ ثابت شده با اسید
acidic fibroblast growth factor	عامل رشد فایبرو پلاست/رشته تنده اسیدی
acidophilic	اسید/ترشا پسندی/خواهی/گرائی، اسیدوفیلیک
acidosis	اسیدوز
acidotropic	اسیدوتروپیک/ترشاپرور
aclacinomycin	آکلاسینومایسین
aconitase	آکونیتاز
aconitin	آکونیتین
aconta	آکونتا
acoustic gene transfer	انتقال ژن آکوستیکی/صوتی
ACP (= acyl carrier protein)	ای.سی.پی.
acquine renal cystic disease	بیماری سیست کلیوی اسبی
Acquired Immune Deficiency Syndrome (AIDS)	سندروم/عارضه کاهش ایمنی اکتسابی (ایدز)
acquired mutation	تحمل/تولرانس/جهش اکتسابی
acral	پایانکی
acrasin	آکرازین
acridine	آکریدین
acridine orange	نارنجی آکریدینی
acriflavine	آکریفلاوین
acrocephalia	مخروط سری

acrocephalic	مخروط سر
acro-cyanosis	آکروسیانوز
acrodermatitis	پوست آماس پایانکی
acrodermatitis enteropathica	آکرودرماتیت آنتروپاتیکا
acrolein	آکرولئین
acromegaly	درشت پایانکی
acromelanism	سیه پایانکی
acromicria	ریزپایانکی
acroneurosis	پی آسیبی پایانکی
acropachia	ستبر استخوانی
acropachy (= acropachia)	
acropachyderma	ستبر پایانکی
acrosin	آکروزین
acrylamide	آکریلامید
acrylamide gel	ژل آکریلامید
acrylic acid	اسید آکریلیک
ACS (= autonomous consensus sequence)	ای.سی.اس.
actidione	آکتیدیون
actin	آکتین
actinin	آکتینین
actinobacillosis	تخته زبانی
actinomycetes	پرتوجلد ها، آکتینومایست ها
actinomycin	آکتینومایسین
actinomyosin	آکتینومیوزین
actinospectacin	آکتینواسپکتاسین
activated biofilter	بیو فیلتر فعال شده، زیست صافی آکتیو شده
activated carbon	کربن آکتیو/فعال شده
activated charcoal	زغال جذب/فعال، زغال فعال شده، زغال چوب اشتار
activated partial thromboplastin time	زمان نیمه ترومبوپلاستین فعال شده
activated sludge process	فرآیند اسلاج فعال شده
activating group	گروه فعال ساز
activation energy	انرژی فعال سازی، کارمایه کنش ور سازی
activation hormone	هورمون فعال کننده
activator	آغاز گر، اشتارکن، عامل محرک، فعال ساز/کننده
activator protein	پروتئین فعال کننده

active biomass	زیست توده فعال/کنش ور، بیوماس آکتیو
active center	مرکز فعال/کنش ور
active immunity	ایمنی فعال/کنش ور
active immunization	ایمنی سازی فعال
active ingredient	ماده موثر
active site	جایگاه کنش ور، ناحیه فعال
active transport	انتقال فعال
active-enzyme centrifugation	سانتریفوژاسیون آنزیم فعال شده
activin	آکتیوین
activity coefficient	ضریب فعالیت
actobindin	آکتوبیندین
actomyosin	آکتومایوزین
actual oxygen transfer efficiency (AOTE)	بازده انتقال اکسیژن واقعی (بازاک واقعی)
Acuron™ gene	ژن آکورون
acute exposure	پرتو گیری حاد، قرار گرفتن شدید/حاد در معرض(چیزی)، اکسپوژر/
acute illness	بیماری حاد
acute lymphocytic leukemia	لوکمی لنفوسیتی حاد
acute myeloblastic leukemia	لوکمی میلوبلاستیک حاد
acute myelomonacytic leukemia	لوکمی میلوموناسیتیک حاد
acute nonlymphocytic leukemia	لوکمی غیر لنفوسیتی حاد
acute promyelocytic leukemia	لوکمی پرومیلوسیتیک حاد
acute sample	نمونه پیشرفته/حاد
acute toxicity	سمیت حاد
acute transfection	ترا آلودگی حاد
acute-phase protein	پروتئین فاز حاد
acycloquanosine	آسیلوکوانوزین
acyclovir	آسیکلوویر
acyl carrier protein	پروتئین حامل آسیل
acylcarnitine transferase	آسیل کارنیتین ترانسفراز
AD (= autosomal dominant)	ای.دی.
ADA (= adenosine deaminase)	ای.دی.ای.
adalimumab	آدالیموماب

adamalysin	آدامالیزین
adamantane	آدامانتان
adaptation	انطباق، سازگاری
adaptation traits	ویژگی های سازگاری
adaptin	آداپتین
adaptive enzyme	آنزیم تطبیقی/سازشی
adaptive immunity	ایمنی تطبیقی/سازشی، مصونیت مزاجی
adaptive radiation	تشعشع سازشی
adaptive response	واکنش تطبیقی/سازشی
adaptive zone	منطقه سازشی
adaptor	آداپتور، تطبیق/سازش دهنده
ADCC (= antibody dependant cellular cytotoxicity)	ای.دی.سی.سی.
ADCC (= apurinic cellular citotoxicity)	ای.دی.سی.سی.
addition polymer	بسپار/پلیمر افزایشی
addition reaction	واکنش افزایشی
additive	افزایشی/افزودنی، جمع پذیر
additive effect	اثر افزایشی/افزاینده/جمعی
additive genes	ژن های افزایشی/فزاینده
additive recombination	باز ترکیبی/رکومبیناسیون/نو ترکیبی افزایشی
address	آدرس
address sequence	توالی آدرس
adducin	آددوسین
adenilate cyclase	آدنیلات سیکلاز
adenin arabinucleoside triphosphate	آدنین آرابینوکلئوزید تریفسفات
adenine	آدنین
adenine phosphoribosyl transferase	آدنین فسفوریبوسیل ترانسفراز
adeno-associated virus	ویروس مرتبط به آدنو
adenoleucodystrophy	آدنولوکودیستروفی
adenomatous polyposis coli (gene)	آدنوماتوس پولیپوزیس کولای (ژن)
adenosine	آدنوزین
adenosine deaminase	آدنوزین دآمیناز
adenosine diphosphate	آدنوزین دی فسفات

(ADP)	
adenosine monophosphate	آدنوزین مونو فسفات
adenosine triphosphatase	آدنوزین تریفسفاتاز
adenosine triphosphate (ATP)	آدنوزین تری فسفات
adenoviridae	آدنوویریدا
adenovirus	آدنو ویروس
adenyl cyclase	آدنیل سیکلاز
adenylate	آدنیلات
adenylate cyclase	آدنیلات سیکلاز
adenylate kinase deficiency	کمبود آدنیلات کیناز
adenylic acid	اسید آدنیلیک
adequate intake	مصرف/درون جذب کافی
ADH (= alcohol dehydrogenase)	آ.د.هاش.، ای.دی.اچ.
ADH (=antidiuretic hormone)	آ.د.هاش.
adhesion molecule	مولکول چسبندگی
adhesion protein	پروتئین چسبندگی
adhesive	چسب
adhesive force	نیروی چسبندگی
adipocyte	آدیپوسیت، سلول چربی (حیوانی)
adipose	آدیپوز، پیه دار، پیه مانند، چرب، چربی دار، روغنی، فربه، مربوط به چربی
adipsin	آدیپزین
adjuvant	آدجووانت، ماده محرک ایمنی، ماده کمکی، یاری کننده، یاور
a-DNA	ای.-دی.ان.ای.، دی.ان.ای.-آلفا/آ.
adolase	آدولاز
adoptive cellular therapy	درمان/معالجه سلولی انتخابی
adoptive immunization	ایمن/مصون سازی انتخابی
ADP (= adenosine diphosphate)	ای.دی.پی.
ADPKD (= autosomal dominant polycystic kidney)	ای.دی.پی.کی.دی.

adrenal cortical hormone	هورمون کورتیکال آدرنال
adrenergic receptor	گیرنده آدرنرژیک
adrenocorticotropic hormone	هورمون آدرنوکورتیکوتروپیک، هورمون قطعه قدامی غده صنوبری
adsorption	برآشامی، برکشی، جذب سطحی، رونشینی
adult hemoglobin	هموگلوبین بالغ
adult stem cell	سلول پایه بالغ، یاخته دودمانی بالغ
advanced informed agreement	ادوانس اینفورم اگریمنت، توافقنامه ادوانس اینفورم/پیش آگاه
adventitious	نابجا، عارضی، غیرطبیعی
adverse effect	اثر معکوس/وارونه، تاثیر زیانبار/مضر
adverse reaction	واکنش معکوس/نا مطلوب
AE (= acrodermatitis enteropathica)	آ.ئی.، ای.ئی.
A-EB (= epidermolysis bullosa acquisita)	آ.-ئی.ب.، ای.-ئی.بی.
aedes albopictus	ادس آلبوپیکتوس
aequorin	اکوئورین
aeration	تهویه، هوا دهی
aeration basin	حوزه تهویه/هوادهی
aeration process	فرآیند تهویه/هوادهی
aerobe	در (حضور) هوا، میکروب هوازی، هوازی
aerobic	ایروبیک، هوازی، هوازیست، هوازیستی، هوایی
aerobic bacteria	ترکیزگان/باکتری های هوازی
aerobic growth	رشد هوازی
aerobic microbe	میکروب هوازی
aerobic reactor	رآکتور ایروبیک
aerobic respiration	تنفس ایروبیک
aerobic treatment	درمان ایروبیکی
aerodynamic	آیروداینامیک، آیرودینامیک
aerolysin	آئرولیزین
aeromonas	آئروموناس
aerosol	آئروسول، اسپری، افشانه، ذره پاش(ی)، هواویز(ه)

aerosol particle	ذره افشانه ای/اسپری شده، ذره جامد/مایع معلق در هوا
aerosol spray	افشانه گردپاش
aerosol vulnerability testing	آزمایش آسیب پذیری ذره پاش
aerosolization	افشانه سازی/آئروسولیزه کردن، تبدیل به افشانه کردن، قابل افشاندن کردن
aerosolize(d)	اسپری/افشانه شده، اسپری کردن، بصورت اسپری در آمده، تعلیق مایع/جامد بصورت گرد/گاز در هوا، ذره پاشیدن
aerosolized biological agent	عامل بیولوژیکی افشانه شده، عامل میکروبی افشانه ای
aerosolizing	افشاندن
AF (= amniotic fluid)	آ.اف.، ای.اف.
afebrile	بدون/بی تب
affected party	شخص/طرف/گروه مبتلا/متاثر
affinity	پیوستگی، پیوند، تمایل، جذب، شباهت، علاقه، کشش، گیرائی، میل ترکیبی، وابستگی
affinity chromatography	کروماتو گرافی تمایلی/کششی/بر اساس میل ترکیبی
affinity tag	برچسب دار، برچسب/تگ/ضمیمه افینیتی، برچسب/تگ/ضمیمه خویشاوندی/قرابت، نشان دار
aflatoxin	آفلاتوکسین
AFP (= α- Fetoprotein)	ای.اف.پی.
aftercrop	کشت دوم
agar	آگار
agarase	آگاراز، آنزیم کافنده آگار
agar-diffusion method	روش آگار دیفیوژن

167

English	Persian
agarose	آگاروز
agarose gel	ژل آگاروز
agarose gel electrophoresis	آگارز ژل الکتروفورز، الکتروفوروز ژل آگاروز
age group	گروه سنی
agent	عامل، عامل بیماری زا، معرف، واسطه
agglomeration	اگلومراسیون، انبار، انباشت، انباشتگی، تجمع، تراکم، توده، دلمه شدن، همجوشی، کلوخه شدن
agglutination	به هم چسبندگی، چسبندگی، هم چسبی
agglutinin	آگلوتینین
agglutinogen	آگلوتینوژن
aggregation	تجمع، تراکم، توده
aggressin	آگرسین
aging	پیر شدن، پیری، قدمت، کهنگی، کهنه شدن
aglycon	آگلیکون
aglycone	آگلیکون
agnotobiotic culture	کشت آگنوتوبیوتیک
agonists	آگونیست ها
agraceutical (= nutraceuticals)	آگراسوتیکال
agranulocyte	تک هسته ای
agricultural	زراعی، کشاورزی
agricultural biodiversity	تنوع زیستی کشاورزی
agricultural biological diversity	تنوع زیست شناختی کشاورزی
agrobacterium	آگروباکتریوم
agrobacterium tumefaciens	آگروباکتریوم تومافاسین
agrobiodiversity	تنوع زیست-کشاورزی
agrobiotechnology	زیست فناوری/فن شناسی کشاورزی
agroecology	اکولوژی زراعی، بوم شناسی کشاورزی
agroforestry	جنگل داری زراعی
agropine	آگروپین
a-helix	مارپیچ ای./آلفا/آ.
AID (= artificial insemination by donor)	ای.آی.دی.
AIDS = (acquired immune deficiency syndrome)	ایدز
air	هوا، هوا دادن
Air America compressors	کمپرسورهای ایر آمریکا
airborne pathogen	عامل بیماری زای هوایی
airborne precaution	اقدام احتیاطی (ذرات) هوابرد، هشدار ذرات معلق در هوا، هشدار (ذرات) هوائی/هوابرد
airbrush	ایر براش، قلم رنگ پاش
airplane	هواپیما
airway	لوله (های) هوا، مسیر تنفس/هوا/هوایی
AK (= adenylate kinase deficiency)	آ.کا.، ای.کی.
Ala (= alanine)	ای.ال.ای.، آلا
alamethicin	آلامتیسین
alanine	آلانین
alanine aminotransferase	آلانین آمینوترانسفراز
alarmone	آلارمون
ALB (= albumin)	ای.ال.بی.
albumin	آلبومین، سفیده تخم مرغ
alcalase	آلکالاز
alcohol dehydrogenase	دئیدروژناز الکل
alcohol oxidase	اکسیداز الکل
ALD (= adenoleucodystrophy)	ای.ال.دی.
aldehyde	آلدهید، آلدئید
aldose	آلدوز
aldosterone	آلدوسترون
aleukia	آلوسمی، فقدان پلاکتهای خون
aleurone	آلورون
alfAFP (= alfalfa antifungal peptide)	آلف ای. اف. پی.
alfalfa antifungal peptide	پپتید ضد قارچ آلف آلفا
algae	جلبک، خزه
algal toxin	سم جلبکی
algicide	آلجیساید، جلبک کش
algin	آلژین

English	Persian
alginate	آلژینات
alginic acid	اسید آلژنیک
algorithm	الگوریتم
alicin	آلیسین، پودر سیر
alien species	گونه خارجی/غیر بومی
aliesterase	آلی استراز
a-linolenic acid	اسید آ.-لینولنیک
alkali	آلکالی، قلیا
alkaline	آلکالاین، قلیایی
alkaline earth metal	فلز قلیلیی خاکی
alkaline hydrolysis	آبکافت قلیایی، آلکالاین هیدرولایزیس، هیدرولیز قلیایی
alkaline lysis	لیز قلیائی/آلکالینی
alkaline phosphatase	آلکالین فسفاتاز، فسفاتاز قلیایی
alkaloid	آلکالوئید، شبه قلیایی
alkannin	آلکانین
alkaptonuria	آلکاپتونوریا، سندرم ادرار چای/کوکاکولا
ALL (= acute lymphocytic leukemia)	ای.ال.ال.
allantoin	آلانتوین
allele	آل، دگره، ژن همردیف، هم ردیف ژنی
allele- specific oligonucleotide hybridization	هم تیرگی اولیگونوکلئوتید مختص به آل
allele-specific oligonucleotide	اولیگونوکلئوتید مختص به آل
allelic exclusion	آللیک اگزکلوژن، حذف آللی
allelopathic	آسیب رسان
allelopathy	دگر آسیبی
allelozyme	آللوزیم
allergy	آلرژی، حساسیت
allicin (= alicin)	آلیسین
alloantigen	آلوآنتی ژن
allogamous	دگر گشن
allogamy	دگر گشنی
allogeneic	آلوژنئیک
allograft	آلوگرافت، پیوند آلو
alloisoleucine	آلوایزولئوسین
allolactose	آلولاکتوز
allometry	آلومتری
allomone	آلومون
allopatric	آلوپاتریک، دگر بوم، دگر نیاک، نا هم بوم
allopatric speciation	گونه زایی دگر بوم
allopurinol	آلوپورینول
allosteric enzyme	آنزیم آلوستریک/دگر ریختار
allosteric site	سایت/موضع آلوستریک/دگر ریختار
allosterism	آلوستریزم/دگر ریختاری
allostery	آلوستری
allotrope	دگرشکل
allotropy	دگرشکلی
allotype	آلوتیپ
allotypic monoclonal antibodies	آنتی بادی های مونوکلونال آلوتیپیک، پادتن های تک بنیانی دگر مونه ای
alloxan	آلوکسان
allozyme (= allosteric anzyme)	آلوزیم
alpha amylase	آلفا آمیلاز، آمیلاز آلفا
alpha amylase inhibitor-1	آلفا آمیلاز اینهیبیتور -1، بازدارنده آلفا آمیلاز نوع اول
alpha DNA	دی.ان.ای نوع/شکل آلفا
alpha galactoside	گالاکتوزید آلفا
alpha helix	مارپیچ آلفا
alpha interferon	انترفرون آلفا
alpha linolenic	لینولنیک آلفا
alpha particle	ذره آلفا
alpha-antitrypsin	آلفا-آنتی تریپسین
alpha-chaconine	آلفا شاکونین
alpha-helice	مارپیچهای آلفا
alpha-rumenic acid (= conjugated linoleic acid)	اسید آلفا رومنیک
alpha-solanine	آلفا سولانین، سولانین آلفا
alpha-synuclein	آلفا-ساینوکلئین
alphavirus	آلفا ویروس، ویروس آلفا
Alport syndrome	سندرم آلپورت
ALS (= amyotrophic lateral	ای.ال.اس.

sclerosis)	
ALS gene (= acetolactate synthase gene)	ژن ای.ال.اس.
alternative medicine	پزشکی/طب غیر سنتی، طب جایگزین/مکمل
alternative mRNA splicing	اتصال جایگزین پیک ار.ان.ای
alternative splicing	اتصال جایگزین
alu family	خانواده آلو
alu sequence	توالی آلو در ملکول دی.ان.ای.
alumina	آلومین
aluminum	آلومینیم، آلومینیوم
aluminum hydroxide	هیدروکسید آلومینیم
aluminum resistance	مقاومت به آلومینیم
aluminum tolerance	تحمل آلومینیم
aluminum toxicity	سمیت آلومینیم
ALV (= avian leukemia virus)	ای.ال.وی.
alveolar macrophage	درشت خوار حفره ای
Alzheimer's disease	بیماری آلزایمر
amanitin	آمانیتین
amantadine	آمانتادین
amatoxin	آماتوکسین
amber codon	رمز بی معنی، رمز پایان دهنده، کدون امبر
amber mutation	جهش امبر
amber suppressor	سرکوبگر امبر/عنبری
ambient measurement	اندازه گیری امبیانت/محیط (ی)
AMC (= arthrogryposis multiplex congenita)	ای.ام.سی.
AMD (= acid maltase deficiency)	ای.ام.دی.
American Type Culture Collection (ATCC)	مجموعه کشت تیپ آمریکا
Ames test	آزمون آمز
AMH (= anti-mullerian hormone)	ای.ام.اچ.
amidase	آمیداز
amino acid	اسید آمینه
amino acid profile	نمودار اسید آمینه

amino acid sequence	ترتیب/توالی اسید آمینه
amino peptidase	آمینو پپتیداز
amino purine	آمینو پورین
aminoacyl	آمینوآسیل
aminoacylase	آمینوآسیلاز
aminoacylation	آمینوآسیلاسیون
aminoacyl-tRNA	آمینوآسیل-تی.آر.ان.ای.
aminoacyl-tRNA synthetase	سینتاز آمینوآسیل-تی.آر.ان.ای.
aminocyclopropane carboxylic acid	آمینو سیکلو پروپان کربوکسیلیک اسید
aminocyclopropane carboxylic acid synthase	آمینو سیکلو پروپان کربوکسیلیک اسید سینتاز
aminoethoxyvinylglycine	آمینو اتوکسی وینیل گلیسین
aminoglycoside antibiotic	آنتی بیوتیک آمینوگلیکوزید
aminopterin	آمینوپترین
aminopyridines	آمینوپیریدین ها
aminotransferases	آمینو ترانسفراز ها
amitosis	آمیتوز
AML (= acute myeloblastic leukemia)	ای.ام.ال.
AMMOL (= acute myelomonacytic leukemia)	ای.ام.ام.او.ال.
amniocentesis	آمنیوسنتز
amnion	آمنیون
amniotic fluid	مایع آمنیون
amorph	آمورف، بدون ریخت، بی شکل
amorphous	بی شکل
AMP (= adenosine monophosphate)	ای.ام.پی.
amphetamine	آمفتامین
amphibolic pathway	مسیر دو حرکتی
amphipathic molecules	ملکول های آمفی پاتیک/دو بخشی
amphiphilic molecules	مولکول آمفی فیل
amphitropic virus	ویروس آمفیتروپیک
amphoteric compound	ترکیب آمفوتریک/خنثی
amphotropic retrovirus	رتروویروس آمفوتروپیک

ampicillin	آمپی سیلین
amplicon	آمپلیکون
amplification	افزایش، بسط، تقویت، تکثیر، فزون سازی
amplified fragment length polymorphism (AFLP)	پلی مورفیزم طولی قطعه تقویت شده
amplify	بسط دادن، تقویت/تکثیر کردن
amplimer	آمپلیمر
amygdalin	آمیگدالین
amylase	آمیلاز، دیاستاز
amylo process	فرآیند آمیلو
amyloglucosidase	آمیلوگلوکزیداز
amyloid plaque	پلاک آمیلوئید
amyloid precursor protein	پروتئین پیش درآمد آمیلوئید
amylopectin	آمیلو پکتین
amyloplast	نشادیسه
amyotrophic lateral sclerosis	اسکلروز آمیوتروفیک جنبی
anabolic	آنابولیک، فراگشتی
anabolic pathway	مسیر آنابولیک
anabolic process	پروسه آنابولیکی
anabolism	جذب و ساخت، جذب و هضم، فراساخت، فراگشت، فراگوهرش، متابولیزم سازنده
Anabolis™	آنابولیس
anabolite	آنابولیت، فراگشته
anaerobe	غیر/نا هوازی
anaerobic	بی هوازیستی، بی/غیر هوازی
anaerobic bacteria	باکتری غیر هوازی
anaerobic digestion	هضم آن-ایروبیک/غیر هوازی
anaerobic growth	رشد آن-ایروبیک/غیر هوازی
anaerobic treatment	درمان آن ایروبیک/غیر هوازی
anal	مخرجی، مقعدی
analgesic	ضد درد، مسکن
analog	متجانس، مشابه، همسان
analog gene	ژن متشابه/همانند/همسان
analogous	شبیه، همانند، هم تراز
analogue	قیاسی، مشابه، همسان
analysis of variance	تجزیه/تحلیل واریانس

analyte	آنالیت، جسم تجزیه ای، ماده مورد تجزیه
anandamide	آناندامید
anaphylatoxin	آنافیلاتوکسین
anaphylaxis	آنافیلاکسیس
anaplasia	آن اپلازی
anaplerotic metabolic pathway	مسیر متابولیکی آناپلروتیک
anchorage-dependent cell	سلول متکی به لنگرگاه/آنکوریج
ancrod	آنکرود
andosterone	آندوسترون
androgen	آندروژن
anemia	آنمی، کم خونی
anencephaly	آن انسفالی
anesthesia	بی حسی، بی هوشی، هوش بری
aneuploid	آنوپلوئید
aneuploidy	آنوپلوئیدی
aneurysm	رگ آماس
Angelman syndrome	سندروم آنجلمان
angiogenesis	رگ زایی
angiogenesis factor	عامل رگ زایی
angiogenic growth factor	عامل رشد آنژیوژنیک
angiogenin	آنژیوژنین آنژوژنین
angiostatin	آنژیواستاتین
angiotensin	آنژیوتنزین
angiotensinase	آنژیوتنزیناز
angiotensin-converting enzyme	آنزیم تبدیل آنژیوتنزین
angiotensin-converting enzyme inhibitor	بازدارنده آنزیم تبدیل آنژیوتنزین
angiotensinogen	آنژیوتنزینوژن
angiotensinogenase	آنژیوتنزینوژناز
angstrom (A)	آنگستروم
animal cell line	خط سلولی حیوانی
animal gene bank	مخزن ژنی حیوانی
animal genetic resources databank	مخزن اطلاعاتی/داده های منابع ژنتیکی حیوانی
animal genome (gene) bank	مخزن ژن حیوانی، مخزن ژنوم جانوری

animal model	مدل جانوری/حیوانی	antibody affinity	آنتی بادی افینیتی
animall cell culture	کشت سلول	chromatography	کروماتوگرافی،
	جانوری/حیوانی		کروماتوگرافی خویشاوندی
anion	آنیون، یون منفی		آنتی بادی ها
anion exchanger	مبادله کننده آنیون	antibody array	آرایه پادتن
anionic	آنیونی	antibody dependant cellular	سیتوتوکسیته سلولی
ANLL (= acute	ای.ان.ال.ال.	cytotoxicity	متکی به آنتی بادی
nonlymphocytic		antibody-laced nanotube	غشاء ریز لوله ای با آستر
leukemia)		membrane	آنتی بادی
anneal	تابکاری	antibody-mediated immune	پاسخگویی ایمنی توسط
annealing	باز ریخت، تابکاری، آنیلینگ،	response	آنتی بادی، واکنش ایمنی
	دورگه/هیبرید شدن زنجیره		بواسطه آنتی بادی ها
	دی.ان.ای. یا آر.ان.ای	antibody-secreting cell	سلول تراوش کننده آنتی
annexin	آنکسین		بادی
annotation	تفسیر/توضیح/حاشیه/شرح	anticholinesterase	آنتی کولین استراز
anomer	آنومر	anticoagulant	آنتی کوآگولانت، ضد انعقاد
anonymous DNA marker	نشانگر دی.ان.ای. ناشناس	anticoding strand	رشته آنتی کدون/غیر رمز
anoxic culture	کشت آنوکسیک		کننده/نامفهوم
antagonism	آنتاگونیزم، پادکرداری، تضاد،	anticodon	آنتی کودون، ضد رمزه
	رقابت، مخالفت،	anticomplementary	ضد مکمل
	ناسازگاری، ناهمسازی	anticonvulsant	ضد تشنج
antagonist	پادکردار، رقیب، دشمن،	anticooperativity	آنتی کواپرتیویتی، ضد تعاونی
	متضاد، مخالف	anticrop agent	عامل آنتی کراپ/ضد محصول
anterior pituitary gland	غده زیر وحشه جلویی، غده	antidiuretic hormone (ADH)	هورمون آنتی
	هیپوفیز قدامی		دیورتیک/پادپیشابی
anthocyanidin	آنتو سیانیدین	antidote	پادزهر، درمان، علاج،
anthocyanin	آنتو سیانین		نوشدارو
anthocyanoside	آنتوسیانوزید	antifolate	آنتی/ضد فولات
antiangiogenesis	آنتی آنژیوژنز، ضد رگ زایی	antifreeze protein	پروتئین ضد انجماد
antibacterial	ضد باکتری	antifungal	ضد قارچ (ی)
antibiosis	ضد حیات (ی)	antigen	آنتی ژن، پادگن
antibiotic	آنتی بیوتیک، پادزی، پاد	antigen presenting cell	سلول ارائه دهنده آنتی ژن
	زیست، پادزیو، ضدزندگی	antigen-antibody complex	مجموعه پادگن-پادتن
antibiotic resistance	مقاوم شدن نسبت به آنتی	antigene	آنتی ژن
	بیوتیک، مقاومت در مقابل	antigenic determinant	تعیین کننده/گمارنده پادگنی
	آنتی بیوتیک	antigenic switching	آنتی ژن سوئیچینگ، تغییر
antibiotic therapy	آنتی بیوتیک درمانی، درمان		آنتی ژنی، سوئیچ آنتی
	با آنتی بیوتیک		ژنی، سوئیچ کردن آنتی
antibiotics	آنتی بیوتیک ها		ژنها
antibody	آنتی بادی، آنتی کور، پادتن،	antigen-presenting cell	سلول ارائه دهنده آنتی ژن
	پادزهر	(APC)	

antihemophilic factor	عامل ضد هموفیلی
antihemophilic globulin	گلوبولین ضد هموفیلی
antihistamine	آنتی هیستامین، ضد حساسیت
anti-idiotype	آنتی ایدیوتیپ
anti-idiotype antibody	آنتی ایدیوتیپ آنتی بادی، آنتی بادی آنتی ایدیوتیپ، آنتی بادی ضد ایدیوتیپ، پادتن آنتی ایدیوتیپ، پادتن ضد ایدیوتیپ
anti-infective	ضد عفونت
anti-infective agent	عامل ضد عفونت
anti-interferon	ضد انتر فرون
antimateriel agent	عامل آنتی ماتریال/ضد ماده
anti-messenger DNA	دی.ان.ای.ضد پیک/پیامبر
antimetabolite	آنتی متابولیت
antimicrobial agent	عامل پاد زیوه ای/ضد میکروبی
antimorph	آنتی مورف
anti-mullerian hormone	هورمون ضدمولریان
antimutator gene	ژن آنتی موتاتور، ژن ضد جهشگر
anti-oncogene	آنتی آنکوژن، ضد تومور زا، ضد سرطان زا
anti-o-polysaccharide antibody	آنتی بادی آنتی او.پلی ساکارید، آنتی بادی ضد پلی ساکاریدهای او.، آنتی بادی ضد چند قندی های او، آنتی او. پلی ساکارید آنتی بادی
antioxidant	ضد اکسنده/اکسیدان
antiparallel	آنتی پارالل، غیر موازی، غیر هم جهت، متقابل
antiplatelet	ضد پلاکت/گروه خون
antiporter	آنتی پورتر
antiproliferative	آنتی پرولیفراتیو/ضد رشد و تولد
antipyretic	تب بر/ضد تب
antisense	آنتی/ضد سنس، غیر قابل ترجمه، غیر کد کننده
antisense RNA	آر.ان.ای. آنتی سنس

antisense strand	رشته آنتی سنس/غیر کد کننده
antisense technology(anti-sense technology)	تکنولوژی/فناوری آنتی سنس/ضد سنس/غیر کد کننده/غیر قابل ترجمه
antiseptic	گند زدا، ماده ضد عفونی کننده
antisera	پاد/ضد سرم ها
antiserum	آنتی سرم
antithrombin III	آنتی ترومبین نوع سوم
antithrombogenous polymer	پلی مر آنتی/ضد ترومبوژنوس
antitoxin	آنتی توکسین، پاد زهر، پادزهرابه، ضد زهرابه، ضد سم
antiviral	ضد ویروس
antivirial agent	عامل ضد ویروس
antixenosis	بیگانه گریزی
antrin	آنترین
aorta	آئورت، بزرگ سرخرگ
aortic	آئورتی، بزرگ سرخرگی
aortic stenosis	استنوز آئورتی
AOTE (= actual oxygen transfer efficiency)	بازاک واقعی
AP (= apurinic)	آ.پ.، ای.پی.
AP endonuclease	اندونوکلئاز ای.پی.
apamin	آپامین
APC (= adenomatous polyposis coli (gene))	ای.پی.سی.
APC (= antigen presenting cell)	ای.پی.سی.
aphicidin	افیسیدین
aphidicolin	افیدیکولین
apigenin	اپیژنین
APL (= acute promyelocytic leukemia)	ای.پی.ال.
aplastic anemia	آپلاستیک آنمی، کم خونی ناشی از ناسازی
apoenzyme	آپو انزیم، جدازیما، رهازیما
apogamous	نامیخته
apogamy	نامیزش
apolipoprotein	آپولیپوپروتئین

apomict	نامیخته
apomictic	نامیخته
apomixis	آپو میکسی، رها آمیزی، نا آمیزی، نامیختگی
apomorphine	آپومورفین
apoptosis	محدودیت ژنتیکی طول عمر سلولها، مرگ برنامه ریزی شده سلولها، مرگ سلولی برنامه ریزی شده
aporepressor	آپوریپرسور
apospory	نامیزی
apparent viscosity	ویسکوزیته ظاهری
apple domains	حوزه (های) اپل/سیبی
applicability	قابلیت اجرا/اعمال
application	استفاده، کاربرد، کاربست
applicator	اپلیکاتور، وسیله
approvable letter	تایید نامه نهایی
APRT (= adenine phosphoribosyl transferase)	ای.پی.آر.تی.
aptamer	آپتامر
aptitude	استعداد، توان، قوه
APTT (= activated partial thromboplastin time)	ای.پی.تی.تی.
apurinic	آپورینیک
apurinic acid	اسید آپورینیک
apurinic cellular citotoxicity	سیتوتوکسیته سلولی آپورینیک
apyrimidinic acid	اسید آپیریمیدینیک
aquaculture	آب کشتی، کشت آبی، کشتاب ورزی، کشت بدون خاک
aquaporin	آکوا پورین
aqueous	آبدار، آبکی، آب نهاد، آبی، محلول در آب
AR (= autosomal recessive)	ای.آر.
Ara ATP (= adenin arabinucleoside triphosphate)	
Ara C (= cytosine arabinoside)	

arabidopsis thaliana	آرابیدوپزیس تالیانا
arachidonic acid	اسید آراشیدونیک
arachin	آراشین
ARCD (= acquine renal cystic disease)	ای.آر.سی.دی.
archea	آرشیا
architectural gene	ژن آرکیتکچرال/ساختمانی/معماری
area of release	حوزه آزادسازی، منطقه آزاد کردن
arenavirus	آرناویروس
Arg (= arginine)	
arginase	آرژیناز
arginine	آرژینین
ARMS-PCR	ای.آر.ام.اس-پی.سی.آر.
armyworm	آرمی وورم، بید سرباز
aromatic	آروماتیک، خوشبو، معطر
array	آرایه
arrestin	آرستین
arrhythmia	آریتمی، بی نظمی، بی نواختی، نا موزونی
ARS (= autonomous replicating segment)	ای.آر.اس.
ARS (= autonomous replicating sequence)	ای.آر.اس.
ARS element	عنصر ای.آر.اس.
arteriosclerosis	آرترواسکلرز، تصلب شرائین
arthralgias	دردهای مفاصل
arthritis	آماس/التهاب مفصل، ورم مفاصل
arthrogryposis multiplex congenita	آرتروگریپوز مالتیپلکس کنجنیتا
arthropathy	آرتریت، التهاب مفصل، بیماری مفصلی
artificial gene	ژن مصنوعی
artificial insemination	تلقیح مصنوعی
artificial insemination by donor	تلقیح مصنوعی توسط دهنده
artificial selection	انتخاب/گزینش مصنوعی
artillery mine	مین توپی (توپخانه ای)

AS (= Angelman syndrome)	ای.اس.
AS (= antigene)	ای.اس.
AS (= aortic stenosis)	ای.اس.
ASC (= antibody-secreting cell)	ای.اس.سی.
ascites	آب آوردن شکم، آب شکم، استسقای شکم
ascites fluid	مایع آب شکم
ascopore	آسکوپور
ascorbic acid	اسید اسکوربیک، ویتامین ث
ascus	آزکوس
ase	ـاز
asepsis	آسپسی
aseptic	آسپتیک، بدون آلودگی/میکروب، بی پلشت، پاک، ضد عفونی شده
asexual	غیر جنسی، فاقد اندام جنسی، فاقد میل جنسی
asexual reproduction	تولید مثل غیر جنسی
Asian corn borer	حشره (آفت) ذرت آسیائی
Asn (= asparagine)	
ASO (= allele-specific oligonucleotide)	ای.اس.او.
ASOH (= allele- specific oligonucleotide hybridization)	ای.اس.او.اچ.
Asp (= aspartic acid)	
asparagine	آسپاراژین
aspartate transaminase	آسپارتات ترانس آمیناز
aspartic acid	اسید آسپارتیک
aspartokinase	آسپارتوکیناز
aspergillus flavus	آسپرژیلوس فلاووس
aspergillus fumigatus	آسپرژیلوس فومیگاتوس
aspergillus niger	آسپرژیلوس نایجر
asphyxiating thoracic dysplacia	دیزپلازی توراسیک آسفیکسیتینگ
aspirate	خفه شدن، فرو بردن/دادن، مکیدن
assay	آزمایش، تجزیه شیمیایی، سنجش، عیار

	سنجی/گیری
assessment	ارزیابی، برآورد، تخمین، سنجش
asset	امتیاز، توان، ثروت، موهبت، نعمت
assimilation	ادغام، جذب، جذب و ساخت، همانند سازی، همگون سازی
associate hospital	بیمارستان وابسته
astaxanthin	آستازانتین
asymptomatic seroconversion	سروکانورژن آسیمپتوماتیک
AT (= ataxia telangiectasia)	ای.تی.
AT3 (= antithrombin III)	ای.تی. تری، ای.تی.3.
AT-AC intron	ای.تی.-ای.سی. اینترون
ataxia telangiectasia	آتاکسیا تلانجی اکتازیا
ATD (= asphyxiating thoracic dysplacia)	ای.تی.دی.
atherosclerosis	تصلب شرائین، سخت شدگی سرخرگها
atmospheric	جوی، مربوط به جو
atom	اتم
atomic force microscopy	ریزبینی نیروی هسته ای، میکروسکوپی نیروی هسته ای
atomic weight	وزن اتمی
atomizer	اسپری، افشانه، ذره پاش
atomizing	پودر کردن، تبدیلسازی به ذرات ریز، تجزیه کردن
Atomizing™	اتمایزینگ
atopic	آتوپیک، مربوط به آلرژی
atopic diathesis	دیاتز آتوپیک
ATP (= adenosine triphosphate)	آ.ت.پ.
ATPase (= adenosine triphosphatase)	آ.ت.پ.آز
atrial natriuretic factor	عامل ناتری-اورتیک دهلیزی
atrial peptides	پپتیدهای دهلیزی
atrioactivase	آتریوآکتیواز
atriopeptins	آتریوپپتین ها
atrolysin	آترولایزین

English	Persian
atropine	آتروپین
attached X chromosome	کروموزوم اکس. پیوسته/متصل
attachment protein	پروتئین اتصال
attenuate	از شدت چیزی کاستن، ضعیف شدن/کردن
attenuated	تضعیف/ضعیف/کم توان شده
attenuation	انحطاط، تضعیف کردن، تکیدگی، ضعیف سازی
attenuation coefficient	ضریب انحطاط/تضعیف/خاموشی
attenuator	ضعیف کننده
attractant	جاذب
attributable proportion	نسبت آتریبیوتبل (به چیزی)، نسبت ناشی از (چیزی)
ATY (= atopic diathesis)	ای.تی.وای.
AU4000	ای. یو. 4000
audiogram	شنوایی نگاره
audiologist	شنوایی شناس
audiology	شنوایی شناسی
audiometer	شنوایی سنج
audiometrician	سنجشگر شنوایی
audiometrist	شنوایی آزما
audiometry	شنوایی سنجی
AUG (initiator codon)	ای.یو.جی.
aura virus	اورا ویروس
aureofacin	اورئوفاسین
autacoid	اتاکوئید
autoantibody	اتوآنتی بادی
autoantigen	اتو آنتی ژن
auto-claving	اتو کلاو (کردن)
auto-correlation	اتو کو ریلیشن، خود ارتباطی، خود همبستگی
autogamous	خود گشن
autogamy	خودگشنی
autogenous control	کنترل خودزا
autograft	اتوگرافت، خود پیوند
autoimmune disease	بیماری خود ایمن
autoimmune disorder	اختلال در سیستم خود ایمنی
autoimmunity	خود ایمنی/مصونی
autologous	اتولوگ، خودی
autolysate	اتولیزات
autolysin	اتولیزین
autolysis	اتولیز
autonomous consensus sequence	توالی توافقی خودمختار
autonomous replicating segment (ARS) element	زنجیره/عنصر توالی همانند ساز اتونوموس، عنصر توالی/جزء همانند ساز خودمختار
autonomous(ly) replicating sequence	توالی همانند ساز خودمختار
autophagy	اتوفاژی
autophosphorylation	اتوفسفریلیشن، خود فسفات افزائی
autopolyploid	اتوپلی پلوئید، خود پر/چند لاد
autopsy	کالبد شکافی/گشایی
autoradiography	اتو رادیوگرافی، خود پرتو نگاری
autosomal dominant	غالب اتوزومی
autosomal dominant polycystic kidney	کلیه پلی سیستیک غالب اتوزومی
autosomal recessive	مغلوب اتوزومی
autosome	اتوزوم، خودتن، کروموزوم غیر جنسی
autotroph	خود پرور/غذا
auxins	اکسین، مواد رشد، هورمون های گیاهی
auxotroph	اکسوتروف، رشد پرور، خود پرور/غذا
Ava I	ای.وی.ای. 1
avian	مربوط به پرندگان
avian leukemia virus	ویروس لوکمی طیور
avidin	آویدین
avidity	آزمندی، ولع
avoidance	اجتناب، احتراز، پرهیز، گریز
axenic culture	کشت آزنیک
axillary lymphadenopathy	لنفا دنوپاتی آکزیلاری
axon	آسه
azadirachtin	آزادیراکتین
azaserine	آزاسرین

English	Persian
AZO (= azoospermia)	ای.زد.او.
azoospermia	آ زوئواسپرمی
azotobacter	ازتوباکتر
azurophil-derived bactericidal factor (ADBF)	عامل باکتری کش مشتق از آزوروفیل

B

English	Persian
B cell (= Bursa dependent cell)	یاخته بی.
B lymphocyte	لنفوسیت بی.
b sitostanol	بی. سیتوستانول، سیتوستانول بی.
BAC (= bacterial artificial chromosome)	بی.ای.سی.
bacillus	باسیل، میلیزه
bacillus amilolique facien RNase	آرناز باسیلوس آمیلولیک فاسین
bacillus anthracis	باسیل/باکتری سیاه زخم
bacillus licheniformis	باسیلوس لایکنیفورمیس
bacillus subtilis	باسیلوس سوبتیلیس
bacillus thuringiensis	باسیلوس تورینزنسیس
back mutation	جهش پسرو/معکوس
backcross	بک کروس، تلاقی برگشتی
bacteremia	باکترمی، پیدایش باکتری در خون، ترکیز (ه) خونی
bacteremic	باکترمیائی، باکترمیک
bacteria	باکتری ها، ترکیزگان
bacteria count	شمارش باکتری ها
bacterial agent	عامل باکتریایی/ترکیزه ای
bacterial artificial chromosome	کروموزوم مصنوعی باکتریایی
bacterial expressed sequence tag (= expressed sequence tag)	اکسپرس سیکوئنس تگ باکتریائی، برچسب باکتریائی توالی اکسپرس
bacterial infection	عفونت باکتریائی/میکروبی
bacterial physiology	فیزیولوژی باکتریایی
bacterial toxin	سم باکتری
bacterial transformation	ترانسفورمیشن باکتریائی، تغییر شکل باکتریائی

English	Persian
bacterial two-hybrid system	سیستم دو رگه باکتریایی
bactericidal	باکتری کشانه، قادر به باکتری کشی
bactericide	باکتری/ترکیزه کش
bacteriocide	باکتری کش
bacteriocin	باکتریوسین
bacteriology	باکتری شناسی
bacteriophage (= phage)	باکتری خوار
bacteriostat	باکتریواستات، عامل بازدارنده رشد باکتری، عامل متوقف کننده رشد باکتری
bacterium	باکتری/ترکیزه
bacterium tularense	باکتری تب خرگوشی، باکتری تب مگس آهو، باکتری تولارمی، باکتری تولارنس، باکتری فرانسیسلا تولارنسیس
baculovirus	باکولوویروس
baculovirus expression vector (BEV)	حامل اکسپرشن باکولو ویروس
baculovirus expression vector system	سیستم حامل اکسپرشن باکولو ویروس
bakanae	باکانائه
baker's yeast	مخمر نانوائی
balance	ترازو، موازنه
BALT (= bronchial-associated lymphoid tissue)	بی.ای.ال.تی.
banjo	بانجو
BAPN (= beta-aminoprionitrile)	بی.ای.پی.ان
BAR gene	بار ژن، ژن بار، ژن بی.ای.آر.
barley	جو
BARnase (= bacillus amilolique facien RNase)	آنزیم بی.ای.آر.، بار ناز
Barr body	پیکره بار
base	باز
base analogue	آنالوگ بازی
base excision repair	ترمیم برداشت بازی
base excision sequence scanning	اسکن توالی خارج سازی/برداشت بازی

English	Persian
base pair	جفت/زوج باز
base sequence	توالی بازها
base substitution	جانشین سازی باز، جایگزینی باز
baseline data	اطلاعات مقدماتی، داده های خط مبنا، داده های معیار، داده های مقدماتی
basic fibroblast growth factor (= fibroplast growth factor)	عامل رشد فیبروبلاست بازی
basic medium	واسطه بازی، محیط کشت بازی
basophilic	بازوفیل، بازوفیلی، قلیا خواه
basophils	باز دوست ها، بازوفیل ها
batch culture	کشت دسته جمعی/گروهی
bce4	بی.سی.ئی. 4
b-conglycinin	بی. کونگلیسینین، کونگلیسینین بی.
BCP (= blue cone pigment)	بی.سی.پی.
BCR (= breakpoint cluster region)	بی.سی.آر.
bcr-abl gene	بی.سی.آر.-ای.بی.ال. ژن، ژن بی.سی.آر.-ای.بی.ال.، ژن عامل سرطان خون انسان
bcr-abl genetic marker (= genetic marker)	نشانگر توارثی/ژنتیکی بی.سی.آر.-ای.بی.ال.
BDNA	دی. ان. ای. بی
b-DNA	دی. ان. ای. بی
B-domain	حوزه/قلمرو بی.
beaker	بشر
Becker muscular dystrophy	موسکولار دیستروفی نوع بکر
Beckwith-Weideman syndrome	سندروم/نشانگان بکویت-ویدمان
bed	بستر، پایه، پی، زیر بنا، طبقه، کف، لایه
behavioral epidemic	اپیدمی/همه گیری رفتاری
benchmark concentration	تراکم/غلظت معیار
bench-scale	ترازوی میزی
benign	بی خطر، خوش خیم
benthos	ته، ته زی، کف، کف زی
bequest value	ارزش میراث/واگذاری
berseem (=berseem clover)	برسیم
berseem clover	شبدر برسیم/مصری
BESS method (= base excision sequence scanning)	روش بس، روش بی.ئی.اس.اس.
BESS t-scan method	روش/متد تی. اسکن بس
best linear unbiased prediction (BLUP)	بهترین پیش بینی بی طرفانه خطی
beta carotene	بتا کاروتن، کاروتن بتا
beta cells	سلول/یاخته های بتا
beta conformation	ترکیب/تطابق بتا
beta interferon	انترفرون بتا
beta oxidation	اکسایش بتا
beta sitostanol	سیتوستانول بتا
beta sitosterol	سیتوسترول بتا
beta-aminoprionitrile	آمنیوپریونیتریل بتا
beta-conglycinin	گونکلیسینین بتا
beta-d-glucuronidase	گلوکورونیداز بتا-د
beta-DNA	بتا دی.ان.ای.، دی.ان.ای. نوع بتا
beta-glucan	گلوکان بتا
beta-glucuronidase	گلوکورونیداز بتا
beta-lactam antibiotic	آنتی بیوتیک بتا-لاکتام
beta-Lactamase	بتا لاکتاماز
beta-secretase	بتا سکرتاز
BFGF (= basic fibroblast growth factor)	بی.اف.جی.اف.
bidentate ligand	لیگاند دو دندانه ای
bifidobacteria	باکتری های دو شاخه ای
bifidus	بایفیدوس
bifunctional vector	بردار/حامل/ناقل دو کاربردی/بایفانکشنال
bifurcate	دو شاخه/منشعب شدن
bile	زردآب، زهره، صفرا
bile acid	اسید صفراوی
bilirubin	بیلی روبین
binary compound	ترکیب دوتایی
binding site	جایگاه پیوند، محل اتصال

English	Persian
bio	زیست
bio microelectromechanichal systems	سیستمهای بیو میکروالکترومکانیکی
bio reactor	راکتور زیستی، زیست واکنشگر
bioaccumulant	زیست انباشت
bioaccumulate	زیست انباشتن
bioaccumulation	زیست انباشتی
bioactivity	فعالیت زیستی
bioassay	زیست آزمون/سنجش/عیاریابی، زیست سنجی، مقایسه زیستی
bioastronautics	زیست فضانوردی، فضانوردی زیستی
bioaugmentation	زیست فزونی
bioautography	اتوگرافی زیستی
bioavailability	زیست دستیابی/دسترسی
bio-bar code	بار کد زیستی، زیست بارکد
biocatalyst	زی فروکافنده، کاتالیزور حیاتی
biocenology	بیو سنولوژی
biocheck 60	بیوچک 60
biochemical	بیوشیمیائی، زیست شیمیائی
biochemical oxygen demand (BOD)	خواست اکسیژن زیست شیمیائی (خاز)، نیاز بیو شیمیائی به اکسیژن
biochemist	زیست شیمیدان
biochemistry	بیو شیمی، زیست شیمی
biochip	تراشه زیستی، زیست تراشه
biocide	آفت کش، پادزیست، حیات/زیست/زیوا کش
biocoenosis	زیستگاه نیادی، زیستگاه نیمه طبیعی، هم زیستی جانوران
bioconcentration factor	عامل تراکم/تمرکز زیستی
bioconversion	بیوکانورژن، تبدیل زیستی
biocrat	بیوکرات

English	Persian
biodegradable	زیست تجزیه پذیر، قابل تجزیه بیولوژیکی
biodegradation	تجزیه بیولوژیکی، زیست فروزینگی، فساد بیولوژیکی
biodesulfurization	گوگرد زدایی بیولوژیکی
biodeteroriation	زیست تخریبی
biodisc	بیو دیسک
biodiversity	تنوع زیستی
biodyne	بیودین
bioelectronics	الکترونیک زیستی، بیو الکترونیک
bioenergy	انرژی زیستی، زیست کارمایه
bioengineering	زیست مهندسی، مهندسی زیستی
bioenrichment	غنی سازی زیستی
biofilm	بیوفیلم، زیست لایه، غشاء زیستی
biofilter	بیوفیلتر
biofouling	آلودگی بیولوژیکی
biofuel	بیو سوخت، سوخت بیولوژیکی
biogas	بیوگاز، گاز بیولوژیکی
biogenesis	زیست زایی
biogeochemistry	شیمی زیست زمینی
biogeography	جغرافیای زیستی، زیست گیتا شناسی، گیتا شناسی زیستی
biohazard	خطر زیستی
bioindicator	معرف زیستی
bioinformatics	داده شناسی زیستی، زیست انفورماتیک
bioinorganic	زیست غیر آلی
bioleaching	زیست آبشویی، فرو شست زیستی
biolistics	بیولیستیک
biologic agent	عامل بیولوژیکی/زیست شناختی
biologic indicator of exposure study	مطالعه نشانگر زیستی تماس

179

English	Persian
biologic monitoring	ردیابی زیستی
biologic response modifier therapy	درمان توسط تغییر دهنده پاسخ بیولوژیکی
biologic transmission	انتقال زیست شناختی
biologic uptake	جذب زیستی
biological	بیولوژیکی، زیست شناختی، زیستی، مربوط به زیست
biological activity	فعالیت زیست شناختی
biological agent	عامل بیولوژیکی/زیست شناختی
Biological and Toxin Weapons Convention (BTWC)	معاهده جنگ افزار های بیو لوژیکی و سمی
biological attack	حمله بیولوژیکی
biological contaminant	آلودکننده زیستی
biological contamination	آلودگی بیولوژیکی
biological control	کنترل/مهار بیولوژیکی/زیست شناختی/زیستی
biological control agent	عامل مهار/کنترل بیولوژیکی
biological defense	پدافند/دفاع بیولوژیکی
biological diseases	امراض زیستی
biological disk	دیسک بیولوژیکی
biological diversity	تنوع زیست شناختی
biological effect	اثر زیست شناختی
biological emergency	فوریت زیست شناختی
biological environment	محیط بیولوژیکی/زیست شناختی
biological filter bed	بستر صافی زیستی
biological half-time	نیم (ه) عمر بیولوژیکی/زیستی
biological incident	رویداد بیولوژیکی
biological indicator of exposure study	آزمایش نشانگر بیولوژیکی سطح تماس
biological integrated detection system	سیستم شناسایی منسجم بیولوژیکی
biological marker	علامت/نشانگر بیولوژیکی
biological marker of exposure	نشانگر بیولوژیکی/زیستی سطح تماس
biological measurement	پیمایش/سنجش بیولوژیکی
biological medium	محیط کشت بیولوژیکی
biological molecule	ملکول بیولوژیکی
biological monitoring	ردیابی بیولوژیکی، فرابینی زیستی
biological operation	عملکرد بیولوژیکی فعالیت زیست شناختی
biological oxidation	اکسایش زیست شناختی
biological oxygen demand	نیاز زیستی به اکسیژن
biological resource	منبع زیستی
biological response modifier	تغییر دهنده واکنش بیولوژیکی
biological rotating disc process	فرآیند دیسک چرخان زیستی
biological shield	حفاظ/سپر زیستی
biological stressor	عامل فشار بیولوژیکی/زیستی
biological threat	تهدید زیست شناختی
biological threat agent	عامل تهدید زیست شناختی
biological transport	انتقال بیولوژیکی
biological uptake	جذب بیولوژیکی
biological vector	ناقل بیولوژیکی
biological warfare	جنگ بیولوژیکی/میکروبی
biological warfare agent	عامل جنگ بیولوژیکی/میکروبی
biological warfare agent classification	رده بندی عامل جنگ میکروبی، طبقه بندی سلاحهای میکروبی
biological warfare agent identification method	روش شناسایی عامل جنگ/سلاحهای میکروبی
biological weapon	سلاح بیولوژیکی
biologicals	زیستی ها
biologics	بیولوژیک، زیست شناختی
biology	بیولوژی، زیست شناسی، علم حیات
bioluminescence	زیست شب افروزش/تابی، فسفر افکنی نور تابی زیستی
biomagnification	بزرگ نمایی زیستی
biomarker	شناسه/علامت/نشانه زیستی
biomass	تراکم حیوانات زنده/زیست، زیست توده
biomass energy	انرژی زیست توده

180

English	Persian
biomaterial	بیومواد
biome	اقلیم زیست، زی/زیست بوم
biomedical testing	آزمون زیست پزشکی
biomedicine	بیو پزشکی
biomembrane	بیو غشاء
bioMEMS (= bio microelectromechanichal systems)	بیو ام.ئی.ام.اس.
biomimetic materials	مواد بیومیمتیک
biomodulator	تعدیل کننده زیستی
biomolecular electronics	الکترونیک زیست ملکولی
biomolecular engineering	مهندسی زیست ملکولی
biomonitoring	زیست ردیابی
biomotor	زیست محرکه/موتور
bionics	بیونیک، زیست لگام شناسی، زیستار شناسی
biopesticide	زیست آفت کش
biopharmaceutical	زیست دارویی
biophotolysis of water	بیو فتو لیز آب
biophysics	بیو فیزیک، زیست فیزیک
biopolymer	بیو پولیمر، پلیمر زیستی، زیست پرپار
biopreparat	بیوپرپارات
bioprobe	پروب/ردیاب/کاووشگر/نشانگر زیستی
bioprocess	فرآیند زیستی
bioprocessing	پردازش زیستی
biopsy	بافت/نمونه برداری، بیوپسی، نمونه بافت
bioreactor	رآکتور زیستی، زیست واکنشگر
bioreceptor	زیست گیرنده
biorecovery	زیست بازیافت
bioregion	زیست اقلیم/ناحیه
bioregulator	زیست تنظیم گر
bioremediation	زیست درمانی
biorythm	بیوریتم، زیست آهنگ/نواخت
bios	بیوس
biosafety	زیست ایمنی
biosafety levels	سطوح زیست ایمنی

English	Persian
Biosafety Protocol	پروتکل بیوایمنی، پروتکل زیست ایمنی
bioscience	بیوساینس، زیست علم
bioscrubber	بیواسکرابر
bioscrubbing	بیواسکرابینگ
bioseeds	بذر مهندسی شده، بیوبذر
biosensor	زیست حسگر
biosensor technology	تکنولوژی زیست حسگر
biosilk	بیو ابریشم
biosorbent	بیوسوربنت
biosphere	زیست کره
biosphere reserve	ذخیره زیست کره
biosynthesis	بیو سنتز، زیست آمایی/ساختی
biota	جانداران یک پهنه، زیوگان
biotechnology	بیو تکنولوژی، تکنولوژی زیست، تکنولوژی/فن شناسی/فناوری
biotelemetry	بیو تله متری
bioterrorism	بیو تروریزم، تروریزم میکروبی
bioterrorist	بیو تروریست، تروریست میکروبی
biotherm	بیوترم
biotic	حیاتی، زیستی
biotic resource	منابع زیستی
biotic stress	تنش زیستی
biotin	بیوتین
biotinylated-DNA	دی.ان.ای بیوتینی شده
biotinylatio	وصل بیوتینی
biotinylation	بیوتینیلاسیون
biotope	بیوتوپ، زیست جا (ی) زیستگاه، محل زیست
biotransformation	تغییر شکل/دگرگونی زیستی، زیست تبدیلی/دگرگونی
biotron	بیوترون
biotype	بیوتیپ، جانداران هم نژاد، زیست گروه، زیمون، سنج/نوع زیستی
biowarfare	جنگ میکروبی

bioweapon	جنگ افزار میکروبی، سلاح بیولوژیکی	boat conformation	صورت بندی قایقی
biphasic	دو مرحله ای	BOD (= biochemical oxygen demand)	خاز
bipolar	دوانتهایی، دوقطبی	boiling point	دمای جوش
bivalent	دو ارزشی، دو تایی، دو ظرفیتی	boletic acid	اسید بولتیک
BLA gene	ژن بی.ال.ای.	bollworms	کرم های پنبه/غوزه
black-layered (corn)	لایه سیاه (ذرت)	bomb	بمب، بمباران کردن
black-lined	خط سیاه (ذرت)	Bombay phenotype	فنوتیپ نوع بمبئی
blank test	آزمون شاهد	bond	پیوند
blast cell	تنده یاخته	bond angle	زاویه پیوند
blast furnace	کوره بلند	bone marrow transplantation	پیوند مغز استخوان
blast transformation	تبدیل بلاست	bone morphogenetic protein	پروتئین ریخت زای استخوانی
bleaching	رنگبری	bottle	بطری، بطری کردن
blood cell	گویچه	botulin toxin	سم بوتولین
blood clotting	لخته شدن خون	botulinum	بوتولینوم
blood derivative	محصولات مشتق از خون، مشتقات خونی	botulinum toxin	سم بوتولینوم
		botulism	بوتولیزم
blood plasma	پلاسما، پلاسمای/دشتینه/سرم خون	bovine leukemia virus	ویروس لوکمی گاوی
		bovine papilloma virus (BPV)	ویروس پاپیلومای گاوی
blood platelet	پلاکت خونی، گرده خون، یاخته لخته کننده	bovine somatotropin	سوماتوتروپین گاوی، هورمون رشد گاوی
blood pressure gauge	دستگاه فشار خون	bovine spongiform encephalopathy	انسفالوپاتی اسپونجیفورم گاوی
blood serum	پیماب/سرم خون	bowel	روده، روده ای، شکم، شکمبه
blood-brain barrier	سد مغزی-خونی		
blot	بلات	Bowman's capsule	پوشینه بومن
blower	بلوئر، دمنده، وزنده	Bowman-Birk trypsin inhibitor	بازدارنده تریپسین (نوع) بومن-برک
blue cone pigment	رنگدانه مخروط آبی	bP (= base Pair)	زوج باز
blue vitriol	کات کبود	bP (= base Pair)	جفت باز
blunt-end DNA	دی.ان.ای. انتها صاف/کور	BP (= bP)	بی.پی.
blunt-end ligation	اتصال/پیوند انتهای صاف/کور	Bp = base pair	جفت باز
BLUP (=best linear unbiased prediction)	بی.ال.یو.پی.	BPG (= blood pressure gauge)	بی.پی.جی.
BLV (= bovine leukemia virus)	بی.ال.وی.	bradykinin	برادی کینین
B-lymphocytes	لنفوسیت های بی.	bradyrhizobium japonicum	برادیریزوبیوم جاپونیکوم
BMD (= Becker muscular dystrophy)	بی.ام.دی.	branch	انشعاب، شاخه فرعی، شاخه شاخه/منشعب شدن
BMT (= bone marrow transplantation)	بی.ام.تی.	brassica	شلغم بیابانی (کلم پیچ و

bronchoscopy	نایژه بینی
bronchus	نایژه
broth	آبگوشت (کشت)
brown stem rot	پوسیدگی قهوه ای ساقه
BSE (= bovine spongiform encephalopathy)	بی.اس.ئی.
BSP (= sulphobromo phtahalein)	بی.اس.پی.
Btal (= B-thalassemia)	
B-thalassemia	بتا تالاسمی
BTWC (=Biological and Toxin Weapons Convention)	بی.تی.دبلیو.سی.
bubonic plague	طاعون میمونی
bucket	دلو، سطل
BUdR (= bromodeoxyuridine)	
buffer solution	محلول بافر
buffer zone	منطقه بافر/حائل/ضربه گیر
buffy coat	رویه بافی
bulk	توده (ای)، حجیم، عمده، فله ای، ماده اضافی/بی مصرف
bulking	جذب آب در روده
bunyaviridae	بونیا ویروس ها
bunyavirus	بونیا ویروس
buoyant density	چگالی شناوری
Burkhit lymphoma	لنفوم بورکیت
burn-through range	برد/رنج/طیف برن ترو
Bursa dependent cells	سلول متکی/وابسته به بورسا
BWS (= Beckwith-Weideman syndrome)	بی.دبلیو.اس.
BXN gene	بی.اکس.ان. ژن، ژن بی.اکس.ان.

C

C (= cytosine)	سی.
c value	ارزش سی
CA repeats (repeats of	تکرار های سی.ای.

	(خانواده کلم)
brassica campestre	براسیکا کمپستر
brassica campestris	براسیکا کمپستریس
brassica napus	براسیکا ناپوس، کلزا
brazzein	برازئین
BRCA (= breast cancer gene)	بی.آر.سی.ای.
BRCA 1 gene	ژن بی.آر.سی.ای. نوع 1
BRCA 2 gene	ژن بی.آر.سی.ای. نوع 2
BRCA gene	ژن بی.آر.سی.ای.
breakdown	اختلال، از هم پاشیدن، انقراض، تجزیه/تقسیم شدن/کردن، فروپاشی
breakpoint cluster region	حوزه/ناحیه خوشه شکننده
breast cancer gene	ژن سرطان سینه
breed	اصل، اصلاح نژاد کردن نسل، پروردن، تخم کشی کردن، تولید مثل کردن، زادگیری کردن، نژاد، نوع
breed at risk	گونه ریسکی/در خطر
breed not at risk	گونه امن/غیر ریسکی
breeder's rights	حقوق اصلاحگر/پرورنده
breeding	اصلاح نژاد، به نژادی، پرورش حیوانات، تخم کشی، جفت گیری، زادگیری
brevetoxin	بروه توکسین، سم بروه
bright greenish-yellow fluorescence	پرتو افشانی درخشان زرد مایل به سبز
broad spectrum	طیف وسیع
broad-specturm	وسیع الطیف
bromodeoxyuridine	بروموداکسی اوریدین
bromoxynil	بروموکسی نیل
bronchi	صورت جمع برونکوس به معنی نایژه، نایژه ها
bronchial	مربوط به برونشی/نایژه، نایژه ای
bronchial-associated lymphoid tissue	بافت لنفوئیدی نایژه ای
bronchiole	نایژک
bronchitis	نازه آماس، نایژاماس
bronchoscope	نایژه بین

English	Persian
cytosine and adenine)	
CAAT (= CAAT box)	سی.ای.ای.تی
CAAT box	جعبه سی.ای.ای.تی.
cachectic factor	عامل کاشکتیک
cachectin	کاشکتین
cacinoembryonic antigen	آنتی ژن کاسینو جنینی، پادگن کاسینو جنینی
cadherin	کادهرین
caesium chloride	کلرید سزیم (نمک)
caffeine	جوهر قهوه، کافئین
CAH (= congenital adernal hyperplasia)	سی.ای.اچ.
Cal (= calorie)	
calciphorin	کالسیفورین
calcitonin	کالسیتونین
calcium	کلسیم
calcium channel-blockers	مسدود کننده های مجرای کلسیم
calcium oxalate	کلسیم اکسالات
calcyclin	کلسیکلین
calicol	کالیکول
callipyge	فربهی، کالیپیجی
callus	پینه، خیز، کالوس، کبره
calmodulins	کالمودولین ها
calomys colosus	کالومیس کولوسوس
calorie	کالری
calorimeter	گرماسنج
calpain	کالپین
calpain-10	کالپین 10
caltractin	کالتراکتین
caltrin	کالترین
Calvin cycle	چرخه کالوین
cAMP (= cyclic 3e, 5e-adenosine monophosphate)	
cAMP receptor protein (= catabolite activator protein)	پروتئین گیرنده سی.ای.ام.پی.
Campbell Hausfeld	کامپبل هاوسفلد
campesterol (= campestrol)	کمپسترول
campestrol	کمپسترول

English	Persian
camphor	کافور
campsterol (= campestrol)	کمپسترول
camptothecin	کامپتوتسین
canavanine	کاناوانین
cancer	چنگار، سرطان
cancer epigenetics	اپی ژنتیک سرطان، تحولات پیدایشی سرطان، روزاد شناسی سرطان
cancerocidal	سرطان کش
CANDA (= computer assisted new drug application)	کاندا
candicidin	کاندیسیدین
candida	کاندیدا
cannabinoid	کانابینوئید، مربوط به شاهدانه/کانابیس
cannitracin	کانیتراسین
canola	منداب روغنی کانادایی
canola	کانولا
CAP (= catabolite activator protein)	سی.ای.پی.
Cap (= 7-methylguanosine)	
CAP site	جایگاه سی.ای.پی.
cap structure	ساختار کلاهکی، سرپوش
capable of being transmitted from one person to another	مستعد انتقال از شخصی به شخص دیگر
capacity	استعداد، توانایی، ظرفیت، قابلیت، گنجایش
capacity building	ظرفیت سازی
capillary electrophoresis	الکتروفورروز/برق بری موبرگی
capillary isotachophoresis	ایزوتاکوفورسیس موئین/مویسان/کاپیلاری، کاپیلاری ایزوتاکوفورسیس
capillary isotechophoresis	ایزوتکوفورسیس موئین/مویسان/کاپیلاری، کاپیلاری ایزوتکوفورسیس
capillary zone electrophoresis	الکتروفورروز/برق بری ناحیه موبرگی
capital femoral epiphysis	اپیفیز ران/فمورال اصلی
capping	پوشش دادن، درپوش

گذاشتن، کپینگ

capsid	پوشش پروتئینی، پوشه، کاپسید، کپسول
capsular	پوششی، پوشینه ای
capsular antigen	آنتی ژن کپسولی
capsule	پوشینه، تخمدان، کپسول
captive breeding	تولید مثل در اسارت
capture agent	عامل اتصال
capture molecule	ملکول اتصال
carbetimer	کاربتیمر
carbohydrate	کربو هیدرات، هیدرات کربن
carbohydrate engineering	مهندسی کربوهیدرات
carbon dioxide	انیدرید کربنیک، دی اکسید کربن، گاز کربنیک
carbon nanotube	نانو تیوب کربن(ی)، ابر ریز لوله کربن(ی)
carboxyl group	گروه کاربوکسیل
carboxyl proteinase	کاربوکسیل پروتئیناز
carcinogen	ماده سرطان زا
carcinogenicity	سرطان زایی، قدرت سرطان زایی
carcinoma	کارسینوما
carcinomatosis	کارسینوماتوز
cardenolide	کاردنولید
cardiac glycoside	گلیکوزید کاردیاکی
carnitine	کارنیتین
carnosine	کارنوزین
carotene	کاروتن
carotenoid	شبه کاروتن
carotid artery	خوابرگ، سرخرگ کاروتید
carrier	ـبر، حامل، ناقل
carrying capacity	ظرفیت برد/حمل، ظرفیت زیست محیطی
cartilage link protein	پروتئین اتصال غضروف
cartilage-inducing factor	عامل ایجاد غضروف
caryogram	کاریوگرام
caryokinesis	کاریوکینزیس
caryon	کاریون
caryotype	کاریوتیپ
cascade	آبشار مانند، آبشاره، آبشیب، دوانه، توالی،

case	پوسته، جلد، غلاف، مورد، مورد بیماری، موضوع، نمونه، نیام، وضعیت
case study	بررسی/مطالعه موردی
case-by-case	مورد به مورد
case-fatality rate	میزان کشندگی موردی
case-finding	بیمار/مورد یابی
casein	کازئین
caseinogen	کازئینوژن
caseous necrosis	بافت مردگی پنیری/نکروز
CASP (= CTD-associated SR-like protein)	سی.ای.اس.پی.
caspase	کاسپاز
cassava	ماینوک (گیاه)، کاساوا (نشاسته)
cassette	کاست
cassette mutagenesis	موتاژنز کاستی
castor bean	کرچک، دانه سمی کرچک
castor oil	روغن کرچک
casual contact	تماس/اتفاقی/روزمره/عادی، تماس غیر دائم/غیر مستمر
catabolic	تجزیه ای، کاتابولیک، فروگشتی، فروگوهرشی
catabolism	تجزیه و تخریب مواد، سوخت، فروساخت، کاتابولیسم فروگشت
catabolite	فروگشته
catabolite activator protein	پروتئین فعال کننده فرآورده تجزیه
catabolite repression	بازدارندگی کاتابولیت، سرکوب فرو گوهره
catalase	کاتلاز
catalysis	فروکافت، فروکافتن، کاتالیز
catalyst	عامل شتاب دهنده، فروکافنده، فعال کننده، کاتالیزور، کاتالیست، واکنش یار
catalytic antibody	پادتن فروکافتی/کاتالیزوری
catalytic domain	منطقه کاتالیزوری

catalytic RNA	آر.ان.ای کاتالیزوری
catalytic site	محل کاتالیزوری، جایگاه فروکافتی
catechin	کاتچین، کاتشین
catecholamine	کاته کولامین
catenated DNA	دی. ان.ای. کاتنیتد/متصل
catenin	کاتنین
cathespin	کاتسپین
cation	کاتیون، یون مثبت
cationic	کاتیونی
causative agent	عامل سببی
caveolae	کاویئولا
caveolin	کاویئولین
C-banding	سی. باندینگ
ccc DNA	دی.ان.ای. سی.سی.سی.
CD (= circular dichroism)	سی. دی.
CDA (= congenital dyserythropoietic anaemia)	سی.دی.ای.
Cdc2 (= cyclin-dependent mitotickinase)	
CDH (= congenital dislocation of the hip)	سی.دی.اچ.
Cdk (= cyclin-dependent kinase)	سی.دی.کی.
cDNA (= complementary DNA)	سی دی.ان.ای.
c-DNA (= complementary DNA)	سی دی. ان. ای.
cDNA array	آرایه زنجیره واحد دی.ان.ای
cDNA clone	همسانه زنجیره واحد دی.ان.ای
cDNA library	مخزن زنجیره واحد دی.ان.ای
cDNA microarray	ریز آرایه زنجیره واحد دی.ان.ای
CDP (= cytidine diphosphate)	سی.دی.پی.
cecrophins	سکروفین ها
cecropin	سکروپین
cecropin a	سکروپین ای.
cecropin a peptide	پپتید سکروپین آ/آلفا/ای.

cefazolin	سفازولین
ceftriaxone	سفتریاکزون
ceiling limit	حد/سقف مجاز
ceiling value	ارزش نهائی، حد اکثر ارزش، حد نهائی ارزش
cell	حفره، سلول، کیسه، یاخته
cell affinity chromatography	کروماتوگرافی کشش سلولی
cell culture	کشت سلول(ی)
cell cycle	چرخه سلولی/یاخته ای
cell cytometry	سیتومتری یاخته ای، یاخته شماری
cell death	مرگ سلولی
cell differentiation	تمایز سلولی، جدایش یاخته ای
cell division	تقسیم یاخته/سلول (ی)
cell fusion	ادغام/الحاق/امتزاج/همجوشی سلولی/یاخته ای
cell growth Y chromosome	کروموزوم وای. رشد سلولی
cell hybrid	سل دو رگه/هیبرید
cell line	خط سلولی، سل لاین
cell mediated immunity	ایمنی با واسطه سلولی
cell membrane (cell wall)	دیواره سلول، شامه یاخته، غشاء سلولی
cell membrane structure	ساختمان دیواره/غشاء سلولی
cell nucleus	هسته یاخته
cell recognition	شناسایی یاخته
cell respiration	تنفس سلولی
cell signalling	علامت دهی سلولی
cell size	اندازه یاخته
cell sorting	دسته بندی سلولی
cell turnover	ترن اوور سلولی، سل ترن اوور
cell wall	دیواره سلولی/یاخته
cell-differentiation protein	پروتئین تمایز سلولی
cell-free gene expression system	سیستم تظاهر ژنی بدون سلول
cell-free translation	ترجمه بدون سلولی
cell-mediated immunity	ایمنی با واسطه سلول
cellular adhesion molecule	ملکول چسبندگی سلولی

186

cellular adhesion receptor	سلولار ادهیژن ریسپتور، گیرنده چسبندگی سلولی
cellular affinity	گیرایی سلولی
cellular aging	پیری سلولی
cellular immune response	واکنش ایمنی سلولی
cellular necrosis	بافت مردگی سلولی، مرگ نسج سلولی
cellular oncogene	تومور زایی یاخته ای، ژن توموری سلولی
cellular pathway mapping	شناسائی مسیر سلولی، نقشه کشی مسیر سلولی
cellular response	پاسخ/واکنش سلولی، سلولار ریسپانس
cellular slime mold	قارچ/کفک مخاطی سلولی
cellulase	سلولاز
cellulolytic activity	فعالیت سلولایتیک
cellulose	سلولز
celsius	سانتیگراد، سلسیوس
cen (= centromere)	
center of diversity	مرکز تنوع
centers of genetic diversity	مراکز گوناگونی/تنوع ژنتیکی
centers of origin	مراکز اصالت
centers of origin and diversity	مراکز اصالت و تنوع/گوناگونی
centimorgan	سانتی مورگان
central dogma	دوگم اصلی/بنیادی/مرکزی، حکم اساسی
central metabolic pathway	مسیر متابولیک/سوخت و ساز مرکزی
central nervous system	دستگاه عصبی مرکزی
centric fission	شکافت/فیسیون مرکزی
centric fusion	فوزیون/همجوشی مرکزی
centrifugation	سانتریفوژاسیون، میان گریزش
centrifuge	دستگاه مرکز گریز، سانتریفوژ، گریزانه، گریز از مرکز، گریز دادن، میان گریز کردن
centrin	سنترین
centriole	میانک

centrioles	سانتریول(ها)
centromere	سانترومر، میانهار
centrosome	سانتروزوم، میان تن
cephalosporin	سفالواسپورین
cerebro-oculo-facial skeletal syndrome	سندروم/عارضه اسکلتی سربرو-اوکلو-فاسیال
cerebrose	سربروز
cervical lymphadenitis	التهاب گره های لنفی گردنی، لنفادنیت سرویکال
cessation cassette	زنجیره قطع کننده/جلوگیری کننده، سسیشن کاست
cetylpyridinium	ستیل پیریدینیوم
CF (= cystic fibrosis)	سی.اف.
CFE (= capital femoral epiphysis)	سی.اف.ئی.
CFTR (= cystic fibrosis transmembrane conductance regulator)	سی.اف.تی.آر
CFU (= colony forming units)	سی.اف.یو.
CG island	جزیره سی.جی.
CGD (= chronic granulomatous disease)	سی.جی.دی.
CGY (= cell growth Y chromosome)	سی.جی.وای
chaconine	شاکونین
chain terminating mutant	جهش یافته پایان بخش زنجیره
chain terminating suppressor	سرکوبگر پایان بخش زنجیره
chain termination codon	کدون پایان بخش زنجیره
chain terminator	پایان بخش زنجیره
chakrabarty	قانون چاکرابارتی
chakrabarty decision	حکم چاکرابارتی
chalcone isomerase	کالکون ایزومراز
chalmydospore	کالمیدواسپور
chalone	چالون، کالون
channel-blockers (= calcium channel-blockers)	مسدود کننده مجرا
chaotropic agent	عامل کائوتروپ/بی نظمی دوست
chaperone	شاپرون، مراقب

chaperone molecule	ملکول شاپرون
chaperone protein	پروتئین شاپرون/مراقب (ملکولی)
chaperonins	شاپرونین ها
Chapin plant and rose powder duster	پودر/سم پاش چاپین
character	حرف، رقم، شخصیت
characterization assay	خصلت سنجی، محک خصلت
characterization of animal genetic resource	خصیصه یابی منابع ژنتیکی جانوری
Charon phage	فاژ شارون
CHD (= congenital heart disease)	سی.اچ.دی.
Chediak-Higashi syndrome	سندروم چدیاک-هیگاشی
CHEF (= contour-clamped homogeneous electric fields)	سی.اچ.ئی.اف.
chelating agent	عامل ایجاد کننده کلات
chelation	چنگک سازی، کلات سازی
chemical agent	عامل شیمیائی
chemical attack	حمله شیمیائی
chemical genetics	علم ژنتیک شیمیائی
chemical molecule	ملکول شیمیائی
chemical oxygen demand (COD)	خواست اکسیژن شیمیائی (خاش)
chemical terrorism	تروریزم شیمیائی
chemical warfare	جنگ شیمیائی
chemical warfare agent	عامل جنگ شیمیائی
chemical weapon	سلاح شیمیائی
chemicalization	شیمیایی کردن/سازی
chemiluminescence	نورتابی شیمیائی
chemiluminescent immunoassay (CLIA)	ایمنی سنجی نورافشانی شیمیایی
chemo-autotroph	خود خوار شیمیائی
chemokine	کموکین
chemometrics	کمومتری
chemoorganotroph (heterotroph)	کموارگانوتروف
chemopharmacology	دارو شناسی شیمیائی، شیمی داروشناسی
chemoprophylactic	کیموپروفیلاکتیک
chemoprophylaxis	پیشگیری (از بیماری یا عفونت) با دارو، پیشگیری شیمیائی، کیموپروفیلاکسی
chemotactics factor	عامل کیموتاکتیک/جذب شیمیائی
chemotaxis	شیمی گرایی، کشش/گرایش شیمیائی
chemotherapy	دارو/شیمی درمانی
chi (= chimera)	
chiasma	کیاسما
chiasmata	جمع کیاسما
chill	لرزیدن
chills	لرز
chimera	بافت ناهمسان، شیمر، کژزاد
chimeraplasty	کیمرا پلاستی
chimeric antibody	آنتی بادی/پادگن نوترکیب، کیمریک آنتی بادی
chimeric DNA	ملکول دی.ان.ای نوترکیب/کیمریک
chimeric protein (chimera)	پروتئین کیمریک/نوترکیب
chip DNA technology	فناوری تراشه دی.ان.ای.
chiral compound	ترکیب کایرال
chiselplow	گاو آهن قلمی
chitin	کیتین
chitinase	کیتیناز
Chl (= chlorophyll)	
chloramphenicol amplification	تقویت کلرآمفنیکل
chlorination	کلر دار کردن/شدن، کلر زدن، کلر زنی
chlorine	کلر
chlorine demand	مقدار کلر مطلوب/مورد نیاز
chlorine dose	دوز/مقدار مصرف کلر
chlorine residual	پس/ته مانده کلر، رسوبات کلر، مقدار کلر ته نشین شده
chloroform	بیهوش کردن با کلروفورم، کلروفورم
chlorophyll	برگ سبزینه، سبزینه،

	کلروفیل
chloroplast	جسم سبزینه ای، سبز دش، سبزدیسه، کلروپلاست
chloroplast transit peptide	پپتید ناقل کلروپلاست
chlorosis	کلروسیز
chlorosome	کلروزوم
cholecalciferol	کولیکالسیفرول
cholecalcin	کوله کالسین
cholecystokinin	کوله سیستوکینین
cholera toxin	توکسین/سم وبا
cholesterol	کلسترول
cholesterol oxidase	کلسترول اکسیداز
choline	کولین
cholinergic	تراونده استیل کولین، کولینرژیک
cholinesterase	کولین استراز
choriogonadotropin	کوریوگونادوتروپین
choriomamotropin	کوریوماموتروپین
chorionic villus sampling	نمونه برداری از پرزهای جفتی/بافت جنینی (پرده کوریون)
chr (= chromosome)	
chromatid	کروماتید
chromatin	رنگینه، فامینه، کروماتین
chromatin modification	تبدیل/تغییر کروماتینی
chromatin remodeling	تغییر مدل کروماتین
chromatin remodeling element	عامل تغییر مدل کروماتینی
chromatin remodeling machine	دستگاه تغییر آرایه کروماتین
chromatography	رنگ نگاری، فام نگاری، کروماتوگرافی
chromatophore	رنگ دار، فام بر، کروماتوفور
chromogenic substrate	بنیاد رنگ زا، زیر بستر رنگ زا، کروموژنیک سوب استرات
chromomere	کرومومر
chromonema	فام راک، کرومونما
chromoplast	رنگ دیسه
chromosomal	فام تنی، کروموزومی
chromosomal packing unit	واحد بسته بندی کروموزومی

chromosomal translocation	جا به جایی کروموزومی
chromosome	رنگین تن، فام تن، کروموزوم
chromosome jumping library	مخزن پرشی کروموزومی
chromosome map	نقشه کروموزومی
chromosome painting	رنگ آمیزی کروموزومها
chromosome walking	روش راهبری کروموزومی
chronic	تدریجی، کهنه، مزمن
chronic disease	بیماری مزمن
chronic effect	اثر/تاثیر مزمن/حاد/جدی، کرونیک افکت
chronic granulomatous disease	بیماری گرانولوماتوز مزمن
chronic heart disease	بیماری قلبی مزمن
chronic intake	درون جذب مزمن/حاد/جدی
chronic lymphocytic leukemia	لوکمی لنفوسیتی مزمن
chronic myelocytic leukemia	لوکمی میلوسیتی مزمن
chronic-obstructive-pulmonary disease	بیماری انسداد/مسدود کننده ریه مزمن
chronobiochemistry	زمان-بیوشیمی، کرونوبیوشیمی
CHS (= Chediak-Higashi syndrome)	سی.اچ.اس.
cht (= chromatid)	
chymosin	رنین، کیموزین
chymotrypsin	کیموتریپزین، کیموتریپسین
CI repressor	بازدارنده سی.آی.
cidofovir	سیدوفوویر
cilia	سیلیا، مژک ها، مژه ها
ciliary neurotrophic factor (CNTF)	سیلیاری نروتروفیک فاکتور، عامل سیلیاری نروتروفیک، عامل عصب گرای مژکی
circadian rythm	نواخت سیرکادیان/شبانه روزی
circular dichroism	دیکروئیزم دایره ای
circular restriction map	نقشه محدودیت حلقوی/دایره ای
circularization	گردشی کردن
cis	سی.آی.اس.، سیز

English	Persian
cis/trans isomerism	ایزومری سیس-ترانس
cis/trans test	آزمون جذب-دفع، آزمون همسو-دگرسو
cis-acting protein	پروتئین سیز- فعال
Cis-acting sequence in RNA polymerase II promoters	پیش برنده توالی سی.آی.اس.-عملگر در پلیمراز 2 آر.ان.ای.
cisplatin	سیزپلاتین
cistron	بخش فعال ژن، سیسترون
citrate synthase	سنتاز سیترات
citrate synthase gene	ژن سنتاز سیترات
citreoviridin	سیترئوویریدین
citric acid	اسید سیتریک، جوهر لیمو
citric acid cycle	چرخه اسید سیتریک
CKI (= cyclin-dependent kinase inhibitor)	سی.کی.آی.
c-kit genetic marker	نشانگر ژنتیکی کیت سی.
clade	هم نیا
cladistics	شاخه بندی، هم نیایی
clathrin	کلاترین
cleavage and polyadenylation specificity factor	عامل تخصیص پلی آدنیلاسیون و کلیواژ
cleavage stimulation factor	عامل تحریک کلیواژ
cleave	تقسیم اول/دوم سلول تخم، کلیواژ
climate change	تغییر آب و هوا
climax community	اجتماع/جامعه اوج/بالیست
clindamycin	کلیندامایسین
clinic	درمانگاه
clinical	بالینی، درمانگاهی
clinical trial	آزمایش بالینی
clinician	پزشک بالینی
CLL (= chronic lymphocytic leukemia)	سی.ال.ال.
clone	تولید مصنوعی، دودمان سلولی، کلون، مشابه سازی، همسانه
cloning	تاک/دودمان/همسانه سازی
cloning vector	ناقل همسانه سازی
clostridium	کلوستریدیوم
clostridium perfringens toxin	سم کلوستریدیوم پرفرینژن
clotting factor IX	عامل لخته کننده نوع 9
clotting factor VIII	عامل لخته کننده نوع 8
clotting factors	عوامل انعقاد
clubbing	چماقی شدن، کلابینگ
cluster	دسته، خوشه (ای)، سنبله، کلاله، گروه
cluster of differentiation	خوشه جدایش
clustering	خوشه ای شدن
CM (= centimorgan)	سی.ام.
CMD (= congenital myeloperoxidase deficiency)	سی.ام.دی.
CMI (= cell mediated immunity)	سی.ام.آی.
CML (= chronic myelocytic leukemia)	سی.ام.ال.
CMP (= cytidine monophosphate)	سی.ام.پی.
CMV (= cytomegalovirus)	سی.ام.وی.
CNF (= congenital nephrotic syndrome of Finnish type)	سی.ان.اف.
CNO (= congenital nephrotic syndrome of other types)	سی.ان.او.
CNS (= central nervous system)	سی.ان.اس.
CNS (= congenital nephrotic syndrome)	سی.ان.اس.
CoA (= coenzyme A)	
co-adaptation	همسازی
coagulation	انعقاد، بستن، بندایش، لخته شدگی/شدن/کردن، هم بستی
CoA-SH (= coenzyme A)	
coat protein	کت پروتئین، پروتئین پوششی/روکشی
coated vesicle	وزیکول پوشش دار/پوشیده/روپوش دار
cobalamin(e)	کوبالامین

coccus	باکتری کروی شکل، کوییزه	cohesive end	انتهای/پایانه چسبنده
co-chaperonin	کمک چاپرونین	cohesive termini	پایانه های چسبنده
Cockayne syndrome	سندروم/عارضه کوکین	cohort	همزادان، هم گروه
cocking	آماده شلیک کردن، مسلح کردن	cohort study	بررسی/مطالعه همزادگان
cocloning	همسانه سازی اضافی	coiled coil	چنبر (ه) پیچیده
COD (= chemical oxygen demand)	خاش	cointegrate	یکی کردن
		colchicine	کلشیسین
code	به رمز در آوردن/نوشتن، رمز، کد، کد گزاری کردن	cold acclimation	خوگیری با سرما
		cold acclimatization	سازش با محیط سرد
codegenerate codon	کدون کودژنزه	cold hardening	مقاوم سازی در مقابل سرما
codex alimentarius (= codex alimentarius commission)		cold tolerance	تحمل سرما، مقاومت درمقابل سرما
codex alimentarius commission	کمیسیون قوانین غذائی	cold-shock protein	پروتئین (ضد) شوک سرمائی، کولد شوک پروتئین
coding parts of a gene	اجزاء رمز ساز ژن	colectomy	کولون برداری
coding region	منطقه رمز ساز(ی)	colicin	کالیسین، کولیسین
coding sequence	توالی رمز گذار	coliform bacteria	باکتری کولی باسیل شکل
coding strand	رشته رمز گزار	coliform organism	ارگانیزم کولی باسیل شکل
codogenic strand	رشته/شاخه کدوژنیک/کد ساز	colistin	کولیستین
		colitis	کولون آماس
codon	رمز، کد ژنتیکی سه گانه، کدون، کله رمز	collagen	چسب زا، کولاژن
		collagenase	کولاژناز
codon bias	بایاس/تمایل کدون	colligative property	خاصیت جمعیتی
coefficent of coincidence	ضریب انطباق/تصادف/هم رخدادی	colloid	چسب ماند، کلوئید (ی)
		colloidal	چسب سان، چسبی، کلوئیدی
co-enzyme	کوآنزیم، کمک مخمر		
coenzyme A	کوآنزیم آ./ای.	colon	کولون
coenzyme Q	کوآنزیم کیو.	colonial morphology	ریخت شناسی دست جمعی
co-evolution	تکامل توام، کو اوولوشن، هم فرگشت	colonoscopy	کولون بینی
co-factor	عامل/ماده کمکی/مشترک، کوفاکتور	colony	اجتماع، پرگنه، کلونی، کولونی، گروه
co-factor recycle	فرایند بازتولید کوفاکتور، کوفاکتور ریسایکل	colony forming units	واحد های تشکیل دهنده کلونی
		colony hybridization	دورگه سازی کولونی، هم تیرگی کولونی
coffee berry borer	حشره سنجد تلخ		
coformycin	کوفورمایسین	colony lift	برداشت کولونی، کولونی لیفت
COFS (= cerebro-oculo-facial skeletal syndrome)	سی.او.اف.اس.	colony stimulating factor	عامل برانگیزنده کولونی
cognate tRNA	گونه صحیح تی.آر.ان.ای.	column development	کروماتوگرافی

chromatography	پیشرفت/توسعه ستون
co-management	هم گردانی/مدیری
combinatorial	ترکیبی
combinatorial biology	زیست شناسی ترکیبی
combinatorial chemistry	شیمی ترکیبی/تلفیقی
combinatorics	کومبیناتوریکز
(combinatorial chemistry)	
combined gas	اسپکترومتر/طیف سنج انبوه
chromatography-mass	رنگ نگاری گازی ترکیب
spectrometre	شده
combining site	محل ترکیب شدن
commensal	همسفره
commensalism	همسفرگی، هم غذایی،
	همکاری
common property resource	گردانندگی/مدیریت منابع
management	املاک مشترک/عمومی
communicable	قابل انتقال، مسری، واگیردار
communicable disease	بیماری مسری
communicable period	دوره واگیر
communism	کمونیزم
community	اجتماع، جامعه، جمعیت
company	جمع، شرکت، گروه، گروهان،
	همراه
comparative analysis	تحلیل تطبیقی
compensating variation	تغییر جبرانی، گونه جبرانی
compensation	انطباق، ترمیم، جبران،
	سازگاری، موازنه
compensation point	نقطه جبران
competence factor	عامل توانش/شایستگی
competency	استعداد، توانائی،
	شایستگی، قابلیت
competent authority	مرجع ذیصلاح
competent cell	سلول مستعد/توانمند
competition	رقابت
competitive exclusion	طرد رقابتی
complement	کمپلمان، مکمل
complement acvtivation	فعال شدن/کردن/سازی
	مکمل
complement cascade	آبشیب مکمل
complement receptor	گیرنده مکمل
complementary	مکمل (همدیگر)

complementary and	پزشکی مکمل و غیر متعارف،
alternative medicine	پزشکی/طب غیر
	سنتی/جایگزین
complementary DNA	دی.ان.ای. مکمل
complementary medicine	پزشکی/طب مکمل و غیر
	متعارف
complementary nucleotide	نوکلئوتید مکمل
complementary RNA	آر.ان.ای مکمل
complementation	تکمیل/کمپلیمانتاسیون/کمپل
	یمنتیشن
complete digestion	گوارش/هضم کامل
complexation	کمپلکسیشن، مرکب شدن
composting	کمپوست/کودآلی/هم‌پوده
	درست کردن
compound heterozygote	سلول/هتروزیگوت مرکب
compressor	کمپرسور
computational biology	زیست شناسی محاسبه ای
computer assisted new drug	کاربری داروی جدید به کمک
application	کامپیوتر
computer-assisted drug	طراحی دارو به کمک کامپیوتر
design	
comutagen	کوموتاژن، هم جهش زا
concatemer	زنجیره کنکاتامر/چند واحدی
concatemeric DNA	دی.ان.ای. کنکاتمریک
concatermeric DNA	دی.ان.ای. کنکاترمریک
(concatemeric DNA)	
concentration	تجمع، تراکم، تمرکز، غلظت
concomitant	توام، ملزوم، همانند، همراه
condenser	چگالنده
conditional lethal mutant	جهش یافته کشنده شرطی
configuration	آرایش، پیکر بندی، شکل،
	هیئت
confined field testing	آزمایش میدانی محدود
confocal microscopy	ریزبینی دوچشمی/دو
	کانونی/کنفوکال
conformation	انطباق، ترکیب، تطبیق،
	ساخت، شکل، صورت
	بندی، همتاشی، هم
	پیکری
congenital adernal	هایپرپلازی آدرنال مادرزادی
hyperplasia	

congenital dislocation of the hip	در رفتگی مفصل ران مادرزادی
congenital dyserythropoietic anaemia	کم خونی دیسرتروپوئیتیک مادر زادی
congenital heart disease	بیماری قلبی مادر زادی
congenital myeloperoxidase deficiency	کمبود میلو پراکسیداز مادر زادی
congenital nephrotic syndrome	سندروم/عارضه نفروتیک مادر زاد
congenital nephrotic syndrome of Finnish type	سندروم/عارضه نفروتیک مادر زاد نوع فنلاندی
congenital nephrotic syndrome of other types	سندروم/عارضه نفروتیک مادر زاد نوع دیگر
congo red	قرمز کنگوئی
conjugate	جفت/مزدوج/هم یوغ شدن
conjugated linoleic acid (CLA)	اسید لینولئیک هم یوغ/مزدوج
conjugated protein	پروتئین مزدوج
conjugating plasmid	پلاسمید هم یوغ/مزدوج
conjugation	آمیختگی، الحاق، امتزاج، جفت شدگی، هم آوری، هم یوغی
connective tissue	بافت پیوندی
consanguineous	خویش، هم خون
consensus sequence	توالی توافقی
consequence management	گرداندگی/مدیریت پی آمد(ی)
conservation	بقاء، پایندگی، محافظت، حفاظت منابع طبیعی/محیط زیست
conservation of biodiversity	حفاظت تنوع زیستی
conservation of farm animal genetic resource	حفاظت ژنتیکی دام و طیور
conservation tillage	کشت و زرع حفاظتی
conservation value	ارزش حفاظت
conserved	باقی، بایسته، حفظ شده
consistency	تداوم، ثبات، سازگاری،غلظت، همآهنگی، همخوانی
consortia	کنسرسیوم ها
consortium	کنسرسیوم

constant region	منطقه ثبات
constitutive enzyme	آنزیم ساختمانی/نهادی
constitutive gene	ژن نهادی
constitutive heterochromatin	ناجور رنگینه ساختمانی/نهادی، هترو کروماتین ساختمانی/نهادی
constitutive mutation	جهش نهادی
constitutive promoter	فعال کننده نهادی
construct	بنا، ساخت، کنستراکت
consultation	رایزنی، مشورت
consumption	مصرف
contact precaution	اقدام احتیاطی تماسی
contact rate	نرخ/میزان تماس
contact tracing	پیگیری اشاعه دهندگان/افراد ساری/ناقل، ردیابی تماسها
contact zone	منطقه تماس
contact zone element	عامل/عنصر منطقه تماس
contact zone thickness	ضخامت/قطر منطقه تماس
contagion	بیماری واگیر دار، سرایت، واگیری
contagious	مسری، ناقل، واگیر
contagious disease	بیماری مسری/واگیر
contained casualty setting	وضعیت تلفات مهار شده/تحت کنترل
contained use	کاربری کنترل شده
contained work	عمل کنترل شده
containment facility	مرکز محدود کننده/باز دارنده
containment level	سطح تنگداشت/محدودیت
contaminant	ماده آلوده کننده، نا سالم کننده، عفونی
contaminate	آلودن، آلوده کردن، سرایت کردن، واگیر داشتن
contamination	آلایش، آلودگی، عامل آلودگی، کثافت، نا پاکی/سالمی
contig map (= contiguous map)	نقشه کن-تیگ
contiguous gene	ژن پیاپی/هم جوار

English	Persian
contiguous map	نقش(ه) پیوسته/هم جوار/پیاپی
con-till (= conservation tillage)	کان-تیل، زراعت/کشاورزی حفاظتی، کشت و زرع استحفاظی
continuous perfusion	ریزش/وامیختگی پیوسته
continuous sample	نمونه پیوسته
contour-clamped homogeneous electric fields	میدانهای الکتریکی همگن کنتور-کلامپ
contraindication	مورد عدم استعمال
control measure	اقدام مهار کننده
control sequence	توالی/زنجیره بازبینی/کنترل/نظارت/کار بری/تنظیم/بازرسی
controlled release	آزادسازی مهار شده
convalescence	نقاهت، دوره نقاهت
convention	پیمان، توافق نامه، قرارداد، کنوانسیون، معاهده
Convention on Biological Diversity (CBD)	معاهده بین المللی حفظ و بهره برداری از منابع بیولوژیکی جهان
conventional pseudogene	شبه ژن قراردادی
convergent evolution	فرگشت همگرا
convergent improvement	اصلاح همگرا
conversion	بازگشت، تبدیل، تغییر، وارونگی، واگردانی
COOH (= carboxyl group)	ث.او.او.هاش.
coordinated framework for regulation of biotechnology	چارچوب هماهنگ برای تنظیم زیست فن شناسی
coordination chemistry	شیمی کو ئوردیناسیون/هم آهنگی
coordination compound	ترکیب هم آرا
coordination number	عدد هم آرایی
cooxidation	هم اکسایش
COPD (= chronic-obstructive-pulm onary disease)	سی.او.پی.دی.
copolymer	کوپلیمر
copy DNA	نسخه دی.ان.ای.
copy number	شماره نسخه (ای)، کپی نامبر
CoQ (= coenzyme Q)	
cordyceps simensis	کوردیسپ سایمنزیس/زیمنزی
core DNA	دی.ان.ای. هسته/اصلی، کور دی.ان.ای.
core enzyme	آنزیم هسته/اصلی، کور آنزیم
core octamer	هسته هشت واحدی
core promoter	راه انداز اصلی
co-repressor	کمک بازدارنده، ممانعت کننده
Cori cycle	چرخه کوری
corn	ذرت
corn borer	حشره ذرت
corn earworm (helicoverpazea = H.zea)	کرم ذرت
corn rootworm	کرم ریشه ذرت
corona	تاج، تاج خورشید، هاله ماه
coronary heart disease	بیماری قلبی کرونر
coronary thrombosis	انسداد شریان های اکلیلی، تشکیل لخته درون سرخرگهای قلب، ترومبوز کرونر، سکته قلبی
correlation coefficient	ضریب همبستگی
cortical cytoplasm	سیتوپلاسم پوسته ای/کورتیکی
corticotrophin (= corticotropin)	کورتیکوتروفین
corticotropin	کورتیکوتروپین
cortisol	کورتیزول
Cos site	جایگاه کاس
Cosmid	کازمید، کاسمید
cosuppression	هم سرکوبی
co-transduction	هم تبدیلی
co-transformation	هم تبدیلی
co-translational modification	کوترانسلیشنال مودیفیکیشن
counterterrorism	ضد تروریزم
country of origin of genetic	کشور مبداء منابع ژنتیکی

English	Persian
resource	
country providing genetic resource	کشور تامین کننده منابع ژنتیکی
covalent bond	پیوند اشتراکی
covalent coordinate bond	پیوند اشتراکی یکسویه
covariation	هم پراشی، هم وروائی، کو واریانس، واریانس مشترک
covarion	کو واریون
cover slip	روکش
covert	پنهان(ی)، نهان(ی)، نهفته
covert release	آزادسازی/انتشار/رهاسازی پنهان/مخفی
cowpea mosaic virus (CpMV)	ویروس آفت لوبیای چشم بلبلی
cowpea trypsin inhibitor	بازدارنده تریپسین از نوع لوبیا چشم بلبلی
cowpox virus	ویروس آبله گاوی
CPSF (= cleavage and polyadenylation specificity factor)	سی.پی.اس.اف.
CR (= complement receptor)	سی.آر.
C-reactive protein	پروتئین سی.-رآکتیو، پروتئین واکنش پذیر سی.
creatine	کریاتین
creatinine	کراتینین
credible threat	تهدید قابل قبول/موثق
critical breed	نژاد در خطر وخیم/بحرانی
critical micelle concentration	تجمع بحرانی میسل ها/ذرات کلوئیدی
critical value of dissolved oxygen concentration	ارزش حیاتی غلظت اکسیژن محلول
critical-maintained breed and endangered-maintained breed	نژاد در خطر بحرانی نگه داشته شده و نژاد در معرض انقراض نگه داشته شده
CRM (= chromatin remodeling machine)	سی.آر.ام.
CRM (= cross reacting material)	سی.آر.ام.
Cro repressor	بازدارنده/مهارکننده کرو
crop	چینه دان، محصول
Croplands equipment	تجهیزات/لوازم کراپلند
cross reacting material	ماده واکنشگر متقاطع
cross reaction	واکنش چلیپائی/متقابل/متقاطع
cross reactivity	واکنش گری/پذیری متقاطع
cross tolerance	تحمل متقاطع
cross-hybridization	آمیختگی/هم تیرگی متقاطع
cross-infection	عفونت متقاطع
crossing of sibling	تقاطع هم نیا ئی
crossing-over	تبادل، تبادل ژنتیکی، تقاطع کروموزومی، تقاطع و تبادل، کراسینگ اور
crossover region	منطقه کراس اوور
crossover suppressor	سرکوبگر کراس اوور
cross-pollination	گرده افشانی چلیپائی/متقاطع
cross-reacting antibody	آنتی بادی واکنشگر متقاطع
crown gall	ساییدگی تاج، کراون گال
CRP (= catabolite activator protein)	سی.آر.پی.
cruciferae	چلیپائیان، گلمیان
Crusader Garden Powder	کود/گرد باغچه کروسیدر
Cry Proteins	پروتئین(های) کرای
cryobiology	زیست شناسی سرما
cryogenic storage	نگه داری سرد خانه ای/کرایوژنیک
cryptic prophage	پروفاژ/پیش باکتری خوار نهان ساز
cryptobiosis	کریپتوبایوسیز، کریپتوبیوز
crystallization	بلور شدگی، تبلور، متبلور سازی
crystalloid	بلور مانند/نما
CS (= Cockayne syndrome)	سی.اس.
CstF (= cleavage stimulation factor)	
CTD-associated SR-like protein	پروتئین شبه اس.آر. وابسته به سی.تی.دی.
CTL (= cytotoxic T lymphocytes)	سی.تی.ال.
CTP (= cytidine triphosphate)	سی.تی.پی.

cultivar	پرورده، پروره، رقم، زراعی/کشته شده، کشته، واریته
cultivated species	گونه زراعی/کشت شده
cultural diversity	تنوع کشت/مزروعی
culture	پرورش، رویاندن، کشت، کشت دادن، محیط کشت
culture medium	محیط کشت
culture preservation	نگهداری کشت
cultured cell	سلول کشتی/کشت شده
cumulative feedback inhibition	سرکوب فیدبک تراکمی، مهار بازخورد تراکمی
cumulative radiation dose	دوز پرتوئی تراکمی
curative	درمان، درمان بخش، درمانی
curcumin	کورکومین
curing agent	عامل درمان کننده
current good manufacturing practices	شیوه های عمل مناسب کنونی
cutaneous anthrax	سیاه زخم پوستی
CVS (= chorionic villus sampling)	سی.وی.اس.
cyanobacteria	آبی ترکیزگان، جلبکهای سبز-آبی، سیانوباکتری
cyanogen	سیانوژن
cyanogen bromide	برومید سیانوژن
cyanosis	سیانوز، کبودی پوست
cyanotic	سیانوزی، مربوط به کبودی پوست
cybrid	سایبرید
cyclic	ادواری، چرخه ای، دوره ای
cyclic 3e, 5e-adenosine monophosphate	3ئی. 5ئی.-آدنوزین مونوفسفات چرخه ای
cyclic adenosine monophosphate	ادنوزین مونو فسفات چرخه ای/متناوب/سیکلیک/گرد شی
cyclic AMP (= cAMP)	آدنوزین مونوفسفات حلقوی
cyclic phosphorylation	فسفری شدن چرخه ای
cyclin	سیکلین
cyclin dependent protein kinase	آنزیم/کیناز پروتئین متکی به سیکلین
cyclin-dependent kinase	کیناز وابسته به سایکلین

cyclin-dependent kinase inhibitor	بازدارنده کیناز وابسته به سیکلین
cyclobutyl dimer	دیمر سیکلو بوتیل
cyclodextrin	دکسترین چرخه ای، سیکلودکسترین
cycloheximide	سایکلوهگزیمید، سیکلوهگزیمید
cyclooxygenase	آنزیم سایکلو اکسیژن، سایکلو اکسیژناز
cyclophosphamide	سایکلوفسفامید
cycloserine	سایکلوسرین
cycloserine enrichment	تقویت سایکلوسرین
cyclosporin	سایکلو اسپورین
cyclosporine (=cyclosporin)	
Cys (= cysteine)	
cyst	حباب، سیست، کیست، کیسه
cystatin	سیستاتین
cysteine	سیستئین
cystic fibrosis	تصلب کیستی بافتها، سیستیک فیبروز، فساد کیستی الیاف، فیبروز سیستی
cystic fibrosis transmembrane conductance regulator	تنظیم کننده رسانائی بین غشائی فیبروز سیستی
cystine	سیستین
cystinosis	سیستینوز
cystinuria	سیستینوری
cystitis	التهاب مثانه، سیستیت
CystX	سیست ایکس
Cyt C (= cytochrome C)	
cytidine diphosphate	سایتیدین دیفسفات
cytidine monophosphate	سایتیدین مونوفسفات
cytidine triphosphate	تریفسفات سایتیدین
cytochalasin	سیتوکالازین
cytochemistry	سیتوشیمی، شیمی سلولی
cytochrome	سیتوکروم، یاخته رنگ
cytochrome C	سیتوکروم سی.
cytogenetics	توارث یاخته، ژنتیک یاخته

196

	ای/سلولی، سیتوژنتیک
cytokines	سیتوکین ها
cytokinesis	سیتوکینزیس
cytokinetics	سیتوکینتیک
cytology	سلول/یاخته شناسی، سیتولوژی
cytolysin	سیتولایزین، سیتولیزین
cytolysis	سیتولیز، یاخته کافت (ی)
cytomegalovirus	یاخته کلان ویروس
cytopathic	آسیب سلولی
cytoplasm	سیتوپلاسم، میان مایه/یاخته
cytoplasmic	میان یاخته ای
cytoplasmic DNA	دی.ان.ای. سیتوپلاسمی
cytoplasmic inheritance	وراثت سیتوپلاسمی، وراثت غیر مندلی/غیر هسته ای، وراثت مادری
cytoplasmic membrane	غشاء سیتوپلاسمی
cytoplasmic vesicle	ریزکیسه سیتوپلاسمی
cytoplasmically (= cytoplasmic)	
cytoplast	سیتوپلاست
cytosine	سیتوزین
cytosine arabinoside	سیتوزین آرابینوزید
cytoskeleton	اسکلت سلولی
cytosol	سیتوزول
cytostatic agent	عامل سیتواستاتیکی
cytotoxic	سم سلولی، سیتوتوکسیک، سیتوتوگزیک، مسموم کننده سلول، یاخته آزار
cytotoxic cell	سلول سیتوتوکسیک
cytotoxic killer lymphocyte	لنفوسیت کشنده سم سلولی
cytotoxic T lymphocytes	لنفوسیت های سیتوتوگزیک نوع تی.
cytotoxicity	سیتوتوکسیسیته، یاخته آزاری

D

D (= Dalton)	دی.
d (= deoxy)	
D.V (= DNA vaccine)	دی.وی.
DA (= Dalton)	دی.ای.
dAdo (= deoxy adenosine)	
daffodil	گل نرگس (زینتی)
daffodil rice	برنج دافودیل، دافودیل رایس
DAG (= diacylglycerol)	دی.ای.جی.
daidzein	دایدزین
daidzen	دایدزن
daidzin	دایدزین
Dalton	دالتون
DASH (= dynamic allele-specific hybridization)	دی.ای.اس.اچ.
data	اطلاعات، داده (ها)
data collection	گرد آوری داده ها
data mining	داتا ماینینگ، داده‌کاوی
daughter	دختر، دخترانه، دختری
daunomycin	دانومایسین
daunorubicin	دانوروبیسین
DB (= dot blot)	دی.بی.
DBC (= double-stranded DNA binding site)	دی.بی.سی.
DBP (= DNA-binding-protein)	دی.بی.پی.
DCR (= dominant control regions)	دی. سی.آر.
dCTP (= deoxy cytosine triphosphate)	
dd (= dideoxy)	
dd NTP (= dideoxynucleoside triphosphate)	
DE (= DNA extraction and quality control)	دی.ئی.
de novo	نو پدید
de novo sequencing	شناسایی فی البداهه/بدون پیش آگاهی/اطلاع قبلی

de novo synthesis	سنتز نوپدید/دینوو
DEAE (= diethylamino-D ethanol)	دی.ئی.ای.ئی.
deafness	کری، ناشنوایی
deagglomeration	تراکم زدایی
deamidation	آمید زدایی، بی آمید شدن
deaminase	دآمیناز
deamination	آمین زدایی، دآمیناسیون
DEB (= mutagen dipoxybutane)	دی.ئی.بی.
decantation	سرریزکنی
decanter	سرریزکن
decanting	سرریزکردن
decarboxylase	دکربوکسیلاز
decay	پوسیدگی، فاسد شدن، گندیدن
decontamination	آلودگی زدایی، پاکیزه سازی، رفع آلودگی، ضد عفونی
decontamination kit	کیت ضد عفونی، وسایل رفع آلودگی
defective virus	ویروس معیوب/ناقص
defensins	دفنسین/دیفنزین ها
deficiency	عیب، کمبود، نارسایی، نقص، نقصان
defined medium	واسطه/محیط کشت تعریف شده
degenerate codon	کدون دژنره
degeneration	دژنراسیون
degeracy	دژراسی
dehydration	آب برداری، آب زدایی، از دست رفتن آب بدن، پسابش
dehydrogenases	دهیدروژنازها
dehydrogenation	هیدروژن/ئیدروژن زدایی/گیری، دهیدروژناسیون
dehydroxyphenylalanine	د هیدروکسی فنیل آلانین
deinococcus radiodurans	داینوکوکوس رادیودورانس، رادیو دورانس د اینوکوک
de-ionized water	آب دیونیزه شده
Del (= deletion)	

Delaney clause	بند/ماده دیلانی
delayed hypersensitivity	پر حساسیتی عقب/به تاخیر افتاده، هایپر سنسیتیویته با تاخیر
delayed-type hypersensitivity reaction	واکنش فوق حساسیتی تاخیر دار
deletion	حذف
deletion mutation	جهش حذفی
deliberate release	آزاد سازی عمدی
delivery	پرتاب، تحویل، رهایی، زایمان، نجات
delta endotoxin	اندوتوکسین دلتا، درون زهرابه دلتا
demand	تقاضا، نیاز
demography	آمار نگاری، جمعیت شناسی
denaturant gradient gel electrophoresis	الکتروفورز ژل با شیب تقلیب کننده، دناتورانت گرادیان ژل الکتروفوروز
denaturation	تخریب، تقلیب، قلب ماهیت، واسرشتی
denaturation DNA	دی.ان.ای. تقلیب، دی.ان.ای. دناتوره
denature	قلب ماهیت کردن، واسرشتن
denatured DNA	دی.ان.ای. واسرشته/تغییر یافته/تقلیب شده
denaturing gradient gel electrophoresis	الکتروفوروز ژل تغییر ماهیت دهنده گرادیان، دینیچرینگ گریدینت ژل الکتروفورسیز
denaturing polyacrylamide gel electrophoresis	الکتروفوروز ژل تغییر ماهیت دهنده پلی آکریلامید
dendrimer	درختپار، دندرایمر
dendrite	دارینه، دندانه، دندریت، شجری
dendritic	دارینه ای، درختواره، دندریتی، شاخه دار، شجری
dendritic cell	سلول دندرایتی/دندریتی
dendritic evolution	فرگشت دندریتی
dendritic langerhans cell	سلول لانگر هانس دندریتی/دندرایتی

dendritic polymer	پلی مر های دندریتی/دندرایتی
dengue fever virus	ویروس تب دانگ
denitrification	احیای نیترات، شوره برداری
denitrifying bacteria	باکتری های شوره زدا
densitometer	تراکم/چگالی/غلظت سنج
density	تراکم، چگالی، غلظت
density gradient centrifugation	سانتریفوژ با شیب چگالی، مرکز گریزش به روش تغییرات چگالی، میان گریزش شیب چگالی
dentato rubral-pallidolusyian atrophy	آتروفی دنتاتو روبرال-پالائیدولوزیان
dentifrice	خمیر دندان، دندان شوی
dentinogenesis imperfecta	دنتینوژنسیس ایمپرفکتا
dentogram	دنتوگرام، نقشه دندانها/دهان
dentrite	دارینه
deoxy	داکسی
deoxy adenosine	داکسی آدنوزین
deoxy cytosine triphosphate	داکسی سیتوزین تری فسفات
deoxy thymidine di phosphate	داکسی تیرامیدین دی فسفات
deoxy thymidine mono phosphate	داکسی تیرامیدین مونو فسفات
deoxy thymidine tri phosphate	داکسی تیرامیدین تری فسفات
deoxynivalenol	دزوکسی نیوا لنول
deoxynuclotid	د اکسی نوکلوتید
deoxyribo nucleic acid	اسید داکسی ریبونوکلئیک
deoxyribonuclease	داکسی ریبونوکلئاز
deoxyribonucleic acid (DNA)	دزوکسی ریبونوکلئیک اسید
deoxyribonucleoprotein	پروتئین داکسی ریبونوکلئیک
deoxyribonucleotide	دزوکسی ریبو نوکلئوتید
deoxyribose	دزوکسی ریبوز
deoxyribovirus	داکسی ریبوویروس
depolarization	واقطبش
depolarize	واقطبیدن
depolarized	واقطبیده
depolarizer	واقطبنده
depression	افت، افسردگی، تنزل، ضعف، فرورفتگی، گودی
deprotection	بی پناه کردن، بی حفاظ کردن دیپروتکشن
deproteinization	دپروتئینیزاسیون
Der (= derivative chromosome)	
derepression	دی ریپرس کردن، دی ریپرشن، رفع سرکوب
derivative chromosome	کروموزوم اشتقاقی
derived	گرفته/مشتق شده، ماخوذ
dermal absorption	جذب از راه پوست، جذب جلدی
dermal adsorption	جذب سطحی پوستی، رو نشینی پوستی
dermal contact	تماس پوستی/جلدی
dermal penetration	نفوذ جلدی/از راه پوست
desaturase	دساچوراز
desensitization	دسنسیتیزاسیون، حساسیت زدائی
Desert Hedgehog Protein (Dhh)	پروتئین خارپشت بیابانی/کویری
desferroxamine manganese	دزفرواکسامین منگنز
desiccator	خشکانه
desulfovibrio	دسولفو ویبریو
determinant	تعیین کننده، دترمینان، گمارنده
deterministic analysis	تحلیل جبرباورانه/گمارشی
deterministic effect	تاثیر جبری/گمارشی
detoxication	دتوکسیکیشن، سم زدائی
development value	ارزش پیشبرد/توسعه
device	ابزار، تدبیر، دستگاه، علامت، وسیله
Devilbiss	دویل بیس
dextran	دکستران
dextrorotary	راست بر، راست گرد، هم جهت عقربه ساعت
DGI1 (= dentinogenesis imperfecta)	دی.جی.آی. 1
DHAP (= di hydoxy acetone	دی.اچ.ای.پی.

English	Persian
phosphate)	
DHEA (= di-hydro epiandrosterone)	دی.اچ.ئی.ای.
DHF (= di-hydro folate)	دی.اچ.اف.
DHFR (= di-hydro folate reductase)	دی.اچ.اف.آر.
Dhh (= Desert Hedgehog Protein)	دی.اچ.اچ.
DHPR (= di-hydro pteridine reductase)	دی.اچ.پی.آر.
DHT (= di-hydro testosterone)	دی. اچ. تی.
DHTR (= di-hydro testosterone receptor)	دی.اچ.تی.آر.
di hydoxy acetone phosphate	دی هیدروکسی استون فسفات
Dia (= diakinesis)	
diabetes	بیماری قند، دیابت
diabetes mellitus optic atrophy deafness	بیماری/سندروم/عارضه ولفرام
diacylglycerol (DAG)	دیاسیلگلیسرول
diadzein (= daidzein)	دیادزین
diagnostic procedure	فرآیند تشخیصی (طبی)
diagnostis	دیاگنوستیز، مربوط به تشخیص طبی
diakinesis	دیاکینزیس
dialysis	تراکافت (ی)، تجزیه، دیالیز
dialysis culture	کشت دیالیزی
diamond	الماس، لوزی
diapedesis	تراگذری
diastereoisomer	دیاستریوایزومر
diauxy	دیا اوکسی
Dic (= dicentric)	
dicentric	دیسنتریک
dicer enzyme	آنزیم خردکننده/دایسر
dideoxy	دیداکسی
dideoxynucleoside triphosphate (dd NTP)	دیداکسی نوکلئوزید تریفسفات
dideoxynucleotide	دی داکسی نوکلئوتید
didN (= dideoxynucleotide)	
dielectrophoresis	دی الکتروفوروز

English	Persian
diethylamino-D ethanol	دی اتیل آمینو-دی. اتانول
differential centrifugation	میان گریزش/سانتریفوژاسیون دیفرانسیلی
differential display	نمایش دیفرانسیلی
differential labeling	برچسب گذاری دیفرانسیلی
differential medium	محیط کشت/واسطه دیفرانسیلی
differential splicing	بهم تابیدگی دیفرانسیلی
differentiation	تفکیک، تمایز، تمایز یابی، جدایش
diffusion	اختلاط، انتشار، پخش، پراکندگی، نفوذ، نشت
digest	گواریدن، هضم کردن/شدن
digestion	گوارش، هضم
diglyceride	دی گلیسرید
dihybrid	دی هیبرید
dihybrid cross	آمیزش دی هیبرید
di-hydro epiandrosterone	دی-هیدرواپیآندروسترون
di-hydro folate	دی.- هایدروفولات
di-hydro folate reductase	دی.- هایدروفولات رداکتاز
di-hydro pteridine reductase	دی.- هایدرو پتریدین رداکتاز
dihydro pyrimidine dehydrogenase	دی هایدروپیریمیدین دهیدروژناز
di-hydro testosterone	دی.-هایدروتستوسترون
di-hydro testosterone receptor	گیرنده دی.-هایدروتستوسترون
dihydrofolate reductase	دی هایدروفولات رداکتاز
dihydroxyphenylalanine	دی هیدروکسی فنیل آلانین
dilatation	اتساع، انبساط
dilution	تضعیف، رقت، رقیق سازی/شدگی/کردن
dimer	دیمر
dinitrophenol	دینیتروفنول
dioctadectylamidoglycylsper mine tetra fluoroacetic acid	اسید دایوکتادکتیلآمیدوگلیسیل سپرمین تترا فلوئوروواستیک
Dip (= diplotene)	
Dipel	دیپل
dipeptidyl-peptidase	دی پپتیدیل-پپتیداز

disk	دیسک، صفحه، صفحه گرد	diphtheria antitoxin	پادزهر دیفتری
disomy	دایزومی	diploid	تمام دانه، توام، دو برابر،
dispenser	شخص/دستگاه پخش/توزیع		دوتایی، دولا، دولاد،
	(کننده)		دیپلوئید، زوجی
dispensing	توزیع	diploid cell	سلول دوتایی/دیپلوئید
dispersal	انتشار، پخش، پراکندگی	diplophase	دولاد چهر، دیپلوفاز
dispersion	انتشار، پراکندگی، پراکنش	diplotene	دیپلوتن
displacement loop	دیسپلیسمنت لوپ، حلقه	dip-pen lithography	دیپ-پن لیتوگرافی، نقش
	جابجائی		زدن/نگارش نوک قلمی
disposal	انهدام، خنثی سازی،	dip-pen nanolithography	دیپ-پن نانولیتوگرافی، نانو
	دسترسی، دفع، دور ریزی		نقش زدن/نگارش نوک
disruptive selection	انتخاب گسلنده، گزینش		قلمی
	گسسته	diptheria toxin	زهر/سم دیفتری، دیفتریا
dissection	تشریح، کالبد شکافی		توکسین
disseminated intravascular	انعقاد درون آوندی/رگی		
coagulation	گسترده	Dir (= direct)	
disseminating	اشاعه دادن، پخش/منتشر	direct fluorescent antibody	پادتن پرتو افشان مستقیم
	کردن	direct repeat	تکرار مستقیم
dissemination	اشاعه، انتشار، پخش، ترویج	direct transfer	انتقال مستقیم
disseminator	اشاعه دهنده، ترویج/منتشر	direct use value	ارزش استفاده مستقیم
	کننده	directed evolution	تکامل/فرگشت جهت
dissimilation	دگرگون سازی،		دار/هدایت شده
	دیسسیمیلیشن، ناشبیه	directed mutagenesis	جهش زایی هدایت شده
	ساختن، ناهمانند سازی،	directed self-assembly	خود گردایش هدایت شده
	ناهمگون سازی	directional cloning	تاک سازی جهت یافته،
dissociating enzyme	آنزیم جدا ساز/انفکاکی		همسانه سازی جهت دار
dissociation	تفکیک، جدایش، جدایی،	directional selection	انتخاب جهت دار
	شکند، واپاشی	dirty	کثیف، کثیف شدن/کردنکننده
dissolved oxygen probe	پرب/پروب/ردیاب/کاووشگر/ز	dirty bomb	بمب کثیف
	شانگر اکسیژن محلول	Dis (= distal)	
distal	انتهایی/دور	disabled strain	جوره/رقم/سویه/واریته
distribution	پخش، تقسیم، توزیع		خنثی شده
disulfide bond	پیوند دی سولفید	disaccharide	دوقندی، دی ساکارید
disulphide bond	پیوند دو گوگردی	disarmed DNA	دی.ان.ای. خنثی شده
disulphydryl bridge	پل دی سولفیدریل	disaster planning	برنامه ریزی بلایا/حوادث
diversity	اختلاف، تفاوت، تنوع،	discriminator	تبعیض کننده، فرق گذار
	گوناگونی	disease outbreak	شیوع بیماری
diversity biotechnology	کنسرسیوم تنوع زیست	disease transmision	انتقال بیماری
consortium	فناوری/بیوتکنولوژی	disinfectant	گند زدا، ماده ضد عفونی
diversity estimation	برآورد تنوع		کننده
DM (= myotonic dystrophy)	دی.ام.	disinfection	ضد عفونی، گندزدایی
		disjunction	دیسجانکشن

English	Persian
DMAD (= diabetes mellitus optic atrophy deafness)	دی.ام.ای.دی.
DMD (= Duchenne muscular dystrophy)	دی.ام.دی.
Dmin (= double minute)	
DNA (= deoxyribo nucleic acid)	دی.ان.ای.
DNA analysis	تحلیل دی.ان.ای.
DNA b helicase	هلیکاز بی. دی.ان.ای.
DNA backbone	بکبون/رکن/محور دی.ان.ای.
DNA bank	مخزن دی.ان.ای.
DNA bending	تاشدگی دی.ان.ای.
DNA bridges	پل های دی.ان.ای.
DNA chimera	بافت نا همسان دی.ان.ای.، کژزاد دی.ان.ای.
DNA chip	تراشه دی.ان.ای.
DNA diagnosis	آزمایش/تشخیص دی.ان.ای.
DNA extraction and quality control	نظارت بر کیفیت و استخراج دی.ان.ای.
DNA fingerprint	دی.ان.ای. فینگرپرینت
DNA fingerprinting (DNA profiling)	تشخیص اثر انگشت دی.ان.ای. ئی، دی.ان.ای. فینگرپرینتینگ
DNA fragmentation	خرد/ریز/قطعه قطعه کردن دی.ان.ای.
DNA glycosylase	آنزیم گلیکوسیل دی.ان.ای.، دی.ان.ای. گلیکوسیلاز، گلیکوسیل دی.ان.ای.آنزیم
DNA gyrase	دی.ان.ای. ژیراز
DNA helicase (gyrase)	دی.ان.ای. هلیکاز
DNA ligase	دی.ان.ای. لیگاز
DNA marker	نشانگر دی.ان.ای.
DNA melting temperature	دمای ذوب دی.ان.ای.
DNA methylase	آنزیم متیله کننده دی.ان.ای.، دی.ان.ای. متیلاز
DNA methylation	متیلاسیون دی.ان.ای.، متیل دار کردن/شدن دی.ان.ای.، متیله شدن دی.ان.ای.
DNA microarray	ریز آرایه دی.ان.ای.
DNA polymerase	دی.ان.ای. پلیمراز
DNA polymerase I	دی.ان.ای. پلیمراز 1
DNA polymerase III	دی.ان.ای. پلیمراز3
DNA polymorphism	چند ریختی/شکلی شدن دی.ان.ای.
DNA primase	پریماز دی.ان.ای.
DNA probe	پرب/پروب/ردیاب/کاووشگر/ز شانگر دی.ان.ای.
DNA profiling (DNA fingerprinting)	تهیه نما/طرح دی.ان.ای.
DNA repair	ترمیم دی.ان.ای.
DNA replication	رپلیکاسیون دی.ان.ای.، همانند سازی دی.ان.ای.
DNA sequence	توالی دی.ان.ای.
DNA sequencing	توالی یابی دی.ان.ای.
DNA shuffling	بر خوردن/زدن دی.ان.ای.، شافل دی.ان.ای.
DNA synthesis	سنتز دی. ان.ای.
DNA topoisomerase	دی.ان.ای.-توپوایزومراز
DNA typing	تشخیص تیپ دی.ان.ای.، تشخیص هویت از راه آنالیز دی.ان.ای.، شناسائی از روی تیپ دی.ان.ای.
DNA vaccine	واکسن دی.ان.ای.
DNA vector	حامل دی.ان.ای.
DNA-binding protein	پروتئین چسبنده/اتصال دی.ان.ای.
DNA-dependent rna polymerase	پلیمراز آر.ان.ای. وابسته به دی.ان.ای.
DNase (= deoxyribonuclease)	دی. ان.آز
DNAse (= deoxyribonuclease)	دی.ان.آز
DNP (= dinitrophenol)	دی.ان.پی.
docosahexanoic acid (DHA)	اسید دوکوساهگزانوئیک
DOGS (= dioctadectylamidoglycyls permine tetra fluoroacetic acid)	دی.او.جی.اس.

domain	حوزه، زمینه، قلمرو، محدوده، ناحیه	complementary DNA (dscDNA)	ای
domain of protein	دامین/دومین/قلمرو پروتئین	double-stranded DNA binding site	محل اتصال دی.ان.ای. دو رشته ای
domestic animal diversity	تنوع حیوانات اهلی	double-stranded RNA adenosine deaminase	آدنوزین دآمیناز آر.ان.ای. دو رشته ای
domestic biodiversity	تنوع زیستی حیوانات اهلی	down processing	روند کاهشی،
domesticate	اهلی کردن		بازداری/سرکوب واکنش
domesticated species	گونه های اهلی شده		عادی
domestication	اهلی سازی، اهلی/خانگی/رام کردن	down promoter mutation	جهش/موتاسیون داون پروموتر، جهش کم کننده نسخه برداری
dominance	قانون بارزیت/برتری/غلبه	down regulating	سرکوب واکنش عادی یک اندام یا سیستم
dominant	بارز، برتر، غالب		
dominant allele	آلل غالب، دگره بارز	down regulation	سرکوب واکنش عادی
dominant control regions	مناطق کنترل/مهار غالب	Down's syndrome	سندروم داون
dominant gene	ژن غالب	downstream	پائین/فرو دست، داون استریم
dominant(-acting) oncogene	ژن توموری (-فعال) غالب	doxycycline	دوکسی سیکلین
donor	اهدا کننده، بخشنده، دهنده	DPD (= dihydro pyrimidine dehydrogenase)	دی.پی.دی.
donor DNA	دی.ان.ای. دهنده/دونر		
donor junction (acceptor junction site)	اتصال دهنده	DPI	دی.پی.آی.
donor splicing site	ناحیه اسپلایس دونر/دهنده	drift	بی مقصد حرکت کردن، جریان
DOPA (= dihydroxyphenylalanine)	دی.او.پی.ای.	droplet	چکه، ریز قطره، قطرک
dopastin	دوپازتین	drosophila	مگس سرکه/میوه
dormancy	خاموشی، عدم فعالیت، مغلوبیت، نهفتگی	drought tolerance	تحمل خشکسالی
dormant seeding	کشت خفته	drought tolerance trait	ویژگی تحمل خشکسالی
dosage compensation	جبران دوز(ی)	DRPA (= dentato rubral-pallidolusyian atrophy)	دی.آر.پی.ای.
dose	دوز، وعده، مقدار، مقدار مصرف		
dosimetry	دوزیمتری	drug design	طرح دارو
dot blot	دات بلات	drug interaction	تاثیر متقابل دارو
dot blotting	دات بلاتینگ	drug resistance	مقاومت به دارو
dot hybridization	هم تیرگی داتی	drug tolerance	تحمل دارو
dot-blot	دات بلات	drum dryer	خشک کننده بشکه ای/طبلی
double cropping	کشت دوگانه		
double helix	مارپیچ دوگانه/مضاعف	drum filter	فیلتر بشکه ای/طبلی
double minute	دابل ماینیوت، ماینیوت دوبل/مضاعف	dsRAD (= double-stranded RNA adenosine deaminase)	
double- stranded RNA binding domain	منطقه اتصال آر.ان.ای. دو رشته ای		
double-stranded	دی.ان.ای. مکمل دو رشته		

dsRBD (= double- stranded RNA binding domain)	
dTDP (= deoxy thymidine di phosphate)	
DTH (= delayed-type hypersensitivity reaction)	دی.تی.اچ.
dTMP (= deoxy thymidine mono phosphate)	
dTTP (= deoxy thymidine tri phosphate)	
Duchenne muscular dystrophy (DMD)	دیستروفی عضلانی نوع دوشن
Dup (= duplication)	
duplex	دو بخشی، دوتایی، دوقلو
duplex DNA	دی.ان.ای دوبخشی
duplication	دوپلیکاسیون، مضاعف سازی
dura mater	سخت شامه
dust	سم پاشیدن، غبار، گرد، گرد گیری
duster	داستر، سم پاش
dyad	دایاد
Dynafog	داینافاگ
dynamic allele-specific hybridization	هم تیرگی دینامیکی مختص آلل
dynamics	پویایی، پویایی شناسی، دینامیک
dynein	داینئین
dynorphin	داینورفین
dysgenic	دیسژنیک
dystrophication	دیستروفی شدن، دیستروفیکیشن

E

e. coli (= Escherichia coli)	ئی.کولای
e. coli k12	ئی. کولای کی.12
E1B (= E1B protein of adenovirus)	ئی.1 بی.
E1B protein of adenovirus	پروتئین ئی. 1 بی. آدنوویروس
E2F (= transcription factor)	ئی.2. اف.

early development	رشد زودهنگام
early gene	ژن پیشین
early protein	پروتئین پیشین
earthworm	کرم خاکی
EBV (= Epstein Barr virus)	ئی.بی.وی.
ECD (= endocardial cushion defect)	ئی.سی دی.
ecdysone	اکدیزون
ecological resilience	برگشت پذیری بوم شناختی
ecological succession	انعطاف پذیری بوم شناختی، توالی بوم شناختی
ecology	اکولوژی، بوم سازگاری/شناخت/شناس ی
ecosystem	اکوسیستم، بوم سازگان/سامانه/نظام، سازگان بوم شناختی،
ecosystem rehabilitation	باز پروری بوم سازگان/زیستگاه، بازیابی اکوسیستم
ecosystem restoration	باز سازی بوم سازگان
ecosystem service	مراقبت از بوم سازگان
ecotourism	اکوتوریزم، بوم گردشگری/جهانگردی
ecotoxicology	بوم سم شناسی
ectoderm	برون پوست
ectodermal	برون پوستی
ectodermic	برون پوستی
ectoplasm	اکتوپلاسم
ectotrophic mycorrhiza	میکوریزای اکتوتروفیک
ectrodactyly-ectodermal dysplasia-clefting	دیسپلازی-شکاف اکتروداکتیلی-اکتودرمال
ectromelia	اکتروملی، سقط اندام
edaphic factor	عامل ادافیک
edge effect	اثر حاشیه ای
edible vaccine	واکسن خوراکی
EDS (= Ehlers-Danlos syndrome)	ئی.دی.اس.
educator	آموزگار، مربی، معلم
Edward syndrome	سندروم ادوارد
EEC (=	ئی.ئی.سی.

ectrodactyly-ectodermal dysplasia-clefting)	
EEO (= electro-end-osmosis value)	ئی.ئی.او.
EF (= elongation factor)	ئی.اف.
EF1 (= elongation factor 1)	ئی.اف.1.
effector	اثر کننده، موثر
effector molecule	مولکول اثر کننده
efficacy	اثر/فایده بخشی، قدرت تاثیر
EGF (= epidermal growth factor)	ئی.جی.اف.
Ehlers-Danlos syndrome	سندروم اهلرز-دانلوس
eicosanoid	ایکوزانوئید
eicosapentaenoic acid	اسید ایکوزاپنتا انوئیک
eicosapentanoic acid	اسید ایکوزاپنتانوئیک
eicosatetraenoic acid	اسید ایکوزاتترا انوئیک
elastase	الاستاز
elastin	الاستین
electrochemical	الکتروشیمیائی، برق شیمیائی
electrochemistry	الکتروشیمی، برق شیمی
electrodialysis	الکترودیالیز
electro-end-osmosis value	ارزش الکترو-اند-اسمز(ی)
electrofusion	الکتروفیوژن
electrolyte	الکترولیت، برق کافته
electron carrier	حامل/ناقل الکترون (ی)
electron microscopy	ریزبینی/میکروسکوپی الکترونی
electron paramagnetic resonance	رزونانس پارامغناطیسی الکترونی
electron transport chain	زنجیره انتقال الکترون
electronegative	الکترومنفی، الکترون کشان
electronegativity	الکترون کشانی
electropermeabilization	ریزتراواسازی
electrophoresis	الکتروفورز، برق بری
electroporation	الکتروپوریشن
electroporesis	الکتروپورزیس
electrostatic	الکترواستاتیک، ساکن الکتریسیته ای
ELISA (= enzyme-linked-immunoso	ئی.ال.آی.اس.ای.
rbent assay)	
elite germplasm	جرم پلاسم نخبه/ممتاز/الیت/برگزیده
ellagic acid	الاژیک اسید
ellagic tannin	الاژیک تانن
elongation factor	عامل تداوم
elongation factor 1	عامل دراز شدن 1
elution	الوشن
EM (= electron microscopy)	ئی.ام.
embedding	جا دادن، جاسازی کردن، فرو کردن، نشاندن
embryo	جنین، رویان
embryo rescue	نجات جنین
embryology	جنین/رویان شناسی
embryonic stem cell	سلول پایه جنینی، یاخته بنیادی/دودمانی جنینی
emergency	اضطرار، حالت فوق العاده، سانحه، فوری، وضع اضطراری
empricial	تجربی
EMS (= ethyl methane sulfonate)	ئی.ام.اس.
emulsion	امولسیون، پیمایه، شیرمایه
enantiomer	انانتیومر، ایزومر نوری
enantiopure	انانتیوپیور، خالص انانتیومری
encapsidation	قرار دادن/گرفتن در روپوش/روکش/پوسته
encapsulated	چکیده، در بر گرفته، کپسولی
encephalopathic	انسفالوپاتیک
end product	محصول نهائی
endangered breed	نژاد به خطر افتاده
endangered-maintained breed	نژاد در معرض انقراض نگه داشته شده
endemic	بیماری بومی، شایع/رایج در محیطی خاص، محلی
endergonic reaction	واکنش انرژی خواه/اندرگونیک/اندوترمیک
end-labelled nucleic acid	اسید نوکلئیک برچسب خورده در انتها
endocardial cushion defect	عارضه بالشتک آندوکاریال

endocardium	درون شامه قلب
endocrine gland	غده درون ریز/اندوکرین
endocrine hormone	هورمون درون ریز
endocrinology	درون ریز شناسی، هورمون شناسی
endocytosis	درون بری/یاختگی
endoderm	درون پوست
endodermal	درون پوستی
endodermic	درون پوستی
endoenzyme	اندوآنزیم
endogamy	اندوگامی
endogenous retro-virus	رتروویروس آندوژنوس
endoglycosidase	اندوگلیکوسیداز
endometrium	آندومتر، جدار زهدان/رحم
endonuclease	اندونوکلئاز
endopeptidase	اندوپپتیداز
endophyte	درون رست/روی
endophytic	درون رست
endophytism	درون رستی
endoplasmic reticulum (ER)	شبکه درونی سیتوپلاسم
endorphin	اندورفین
endosome	اندوزوم، درون تن
endosperm	اندوسپرم، خورش، درون دانه
endospore	اسپور داخلی، اندوسپور، درون هاگ
endostatin	اندوستاتین
endothelial cell	یاخته درون پوشه ای/اندوتلیال
endothelial nitric oxide synthase	سینتاز اکسید نیتریک اندوتلیال
endothelin	اندوتلین
endothelium	اندوتلیوم، بافت توپوشی، توپوش، درون پوش
endothelium-derived	مشتق شده از بافت توپوشی/اندوتلیوم
endotoxic	اندوتوکسیک
endotoxin	اندوتوکسین، درون زهر/زهرابه، زهرابه داخلی
engineered antibody	آنتی بادی/کور مهندسی شده، پادتن/زهر مهندسی شده

engineering	مهندسی کردن
enhanced nutrition crop	محصول غذائی تقویت شده
enhancement	افزایش، بهبود، تقویت
enhancer	تقویت گر، جلوبرنده
enhancer element	عنصر تقویت گر
enkephalin	انکفالین
enolpiruvil shikimate	انولپایروویل/انولپیروویل شیکیمیت
enolpyruvil shikimate	انولپایروویل/انولپیروویل شیکیمیت
enoyl-acyl protein reductase	انول-آسیل پروتئین رداکتاز
enriched medium	محیط کشت غنی شده
ensiling	آنسیل کردن، انسایل کردن، انسایلینگ
enterocyte	آنتروسیت، انتروسیت
enterotoxin	انتروتوکسین، روده زهرابه
entrainment	انترینمنت، با خود بردن/بری
entrapped cells	سلول/یاخته های در تله/گیر افتاده
entro-toxigenic E.coli	ئی. کولای آنتروتوکسیژنیک
Enviromist™	انوایرومیست
environment	محیط زیست
environment impact assessment	پیامد سنجی
environmental biotechnology	بیو تکنولوژی محیط زیست، زیست فناوری زیست محیطی
environmental etiological agent	عامل سبب شناختی محیطی
environmental factor	عامل زیست محیطی
environmental fate	سرنوشت محیط زیست
environmental fate model	مدل سرنوشت محیط زیست
environmental health	بهداشت محیط
environmental media	محیط کشت/مدیوم محیطی
environmental media and transport mechanism	مکانیسم انتقال و واسطه گری محیطی
environmental monitoring	فرابینی محیط زیست
environmental pathway	گذرگاه محیط زیست
environmental pollutant	آلاینده محیط زیست
environmental protection	حفاظت محیط زیست
environmental sample	نمونه محیطی

environmentalism	دفاع از محیط زیست		شناسی، پس زایشی
environmentalist	مدافع محیط زیست	epiglottis	برچاکنای
enzootic	آفت در حیوانات، بیماری	epimer	اپیمر، دوپار
	حیوانی، جانورگیر	epimerase	اپی مراز
enzymatic	آنزیمی	epiphysis	رومغزی
enzyme	آنزیم، زیمایه	epiphysitis	اپی فیزیت
enzyme activation	فعال سازی آنزیم	epiphyte	رورست
enzyme activity	فعالیت آنزیم	epiphytic	رورست
enzyme analysis	تجزیه آنزیم	epiphytism	رورستی
enzyme assay	سنجش آنزیم	episome	اپیزوم، روتن
enzyme denaturation	تبه گونی/واسرشتی آنزیم	epistasis	اپیستازی، اثر متقابل ژن ها
enzyme derepression	دی ریپرس/دی ریپرشن/رفع	epithelial	پوششی
	کردن سرکوب	epithelial cell	یاخته روپوشه ای
	زیمایه/دیاستاز/آنزیم	epithelial projection	اپیتلیال پروجکشن،
enzyme immobilization	متوقف کردن آنزیم		برجستگی
enzyme immunoassay	سنجش ایمنی آنزیم		پوششی/روپوشه ای
enzyme inhibitor	بازدارنده آنزیم (ی)	epithelium	اپی تلیوم، بافت پوششی،
enzyme kinetics	جنبش شناسی/کینتیک		برون پوش، بشره غشاء
	آنزیمی		مخاطی، پوشش
enzyme regulation	نظم آنزیم		سنگفرشی
enzyme repression	سرکوب/ریپرس/ریپرشن	epitope	اپی توپ، تعیین کننده، تعیین
	کردن		کننده پادگنی
	زیمایه/دیاستاز/آنزیم،	epizootic	اپیزوئوتیک، ناخوشی همه
	متوقف شدن فعالیت آنزیم		گیر حیوانی، همه گیری
enzyme-linked	سنجش ایمنی وابسته به		دامی
immunoassay	آنزیم	EPR (= electron	ئی.پی.آر.
enzyme-linked	سنجش ایمونوسوربنت	paramagnetic resonance)	
immunosorbent assay	متصل/وابسته به آنزیم	epsilon toxin	زهر سم اپسیلون
eosinophils	ائوزین دوست ها، ائوزینوفیل	Epstein Barr virus	ویروس اپشتاین بار
	ها	equilibrium theory	فرضیه تعادل
epidemic pneumonia	ذات الریه همه گیر	equipment	تجهیزات، دستگاه، لوازم،
epiderm	روپوست		وسایل
epidermal	روپوستی	equivalent weight	وزن هم ارز
epidermal growth factor	عامل رشد رو پوستی	ER (= endoplasmic	ئی.آر.
epidermal growth factor	پذیرنده عامل رشد روپوستی	reticulum)	
receptor		ergotamine	ارگوتامین
epidermic	روپوستی	ERV (= endogenous	ئی.آر.وی.
epidermolysis bullosa	اپیدرمولیز بولوسا آکویزیتا	retro-virus)	
acquisita		erwinia caratovora	باکتری اروینیا کاراتوورا
epididymo-orchitis	التهاب بیضه و بریخ	erwinia uredovora	باکتری اروینیا اوره دوورا
epigenetic	اپی ژنتیک، پس زاد	erythritol	اریتریول

erythrocyte	اریتروسیت، گلبول قرمز
erythropoiesis	اریتروپوئیزیس، خونسازی، گویچه قرمز سازی
erythropoietin	اریتروپوئیتین
ES (= embryonic stem cell)	ئی.اس.
ES cells (= embryonic stem cells)	سلول/یاخته های ئی.اس.
eschar	اثر/زخم، اسکار
Escherichia coli (E. coli)	اشریشیاکولای
Escherichia coliform	باکتری اشرشیا کولیفورم
essential amino acid	آمینو اسید اساسی/ضروری، اسید آمینه اساسی/ضروری
essential fatty acid	اسید چرب اساسی/ضروری
essential nutrient	ماده غذایی اساسی/ضروری
essential polyunsaturated fatty acid	اسید چرب پر اشباع نشده اساسی
EST (= expressed sequence tag)	ئی.اس.تی.
establishment potential	استعداد/امکان/پتانسیل/توان استقرار
ester	استر
esterification	استری شدن
estrogen	استروژن، فحل زا
ETEC (= entro-toxigenic E.coli)	ئی.تی.ئی.سی.
ethanol	اتانول
ethical values	ارزش های اخلاقی
ethidium bromide	اتیدیوم بروماید، برومید اتیدیوم
ethnobiology	رامه شناسی، قوم زیست شناسی
ethyl acetate	اتیل استات
ethyl methane sulfonate	سولفونات اتیل متان
ethylene	اتیلن
etiological agent	عامل سبب شناختی
etiology	اتیولوژی، سبب شناسی، علت شناسی، مطالعه علل بیماری
etioplast	اتیوپلاست

eucaryote	هوهسته ای، یوکاریوت
euchromatin	کروماتین حقیقی، یوکروماتین
eugenics	به نژاد شناسی، به نژادی
eukaryote	هسته دار، یوکاریوت
eukaryotic	هوبر، یوکاریوتیک
eukaryotic cell	سلول یوکاریوتیک، یاخته هوبر
euploid	بهگان، بهلاد، هولاد، یوپلوئید
European Corn Borer	حشره ذرت اروپایی
eutrophication	اوتروفیکیشن، مردابی شدن، هو پرور شدگی
evacuation	تخلیه
evaluation	ارزشیابی، ارزیابی، برآورد
event	رویداد، حادثه، واقعه، وقوع
evolution	تحول، تکامل، رشد، فرگشت
ex vivo	اکس ویوو
examination body	هیئت بررسی
excipient	اکسیپینت
excision	انقطاع، برداشتن، بریدن، جدا سازی،حذف، قطع
excitation	برانگیزش، تحریک
excitatory	تحریک شونده، تحریکی
excitatory amino acids	اسیدهای آمینه تحریک شونده
exclusion chromatography	کروماتوگرافی تخریبی/طردی
exergonic reaction	واکنش انرژی ده
existence value	ارزش وجودی
exobiology	اگزوبیولوژی، برون زیست شناسی، زیست شناسی کیهانی
exocytosis	اگزوسیتو،ز برون رانی، برون یاختگی
exogenous	اگزوژنی، برون خاست، برون زا، برون زاینده، پیدا زا، دو لپه
exoglycosidase	آنزیم اکسوگلیکوسید، اکسوگلیکوسیداز
exon	اکسون، اگزون، رمزآور
exonuclease	اگزونوکلئاز
exotic germplasm	جرم پلاست نابومی/بیگانه/خارجی

English	Persian
exotic species	گونه های غیر بومی/بیگانه
exotoxin	برون زهر، سم خارجی
expected progeny difference	درجه بندی عددی ژنتیک والدین دام
expiration	بازدم، تاریخ مصرف، دم بر آوری، مرگ
explosion method	روش انفجاری
explosive	انفجاری، قابل انفجار، منفجره
exponential growth phase	مرحله رشد تصاعدی
export	صادرات، صادر کردن، صدور
exporter	صادر کننده
exposure	پرتوگیری، تماس، در معرض چیزی قرارگرفتن، سطح تماس، قرار دادن در معرض چیزی، نور دهی
express	آشکار، اکسپرس، صریح، ظاهر کردن، فوری
expressed sequence tag (EST)	اکسپرس سیکوئنس تگ، برچسب توالی اکسپرس
expression	بیان، تظاهر، نمایانی، هویدایی
expression analysis	تحلیل تظاهر
expression array	آرایه تظاهر
expression library	مخزن نمایانی/تظاهر/هویدائی
expression profiling	دسته بندی نمایانی/تظاهر/هویدائی
expressive dysphasia	دیسفیژیای آشکار/بارز.
expressivity	بروز، درجه تظاهر، رسایی، نمود
ex-situ conservation	حفاظت در خارج از محل
ex-situ conservation of farm animal genetic diversity	حفاظت تنوع ژنتیکی دام و طیور در خارج از محل
extended spectrum penicillin	پنی سیلین دارای طیف گسترده
extension	بازشدگی، کشیدگی
external cost	خسارت ظاهری، هزینه خارجی
external-beam radiation	اکسترنال بیم رادیشن، پرتوافکنی اشعه خارجی، پرتوافکنی اکسترنال بیم، پرتو دهی خارجی

English	Persian
extinct	خاموش، منقرض
extinct breed	نژاد/نسل منقرض
extinction	اضمحلال، انقراض، نابودی
extinguisher	خاموش کننده
extracellular	برون/خارج سلولی/یاخته ای
extracellularly	خارج سلول(ی)
extrachromosomal gene	ژن خارج کروموزومی
extrachromosomal genetic element	عنصر ژنتیکی خارج کروموزومی
extractive reserve	ذخیره استخراج
extranuclear gene	ژن خارج هسته ای، ژن سیتوپلاسمی
extraocular	برون چشمی
extremophilic bacteria	باکتریهای دوست/افراط گرا/اکسترموفیلیک
extremozyme	اکسترموزایم، زیمایه غائی/فرینه/نهائی، فرین زیمایه
exudative	ترواش کننده، ترواشی

F

English	Persian
F (= Faraday constant)	اف.
F (= phenylalanine)	اف.
F Met	اف.مت
F'	سلول اف. پرایم/پریم
F+-cell	سلول اف. مثبت
F1P (= fructose-1-phosphate)	اف. 1 پی.
F6P (= fructose-6-phosphate)	اف.6 پی.
F8C (= clotting factor VIII)	اف.8 سی.
F8VWF (= von Willebrand factor disease)	اف.8 وی.دبلیو.اف.
F9 (= clotting factor IX)	اف.9.
FA (= Fanconi anaemia)	اف.ای.
FAB classification (= French-American-British Cooperative Group)	طبقه بندی اف.ای.بی.

English	Persian
facilitated diffusion	انتشار تسهیل شده
facilitated folding	تاشدگی تسهیل شده
facultative	دو گونه زی
facultative anaerobe	بی هوازی اختیاری
facultative cell	سلول اختیاری
FAD (= flavin adenine dinucleotide)	اف.ای.دی.
FAD gene (= flavin adenine dinucleotide)	ژن اف.ای.دی.
FADH$_2$ (= reduced flavin adenine dinucleotide)	اف.ای.دی.اچ.2
fall armyworm	آرمی وورم پائیزی، بید سرباز پائیزی
fallow	آیش، زمین آیش
FALSE positive	مثبت کاذب
fame	آوازه، شهرت
familial adenomatous polyposis	پلیپوز آدنوماتوسی خانوادگی
familial hypercholesterolaemia	کلسترول بالای ارثی، هایپرکلسترولمی خانوادگی
familiarity	آشنایی، آگاهی، شناخت
Fanconi anaemia	آنمی فانکونی، کم خونی فانکونی
FAP (= familial adenomatous polyposis)	اف.ای.پی.
Faraday constant	ضریب فاراده
farnesoid X receptor (FXR)	گیرنده فارنسوئیدی اکس.
farnesyl transferase	فارنزیل ترانزفراز
fat(s)	پرواری، چاق، چرب، چربی، زمین حاصلخیز
fatigue	خستگی، شکستگی، کوفتگی، فرسودگی
fatty acid	اسیدچرب
fatty acid methyl ester	متیل استر اسید چرب
fatty acid oxydation	اکسیداسیون اسید چرب
fatty acid synthetase	سنتتاز/لیگاز اسید چرب
fauna	جانوران، زایگان، فون، فونا
FBJ osteosarcoma virus (=oncogene FOS)	ویروس استئوسارکومای اف.بی.جی.
FBP (=	اف.بی.پی.
fructose-1,6-bis-phosphate)	
FBPase (= fructose bis-phosphatase)	اف.بی.پی.آز
F-cell	سلول اف.
FCS (= fetal calf stem)	اف.سی.اس.
Fd (= ferrodoxin)	اف.دی.
F-duction	اف.داکشن، اف. رسانی
fecundity	باروری، حاصلخیزی، زایایی، گشنیدگی
fed-batch culture	کشت فد-بچ
feedback inhibition	سرکوب پسخور
feeder	دستگاه تغذیه اتوماتیک، -خور، خورنده
feedstock	علوفه، علیق، فید استاک، مواد خام
FEN (= flap endo nuclease)	اف.ئی.ان.
fermentation	تخمیر
fermentation technology	فناوری تخمیر/رشد موجودات ذره بینی
fermenter	فرمنتر
ferritin	فریتین
ferrobacteria	باکتری های آهن دار، فروباکتری
ferrochelatase	آنزیم فرو کلات، فروکلاتاز
ferrodoxin	فرودوکسین
fertility	باروری، حاصلخیزی، زایایی
fertility factor	عامل باروری
fertilization	بارور سازی، حاصلخیزسازی، کود دهی، لقاح
fetal calf stem	سلول بنیادی جنین گوساله
fetal hemoglobin	هموگلوبین جنینی
Feulgen stain	لکه فئولگن
FGA (= fibrinogen, alpha chain)	اف.جی.ای.
FGF (= fibroblast growth factor)	اف.جی.اف.
FH (= familial hypercholesterolaemia)	اف.اچ.
fibrin	فیبرین
fibrinogen	فیبرینوژن

fibrinogen, alpha chain	زنجیره آلفا فیبرینوژن
fibrinolytic agent	عامل فیبرین کافت، عامل فیبرینولایتیک
fibroblast	رشته تنده، فیبروبلاست، یاخته بافت همبند
fibroblast growth factor	عامل رشد رشته تنده
fibronectin	فیبرونکتین
fibroplast growth factor (= fibroblast growth factor)	عامل رشد فیبروپلاست
ficoll	فیکول
field inversion gel electrophoresis	الکتروفوروز ژلی فیلد اینورژن
field trial	کارآزمایی صحرایی
FIGE (= field inversion gel electrophoresis)	اف.آی.جی.ئی.
fill	پر شدن/کردن
filler epithelial cell	یاخته روپوشه ای پرکننده
film yeast	فیلم ییست
filopodia	پای کاذب
filoviridae	فیلوویروس ها
filovirus	فیلوویروس
filtration	پالایش، تصفیه
fimbria	فیمبریا
finger protein	پروتئین انگشتی
fingerprinting	اثر انگشت، انگشت نگاره/نگاری
firefly luciferase-luciferin system	نظام لوسیفراز-لوسیفرین حشره(کرم) شب تاب
first filial hybrid	آمیخته/آمیزه/دورگه نسل اول
first meiotic anaphase	اولین آنافاز میوتیک
first meiotic metaphase	اولین متافاز میوتی
first meiotic prophase	اولین پروفاز میوتی
FISH (= fluorescence in situ hybridization)	اف.آی.اس.اچ.
fission	فیسیون
fitness	تندرستی
five prime	پنج پریم
fixed suspended solids (FSS)	مواد جامد معلق ثابت (مج مث)
flaccid	چروکیده، سست، شل، نرم

flagella	تاژکها، فلاژلا
flagellin	فلاژلین
flagship species	گونه مرغوب
flanking region	منطقه جانبی/کناری
flanking sequence	توالی جانبی/کناری
flap endo nuclease	فلاپ اندونوکلئاز
flavin	فلاوین
flavin adenine dinucleotide	دی نوکلئوتید فلاوین آدنین
flavin mononucleotide	مونو نوکلئوتید فلاوین
flavin mono-nucleotide	مونونوکلئوتید فلاوین
flavin nucleotide	نوکلئوتید فلاوین
flavin-linked dehydrogenase	دهیدروژناز وابسته به فلاوین
flavinoid	فلاوینوید
flaviviridae	فلاوی ویروس (ها)
flavivirus	فلاوی ویروس
flavonoid	فلاونوئید
flavonols	فلاوونول (ها)
flavoprotein	فلاووپروتئین
flesh-eating infection	عفونت خورنده
floatation	شناورسازی
flocculant	فلوکولانت
flocculation	فلوکولاسیون
flooding	آب انداختن، غرق کردن
flora	فلورا، گیاهان
flotation	شناوری
flourescent dye	رنگینه فلوئورسان
flow	جریان
flow cell	سلول/یاخته فلو/جریان
flow cytometry	سیتومتری فلو/جریان، فلو سایتومتری
fluctuant	بی ثبات، متغیر
fluidized	سیال/شاراینده شده
Fluidizer	فلوئیدایزر
fluorescence	پرتو افشانی، شب رنگی، شب نمایی، فلو ئورسانس
fluorescence activated cell sorter (FACS)	مجزا کننده سلولی فعال شده با پرتوافشانی/شب نمایی/شارندگی/تشعشع ع ماهتابی
fluorescence in situ	دورگه گیری به روش

211

hybridization (FISH)	پرتوافشانی در محل، فلوئوروسنس هم تیرگی در محل
fluorescence mapping	نقشه کشی/برداری فلوئورسانس
fluorescence multiplexing	مالتی پلکسینگ باپرتوافشانی/شب نمایی/شارندگی/تشعشع ماهتابی، مخابره هم زمان چند پیام با پرتوافشانی/شب نمایی/شارندگی/تشعشع ماهتابی
fluorescence polarization	قطبیت فلوئورسانس
fluorescence resonance energy transfer	انتقال انرژی تشدیدی فلوئورسانس
fluorescent dye	رنگینه فلوئوروسنت
fluorogenic probe	پرب/پروب/ردیاب/کاوشگر/ز شانگر فلوئوروژنیک/فلوئورسانس زا/شب نمایی/پرتوافشانی/شارند گی زا
fluorophore	فلوئورسانس ساز، فلئوروفور
fluoroquinolone	فلوئوروکینولون
flush end	پایانه صاف/کور/پخ
flux	جریان، دسته اشعه، ذوب کننده، روان ساز، سیال، سیلان، شار، گداز آور،
flying	پرنده، پرواز کردن
FMN (= flavin mono-nucleotide)	اف.ام.ان.
FMR1 (= fragile X mental retardation)	اف.ام.آر.1
focal point	کانون، نقطه کانونی
focus group	فوکاس گروپ، گروه کانونی
fog	تاری، تیرگی، سردرگمی، مه
fogger	فاگر
follicle stimulating hormone	هورمون تحریک کننده فولیکول
Fontan fogger	فاگر فونتان

food contamination	آلودگی مواد غذایی/غذایی
food microbiology	میکروبیولوژی تغذیه
food web	شبکه غذایی
footprinting	اثر پا گرفتن
foreign DNA	دی.ان.ای. نا بومی/بیگانه
formaldehyde dehydrogenase	دهیدروژناز فورمالدهید
formite	فورمایت، فورمیت
formulation	تدوین، تنظیم، فرمول بندی، فرموله کردن
forward mutation	جهش پیش رو
FOS (= oncogene FOS)	اف.او.اس.
fouling	آلودن، خراب کردن، کثیف کردن
founder effect	اثر بانی، اثر بنیان گذار
fragile X mental retardation	عقب افتادگی ذهنی اکس. شکننده
fragile X syndrome	سندروم اکس شکننده
fragmentation	تجزیه، تقطیع، تکه تکه شدن، تلاشی،جزء به جزء کردن، فروپاشی
frameshift	تغییر چارچوب، دگر قالب
frameshift mutation	جهش تغییر چارچوب، جهش دگر قالب
FRAX (= fragile X syndrome)	اف.آر.ای.اکس.
FRDA (= Freidreich's ataxia)	اف.آر.دی.ای.
free energy	انرژی آزاد
free fatty acid	اسید چرب آزاد
free radical	بنیان/رادیکال/ریشه آزاد
free-rider problem of public goods	مسئله استفاده رایگان کالاهای عمومی
Freidreich's ataxia	آتاکسی نوع فردریک
French-American-British Cooperative Group	گروه تعاون فرانسوی-آمریکائی-انگلی سی
fructan	فروکتان
fructification	بارآوری، فروکتیفیکاسیون، میوه دهی
fructooligosaccharide (FOS)	فروکتوالیگوساکارید

fructose	فروکتوز، قند میوه
fructose bis-phosphatase	بیس-فسفاتاز فروکتوز
fructose oligosaccharide	فروکتوز الیگوساکارید
fructose-1,6-bis-phosphate	فروکتوز-6،1-بیس فسفات
fructose-1-phosphate	فسفات-1-فروکتوز
fructose-6-phosphate	فسفات-6-فروکتوز
FSS (= fixed suspended solids)	مث مج
fulminant	برق آسا، فولمینان
fumarase	فوماراز
fumaric acid	اسید فوماریک
fumonisin	فومونیسین
functional food	مواد غذائی کارکردی/عملی/کاربردی
functional genomics	ژنوم شناسی کارکردی/عملی/کاربردی
functional group	گروه عامل(ی)/عمل کننده
functional plan	طرح کاربردی/عملی
fungal toxin	سم قارچی
fungicide	ضد باکتری/قارچ، قارچ کش
fungus	قارچ (ی)
furanocoumarin	فورانوکومارین
furanose	فورانوز
furocoumarin	فوروکومارین
fusaric acid	اسید فوساریک
fusarium	فوساریوم
fusarium graminearum	فوساریوم گرامینئاروم
fusarium moniliforme	فوساریوم مونیلیفورم
fusion	امتزاج، فوزیون، فیوژن، همجوشی
fusion gene	ژن ممزوج/همجوش
fusion inhibitor	بازدارنده امتزاج/همجوشی
fusion protein	پروتئین امتزاج/همجوشی
fusion toxin	سم امتزاج/همجوشی
fusogenic agent	عامل ادغام/امتزاج/همجوشی زائی
futile cycle	چرخه بی ثمر/عبث، دور بی حاصل

G

G (= Gibbs free energy)	جی.
G (= guanine)	جی.
g+	جی مثبت
G1 (= G1 phase of the cell cycle)	جی.1
G1 phase of the cell cycle	فاز جی.1 از چرخه سلولی
G1P (= glucose-1-phosphate)	جی.1 پی.
G2 (= G2 phase of the cell cycle)	جی.2
G2 phase of the cell cycle	فاز جی.2 از چرخه سلولی
G3P (= glyceraldehyde-3-phosphate)	جی.3پی.
G6P (= glucose-6-phosphate)	جی.6پی.
G6PD (= glucose-6-phosphate dehydrogenase)	جی.6پی.دی.
GABA (= gamma-aminobutyric acid)	جی.ای.بی.ای.
Gaia hypothesis	فرضیه گایا
gain of function mutation	جهش بدست آوردن کاربری
Gal (= galactose)	
galactomannan	گالاکتومانان
galactose	گالاکتوز
galactosemia	گالاکتوسمی
gall	زرداب، زهر، کیسه صفرا، مازو، ورم بافت گیاه
galvanized iron	آهن سفید
gamete	زامه، سلول تناسلی/زایشی، گامت، یاخته جنسی
gametogenesis	تشکیل گامت
gametophyte	گامتوفیت
gamma globulin	گاما گلوبولین
gamma interferon	اینترفرون گاما
gamma-aminobutyric acid	اسید آمینوبوتریک گاما
gamone	گامون
ganglion	عقده/گره عصبی، گانگلیون

GAP (= GTPase activating protein)	جی.ای.پی.
gap gene	ژن بینابینی/گپ
gap junction	اتصال بینابینی/گپ
gap period	زمان بینابینی، وقفه
gap phase	فاز بینابینی
gap repair	تعمیر بینابینی
garden	باغ، باغبانی کردن، باغچه، باغستان، باغی، بوستان، حیاط، گلستان، ناحیه حاصلخیز
gas	گاز، گازی، نفخ
gas chromatography	رنگ نگاری/کروماتوگرافی گازی
gas exchange	مبادله گازی
gas-liquid chromatography	رنگ نگاری گاز-مایع
gastric	شکمی، گاستریکی، معدی
gastrin	گاسترین
gastrointestinal hormone	هورمون معده-روده ای
gated transport	نقل و انتقال مهار شده
gating current	جریان مهار کننده
gauche conformation	ترکیب/تطبیق/کنفورمیشن گاش
G-banding (= Giemsa chromosome banding)	اتصال جی.
GBY (= gonado blastoma Y)	جی.بی.وای.
GC box	جعبه جی.سی.
GC-MS (= combined gas chromatography-mass spectrometre)	جی.سی.-ام.اس.
GCP (= good clinical practice)	جی.سی.پی.
GCPS (= Grieg's cephalo polysyndactyly syndrome)	جی.سی.پی.اس.
G-DNA	جی.- دی.ان.ای.
GDP (= guanosine di-phosphate)	جی.دی.پی.
GE (= gene expression)	جی.ئی.
gel	بستن، ژل، ژله، سفت شدن، محلول ژلاتینی

gel diffusion	انتشار/پراکندگی ژلی
gel electrophoresis	الکترو فورز ژلی
gel filtration	صافی کردن/جداسازی ژلی
gelatinase	ژلاتیناز
gelatinization	تبدیل به ژلاتین شدن/کردن، ژلاتینیزاسیون
gelation	انعقاد، بسته شدن، ژلاسیون، لخته شدن
gelsolin	ژلسولین
gem	جواهر، ژم، گوهر
gene	زاد، ژانه ژن، عامل توارث
gene + chromosome	ژن بعلاوه کروموزوم
gene amplification	افزایش/تکثیر ژن (ی)
gene array system	سیستم آرایه ژن (ی)
gene bank	مخزن ژن (ی)
gene chip	تراشه ژن (ی)
gene cloning	کلون/همانند/همسانه سازی ژن (ی)
gene cluster	خوشه ژن (ی)
gene complex	کمپلکس ژن (ی)
gene conversion	تکثیر بیش از انتظار ژن (ی)، کانورژن ژن (ی)،وارونگی ژن (ی)
gene delivery	تحویل ژن (ی)
gene dosage	دوز ژن (ی)
gene duplication	دوپلیکاسیون ژن (ی)، کپی برداری ژن (ی)،همانند سازی ژن (ی)
gene engineering	مهندسی ژن (ی)
gene expression	بروز/تجلی/تظاهر ژن (ی)
gene expression analysis	تحلیل تظاهر ژن (ی)
gene expression cascade	آبشیب تظاهر ژن (ی)
gene expression marker	علامت تظاهر ژن (ی)
gene expression profiling	نمای تظاهر ژن (ی)
gene family	خانواده ژن (ی)
gene flow	انتشار/جریان ژن (ی)
gene frequency	بسامد/فراوانی ژن (ی)
gene function analysis	تحلیل کارکرد/کار ژن (ی)
gene fusion	امتزاج/همجوشی ژن (ی)
gene imprinting	نقش بندی ژن (ی)
gene insertion	الحاق/دخول ژن (ی)

English	Persian
gene introgression	الحاق یک ژن درون مخزن ژنی یک جمعیت
gene library	مخزن ژن (ی)
gene linkage	جفت شدگی/پیوستگی ژن (ی)
gene machine	ماشین ژن (ی)
gene manipulation	دستکاری ژن (ی)
gene map	نقشه ژن (ی)
gene mapping	نقشه برداری/نقشه کشی/نگاشت ژن (ی)
gene modification	دستکاری ژن (ی)
gene mutation	جهش ژن (ی)
gene pool	خزانه/مخزن ژن (ی)
gene probe	پرب/پروب/ردیاب/کاووشگر/ز شانگر ژن (ی)
gene repair	اصلاح/تعمیر ژن (ی)
gene replacement therapy	درمان توسط جایگزینی ژن (ی)
gene sequencing	شناسائی توالی ژن (ی)
gene silencing	جلوگیری/سرکوب نمایانی ژن (ی)
gene splicing	پیوند ژن (ی)
gene stacking	استک کردن ژن (ی)، ژن استکینگ
gene subtraction	تفریق ژن (ی)
gene switching	ژن سوئیچینگ، سوئیچ کردن ژن (ی)
gene synthesis	سنتز ژن (ی)
gene targeting	ژن تارگتینگ
gene taxi	تاکسی ژن (ی)
gene therapy	درمان ژن (ی)، ژن درمانی
gene transcript	رونوشت/نسخه ژن (ی)
gene translocation	انتقال/تراجایی/جابه جایی ژن (ی)
gene-bank	بانک/مخزن ژن (ی)
genecology	بوم شناسی ژن (ی)، زادبوم شناسی، ژن اکولوژی
general release	آزادسازی کلی
general transcription factor	عامل نسخه برداری کلی
generalized transduction	ترا رسانی کلی
generating	تولید کننده، مولد
generation	پیدایش، تولید، تولید مثل، زادآوری، زادگان، زاد و ولد، زایش، نسل
generation time	زمان تولید/زادآوری/نسل
generator	زایگر، ژنراتور، مولد
generic	ژنریک، عام، عمومی، کلی، نام جنسی، همانی
genestein	ژنستئین
genetic assimilation	جذب و ساخت ژنتیکی، همانند سازی توارثی، همگون سازی ژنتیکی
genetic block	سد/مانع ژنتیکی
genetic code	رمز ژنتیکی/وراثتی
genetic complementation	مکمل سازی ژنتیکی
genetic counseling	مشاوره ژنتیکی
genetic death	مرگ ژنتیکی (جهش)
genetic defect	نقص ژنتیکی
genetic disease	بیماری ژنتیکی
genetic distance	فاصله ژنتیکی
genetic distancing	اندازه گیری فاصله ژنتیکی
genetic diversity	گونه گونی/تنوع ژنتیکی، گوناگونی زاد شناختی
genetic drift	سرگردانی ژنتیکی
genetic effect	اثر ژنتیکی
genetic engineering	مهندسی ژنتیک(ی)
genetic equilibrium	تعادل/ثبات ژنتیکی
genetic erosion	فرسایش ژنتیکی
genetic event	رویداد ژنتیکی
genetic fingerprinting	انگشت نگاری ژنتیکی
genetic heterogeneity	ناهمگنی ژنتیکی/وراثتی
genetic imbalance	عدم توازن ژنتیکی
genetic information	اطلاعات ژنتیکی
genetic inheritance	وراثت ژنتیکی
genetic lethal	کشنده ژنتیکی
genetic linkage	اتصال/پیوند ژنتیکی
genetic linkage map	نقشه اتصال/پیوند ژنتیکی
genetic load	بار ژنتیکی
genetic manipulation	دستکاری ژنتیکی
genetic map	نقشه ژنتیکی
genetic marker	نشانگر توارثی/ژنتیکی
genetic material	ماده ژنتیکی

English	Persian
genetic modification	تغییر ژنتیکی
genetic mutation	جهش ژنتیکی
genetic predisposition	مهیا سازی ژنتیکی
genetic probe	پرب/پروب/ردیاب/کاوشگر/نز شانگر ژنتیکی
genetic profiling	نمای ژنتیکی
genetic recombination	بازپیوستگی/باز ترکیبی/رکومبیناسیون/نوَ رکیبی ژنتیکی
genetic resource	منبع ژنتیکی
genetic screening	غربال ژنتیکی
genetic sensitivity	حساسیت ژنتیکی
genetic susceptibility	استعداد ژنتیکی
genetic targeting	ژنتیک تارگتینگ
genetic testing	آزمایش ژنتیکی
genetically engineered	مهندسی (ژنتیک) شده
genetically engineered interferon	انترفرون مهندسی (ژنتیک) شده
genetically engineered microbial pesticide	آفت کش میکروبی طراحی شده ژنتیکی
genetically modified	تغییر داده ژنتیکی
genetically modified food (GMF)	غذای تغییر داده شده بطریق ژنتیک(ی)
genetics	زادشناسی، ژنتیک، وراثت
genistein	ژنیستئین
genistin	ژنیستین
genitourinary tract	جهاز/دستگاه تناسلی-ادراری
genome	زادان، ژنوم، گنجینه توارثی، ماده ژنتیکی مجموعه ژنها
genome map	نقشه ژنوم
genomic blotting	ژنومیک بلاتینگ
genomic exclusion	استثناء ژنومیکی
genomic imprinting	نقش زدن ژنومیکی
genomic library	مخزن ژنومی
genomic sciences	علوم ژنومی
genomics	ژن شناسی
genonema	ژنونما
genophore	ژنوفور
genosensor	ژنوسنسور
genosensor	حسگر زنی

English	Persian
genotoxic	ژنوتوکسیک
genotoxic carcinogen	سرطان زای/کارسینوژن ژنوتوکسیک
genotype	ریخته ارثی، زادمون، ژنوتیپ، نژادگان، نژادمانه
genus	قسم، جنس، جور، نوع
geometric mean titer	متوسط غلظت هندسی
geomicrobiology	زمین زیست شناسی میکروبی، زمین زیوه شناسی، زمین میکروبشناسی
geoponics	خاک کشت
geotropic	زمین گرا
geotropism	زمین گرایی
germ cell	سلول تخم، یاخته تناسلی/جنسی/زایشی
germ cell gene therapy	ژن درمانی سلول/یاخته تناسلی/جنسی
germ plasm	بافت تولید مثلی
germinate	جوانه زدن، خلق کردن، رویش جرم، روییدن، زاییده شدن، سبز کردن، شکل گرفتن
germplasm	جرم پلاسم
gestation	آبستنی، بارداری، حاملگی، دوره تکوین، دوره جنینی
gestation period	دوره آبستنی
GH (= growth hormone)	جی.اچ.
ghost	روح، شبح، گوست
gibberella ear rot	پوسیدگی ذرت ژیبرلائی/جیبرلائی
gibberella zeae	جیبرلا زیا، ژیبرلا زیا
gibberellin	جیبرلین، ژیبرلین
Gibbs free energy	انرژی آزاد گیبز
Giemsa chromosome banding	اتصال کروموزومی جی امسا
gigantism	غول پیکری
gland	اندام ترشحی، غده
glass electrode	الکترود شیشه ای
GLC (= gas-liquid chromatography)	جی.ال.سی.

216

Glc (= glucose)	
glia	گلیا
gliadin	گلیادین
Gln (= glutamine)	
globin	گلوبین
globular protein	پروتئین گویچه ای
globulin	گلوبولین
glocosyl transferase factor	عامل ترانسفرازگلوکوسیل
glomalin	گلومالین
glomerulus	کلافک
glottis	چاکنای
Glu (= glutamic acid)	
glucagon	گلوکاگون
glucan	گلوکان
glucocerebrosidase	آنزیم گلوکوسربروسید، گلوکوسربروسیداز
glucogenic amino acid	اسید آمینه قندزا
gluconeogenesis	شیرین نوزایی، گلوکونئوژنز
glucose	قند، گلوکز
glucose 6 phosphate dehydrogenase deficiency	کمبود گلوکز 6 فسفات دهیدروژناز
glucose isomerase	گلوکز ایزومراز
glucose oxidase	آنزیم گلوکوز اکسید، گلوکز اکسیداز
glucose phosphate isomerase	گلوکز فسفات ایزومراز
glucose-1-phosphate	فسفات-1-گلوکز
glucose-6-phosphate	فسفات-6-گلوکز
glucose-6-phosphate dehydrogenase	دهیدروژناز فسفات-6-گلوکز
glucosidase	گلوکسیداز
glucosinolates	گلوکوزینولات (ها)
glufosinate	گلوفوزینات
gluphosinate	گلوفوزینات
glutamate	گلوتامات
glutamate dehydrogenase	گلوتامات دهیدروژناز
glutamic acid	اسید گلوتامیک
glutamic acid decarboxylase	دکربوکسیلاز اسید گلوتامیک
glutamine	گلوتامین
glutamine synthetase	گلوتامین سنتتاز
glutathione	گلوتاتیون
glutathione (reduced form)	گلوتاتیون (تضعیف شده)
glutathione disulfide (oxidized form)	دیسولفیدگلوتاتیون (اکسید شده)
glutathione synthetase deficiency	کمبود گلوتاتیون سینتتاز
glutelin	گلوتلین
gluten	گلوتن
glutenin	گلوتنین
Gly (= glycine)	
glyceraldehyde	گلیسرالدهید
glyceraldehyde-3-phosphate	فسفات-3-گلیسرآلدهید
glycetein	گلیستئین
glycine	قند سریشم، گلیسین
glycine max	گلیسین ماکس
glycinin	گلیسینین
glycitein	گلیسیتئین
glycitin	گلیستین
glycoalkaloid	گلیکوالکالوئید
glycobiology	قند-زیست شناسی، گلیکوبیولوژی
glycocalyx	گلیکوکالیکس
glycoform	گلیکوفورم
glycogen	گلیکوژن، نشاسته حیوانی
glycolipid	گلیکولیپید
glycolysis	شیرین کافت (ی)،کافت گلیکوزی، گلیکولیز
glycoprotein	گلیکو پروتئین
glycoprotein remodeling	تغییر طراحی گلیکو پروتئین، تغییر مدل گلیکوپروتئین
glycosidases	گلیکوسیداز ها
glycoside	گلیکوزید
glycosidic	گلیکوزیدی
glycosinolate	گلیکوزینولات
glycosylation	گلیکوزیلاسیون
glycosyltransferase	گلیکوزیل ترانسفراز
glyphosate	گلیفسات
glyphosate isopropylamine salt	نمک گلیفسات ایزوپروپیلامین
glyphosate oxidase	گلیفسات اکسیداز
glyphosate oxidoreductase	گلیفسات اکسیدو ردوکتاز
glyphosate-trimesium	تریمزیوم گلیفوسات

217

English	Persian
GM-CSF (= granulocyte-macrophage colony-stimulating factor)	جی.ام.-اس.اف.
GMF (= genetically modified food)	جی.ام.اف.
GMP (= guanosine mono-phosphate)	جی.ام.پی.
GMT (= geometric mean titer)	جی.ام.تی.
gnotobiosis	گنوتوبیوزیس
gnotobiotic animal	جانور گنوتوبیوتیک
GNRP (= guanine nucleotide releasing protein)	جی.ان.آر.پی.
GO gene	ژن جی.او.
Goldberg-Hogness box	جعبه گلدبرگ-هاگنس
GoldenRice™	گلدن رایس.
Golgi apparatus	دستگاه گلژی
Golgi body	بدنه/پیکره گلژی
Golgi complex	کمپلکس/همتافت گلژی
Golgi's apparatus	دستگاه گلژی
gonado blastoma Y	گونادوبلاستومای نوع وای.
good clinical practice	عملکرد بالینی مفید
Good Manufacturing Practice	کارکرد تولیدی مفید
GPI (= glucose phosphate isomerase)	جی.پی.آی.
grade	خلوص درجه، درجه بندی کردن، رده بندی کردن، رقم، زوج، زینه، طبقه، عیار، کیفیت، میزان، نوع
graduated cylinder	استوانه مدرج
graf	پیوند، پیوند زدن، پیوندک، پیوندی، قلمه
graft rejection	دفع پیوند
gram molecular weight	ملکول گرم، وزن مولکولی بر حسب گرم
gram stain	رنگ آمیزی گرام، رنگ گرام
Gramacidin	گراماسیدین
gram-negative	گرم منفی
gram-negative bacteria	باکتری گرم منفی
gram-positive	گرم مثبت
gram-positive bacteria	باکتری گرم مثبت
granulation tissue	بافت التیامی، جوانه بافتی، گوشت نو، نسج التیامی
granulocidin	گرانولوسیدین
granulocyte	چند هسته ای، گرانولوسیت، گویجه سفید دانه ای، یاخته دانیزه ای
granulocyte colony stimulating factor	عامل برانگیزنده کولونی گویجه سفید دانه ای
granulocyte-macrophage colony-stimulating factor	عامل محرک کولونی گرانولوسیت ماکروفاژ
granuloma	گرانولوما
green fluorescent protein	پروتئین سبز فلوئورسنتی
green leafy volatile	گرین لیفی ولتایل
greenhouse effect	اثر گلخانه ای
greenhouse gas	گاز گلخانه ای
greenleaf technologies	گرین لیف تکنولوجیز
Grieg's cephalo polysyndactyly syndrome	سندروم سفالو پلی سینداکتیلی گریگ
ground state	حالت حداقل انرژی
group-transfer reaction	واکنش انتقال گروهی
growth	افزایش، تومور، رشد، رویش، غده، نمو
growth curve	منحنی رشد
growth factor	عامل رشد
growth hormone	هورمون رشد
growth phase	دوره/مرحله رشد
GS (= glutamine synthetase)	جی.اس.
GSD (= glutathione synthetase deficiency)	جی.اس.دی.
GSH (= glutathione (reduced form))	جی.اس.اچ.
GSSG (= glutathione disulfide (oxidized form))	جی.اس.اس.جی.
GT (= gene therapy)	جی.تی.
GTF (= general transcription factor)	جی.تی.اف.
GTF (= glocosyl transferase factor)	جی.تی.اف.

GTP (= guanosine tri-phosphate)	جی.تی.پی.
GTPase (= guanosine tri-phosphatase)	جی.تی.پی.آز
GTPase activating protein	پروتئین فعال ساز جی.تی.پی.آز
guanine	گوانین
guanine nucleotide releasing protein	پروتئین رها ساز نوکلئوتید گوانین
guanosin	گوانوزین
guanosine	گوانوزین
guanosine di-phosphate	دی-فسفات گوانوزین
guanosine mono-phosphate	گوانوزین مونو فسفات
guanosine tri-phosphatase	گوانوزین تری فسفاتاز
guanosine tri-phosphate	گوانوزین تری فسفات
guanylate	گوانیلات
guanylate cyclase	گوانیلات سیکلاز
guessmer	گسمر
guild	صنف، هم آشیانه
gun	اسلحه، پمپ دستی، تفنگ (ی)
gynandromorph	جنسیت آمیخته، گایناندرومورف، زن-مرد/ماده-نر ریخت
gyrase (= DNA gyrase)	ژیراز

H

h (= hour)	
H.zea (= corn earworm)	اچ. زیا
H_2O_2 (= hydrogen peroxide)	هاش دو او. دو
habitat	بوم، رستنگاه زیستگاه، محل سکونت، مسکن طبیعی، موطن
habitat restoration	آبادسازی زیستگاه، اصلاح بوم
habituated culture	کشت خوگرفته
HAEC (= human artificial episomal chromosome)	اچ.ای.ئی.سی.
haemophilia	هموفیلی
hairpin loop	حلقه سنجاق سری، هیرپین

	لوپ
hallucination	توهم
halophile	آب شور گرای، گیاه آب شور، گیاه شوره خواه، شوره زی، نمک دوست
hand	دست، دستی، عقربه، کارگر
handedness	عدم تقارن (در دستها)
hand-held	دستی
hanging-drop culture	کشت هنگینگ-دراپ
Hank's balanced salt solution	محلول متعادل نمکی هنک
hantavirus	هانتاویروس
HAP gene (= highly available phosphorous gene)	ژن اچ.ای.پی.
haploid (= monoploid)	هاپلوید
haploid cell	سلول تک لاد/هاپلوئید
haplophase	تک لاد چهر، هاپلوفاز
haplotype	تک لاد مونه، نیمگان جور، هاپلوتیپ
haplotype map	نقشه تک لاد مونه/نیمگان جور/هاپلوتیپ
hapmap	هپ مپ
hapten	هپتن
haptene	نیم پادگن
haptoglobin	هاپتوگلوبین
hardening	آبدیده کردن، سخت گردانی، مقاوم سازی
Hardy-Weinberg equation	معادله هاردی-واینبرگ
Hardy-Weinberg law	قانون هاردی-واینبرگ
harpin	هارپین
harvesting	برداشت کردن، خرمن کردن، در آوردن (عضو بدن جهت پیوند)، دروکردن
harvesting enzyme	آنزیم درو/هاروستینگ
HAT (= hypoxanthine–aminopterin-thymidine)	اچ.ای.تی.
hazard	زیان، خطر مخاطره
hazardous substance	ماده خطرناک
Hb (= hemoglobin)	اچ.بی.

219

Hb F (= fetal hemoglobin)	اچ.بی.اف.
HBA (= adult hemoglobin)	اچ.بی.ای.
HbA (= adult hemoglobin)	اچ.بی.ای.
HBD (= hypophosphataemic bone disease)	اچ.بی.دی.
HBS (= heteroduplex binding site)	اچ.بی.اس.
HBV (= hepatitis B virus)	اچ.بی.وی.
HC (= hypertrophic cardiomyopathy)	اچ.سی.
hcG (= human chronic gonadotropin)	اچ.سی.جی.
HD (= Huntington's Disease)	اچ.دی.
HDL (= high-density lipoprotein)	اچ.دی.ال.
HE (= hereditary elliptocyptosis)	اچ.ئی.
health hazard	خطرناک (برای سلامتی)
heat shock gene	ژن (مقاوم به) شوک گرمائی
heat-shock protein	پروتئین شوک حرارتی
heavy-chain variable	متغیر زنجیره سنگین
hedgehog protein	پروتئین خار پشت/جوجه تیغی
hela cell	سلول هلا
helicase	هلیکاز
helicoverpa zea	هلیکووریازیا
helicoverpazea (=corn earworm)	هلیکووریازیا
helix	مارپیچ، مارپیچ استوانه ای، منحنی پیچ، هلیکس
helix breaker	مارپیچ شکن
helix terminator	پایان بخش مارپیچ
helix-destabilizing proteins	پروتئین بی ثبات کننده مارپیچ
helper phage	فاژ کمک کننده
helper T cell	کمک کننده تی. سل/سلول تی
helper virus	ویروس کمک کننده
hemagglutinin	هم آگلوتینین
hematochezia	خون مدفوعی، هماتوکزی
hematocrit	خون بهر/سنجه

hematogenous	خون زا
hematologic growth factor	عامل رشد خون شناختی
hematopietic growth factor	عامل رشدگلبولهای قرمز خون
hematopoietic stem cell	یاخته دودمانی خون ساز
hematopoietin	هماتوپوئیتین
heme	خون، هم
hemicellulase	همی سلولاز
hemicellulose	همی سلولز
hemizygous	نیم جور تخم، همیزایگوس
hemizygous gene	ژن نیم جور تخم، ژن همیزایگوس
hemoglobin	هموگلوبین
hemolymph	خون-لنف، همولنف
hemolysin	همولیزین
hemolysis	همولیز
hemophilia	دیربندآمدن خون، هموفیلی
hemopoiesis	هموپوئیزیس
hemopoietic stem cells	سلول بنیادی خون ساز/هموپوئیتیک
hemopoietin	هموپوئیتین
hemorrhagic mediastinitis	التهاب و خونریزی میان سینه
hemorrhagic pleural effusion	هموراژیک پلورال افیوژن
hemostasis	انعقاد خون، بند آمدگی خون، همو استاز
hemostatic derangement	اختلال درانعقاد خون، بی نظمی در بند آمدن خون
HEPA filter	فیلتر هپا/اچ.ئی.پی.ای.
heparin	هپارین، هورمون کبدی
hepatatrophic	هپاتاتروفیک
hepatic triglyceride-lipase	تریگلیسرید لیپاز هپاتیکی
hepatitis B virus	ویروس هپاتیت بی.
hepato-renal syndrome	عوارض/نشانگان کبدی-کلیوی
herbicide	علف/گیاه کش
herbicide resistance	مقاومت نسبت به علف کش
herbicide-resistant crop	محصول زراعی مقاوم نسبت به علف کش
herbicide-tolerant crop	محصول زراعی تحمل کننده علف کش
hereditary code	کد وراثتی

hereditary disease	بیماری وراثتی	heterogeneous	نامتجانسی، ناهمگنی ناجور، نامتجانس، ناهمگن، ناهمگون، هتروژن
hereditary elliptocyptosis	الیپتوسیپتوز وراثتی	heterogeneous nuclear RNA	آر.ان.ای. هسته (ای) ناهمگون
hereditary fructose intolerance	ناسازگاری فروکتوزی ارثی	heterokaryon	دگر/ناجور هسته، هتروکاریون
hereditary material	ماده وراثتی	heterokaryosis	دگر هستگی، ناجور هسته بودن، هتروکاریوزیس
hereditary non-polyposis colorectal cancer	سرطان کولو-رکتال ارثی غیر پولیپوزی	heterologous	دگر ساخت/سان، نا همساخت
hereditary persistence of fetal hemoglobin	ماندگاری ارثی هموگلوبین جنینی	heterologous DNA	دی.ان.ای. دگر ساخت/ناهمساخت
hereditary spherocytosis	اسفروسیتوز ارثی	heterologous probing	پروب/کاوشگری ناهمگن
heredity	ارثی، برماند (ی)، توارث، وراثت (ی)، علم توارث	heterologous protein	پروتئین دگرساخت/ناهمساخت/ه ترولوگ
heritability	توانائی ارثی، توارث پذیری، قابلیت توارث، وراثت پذیری	heterology	دگرساختی/ناهمگنی
		heteromorphism	دگر ریختی، گوناگونی طبیعی یک کروموزوم، ناجورشکلی، هترومورفیزم
hermaphrodite	دوجنسی، نر-ماده، هرمافرودیت		
herpes simplex virus	ویروس هرپس سیمپلکس	heteroploid (= aneuploid)	هتروپلوئید
herpes virus saimiri	ویروس هرپس سایمیری	heterosis	بنیه دورگه، دگرینگی، رشد غیر مترقبه، هتروز
hetero-	ناجور-		
heteroacryon	هتروآکریون	heterotroph	دگر پرور/غذا، ناخود پرور، هتروتروف
heteroallels	هتروآلل ها، ناجور دگره ها		
heteroantigen	هتروآنتی ژن	heterotrophic	دگر پروری/خواری/غذایی، ناخود پروری
heterochromatin	دگر رنگینه، کروماتین ناجور، ناجور رنگینه، هتروکروماتین	heterozygosity	ناجور تخمی، ناخالصی
		heterozygote	ناجور تخم، نا خالص، هتروزیگوت
heterocyclic	ناجور حلقه ای، هتروسیکلیک		
heterodimer	ناجور دوپار، هترو دیمر	heterozygous	نا خالص
heteroduplex	دوگانه ناجور، ناجور دوتایی	hex A (= hexosaminidase A)	هگز ای.
heteroduplex binding site	محل اتصال هترودوپلکس/مضاعف ناهمگن	hex B (= hexosaminidase B)	هگز بی.
		hexadecyltrimethylammonium bromide	هگزادسیل تری متیل آمونیم برومید
heteroduplex DNA	دی.ان.ای. دوگانه ناجور/ناجور دوتایی	hexosaminidase A	هگزوآمینیداز ای.
heteroecious	هترو اسیوس	hexosaminidase B	هگزوآمینیداز بی.
heterogametic	دوزامه ای، ناجور زامه ای، هتروگامتی	hexose	هگزوز
heterogametic sex	جنسیت ناجور زامه ای/هتروگامتی	hexose mono-phosphate	مونوفسفات هگزوز
heterogamic	هتروگامی	HFI (= hereditary fructose	اچ.اف.آی.
heterogeneity	عدم تجانس، ناجوری،		

intolerance)	
Hfr	اچ.اف.آر.
Hfr cell	سلول اچ.اف.آر.
hGH (= human growth factor)	اچ.جی.اچ.
HGPRT (= hypoxantine-guanine phospho-ribosyl transferase)	اچ.جی.پی.آر.تی.
high blood pressure	فشار خون بالا
high level waste	پسماند پر پرتو
high mobility group protein	پروتئین گروهی شدیدا متحرک
high molecular weight kininogen	کینینوژن با وزن ملکولی بالا
high profilerative potential	استعداد انتشار بالا
high resolution chromosome banding	باندینگ/رنگ آمیزی/نوارگذاری کروموزومی با دقت بالا
high-content screening	غربالگری با حجم بالا
high-density lipoprotein	لیپو پروتئین پر غلظت/غلیظ
high-efficiency particulate air filter	صافی/فیلتر تصفیه ذرات هوا با کارآیی بالا
high-efficiency particulate air filter mask	ماسک تصفیه ذرات هوا با کارآیی بالا
high-frequency recombinant	بازترکیب/ریکومبینان پر بسآمد
high-frequency transduction	ترارسانی پر بسآمد
highly available phosphorous (HAP)	فسفر شدیدا در دسترس
highly available phosphorous gene	ژن فسفر شدیدا در دسترس
highly unsaturated fatty acid	اسید چرب شدیدا اشباع نشده
high-pressure/high performance liquid chromatography	رنگ نگاری مایع فشار و کارکرد بالا
high-throughput identification	شناسایی با ظرفیت پذیرش بالا
high-throughput screening	جدا سازی با ظرفیت بالا
hilar adenopathy	آدنوپاتی هیلار
hirudin	ماده ضد انعقاد خون،

	هیرودین
hisD (= histidinol dehydrogenase)	
histamine	هیستامین
histidine	هیستیدین
histidinol dehydrogenase	دهیدروژناز هیستیدینول
histiocyte	بافت یاخته
histoblast	بافت تنده، یاخته متشکل بافت
histocompatibility	بافت سازگاری، سازگاری بافتی/نسجی
histocompatibility antigen	آنتی ژن سازگاری بافتی
histone	هیستون
histopathologic	آسیب شناسی بافتی، بافت آسیب شناسی
HIV (= human immunodeficiency virus)	اچ.آی.وی.
HLA (= human leucocyte antigens)	اچ.ال.ای.
HMG coA reductase (= 3-hydroxy 3-methyl glutaryl coenzyme A reductase)	
HMP (= hexose mono-phosphate)	اچ.ام.پی.
HMWK (= high molecular weight kininogen)	اچ.ام.دبلیو.کی.
HNPCC (= hereditary non-polyposis colorectal cancer)	اچ.ان.پی.سی.سی.
hnRNA (= heterogenous nuclear RNA)	
Hogness box	جعبه هاگنس
holandric gene	ژن هولاندریک (پدر به پسر)
holin	هولین
Holliday mode	حالت هالیدی
Holliday structure	ساختار هالیدی
hollow fiber separation	تفکیک فیبر میان تهی
hollow fibre reactor	راکتور فیبر میان تهی
holoenzyme	آنزیم کامل، هام زیمایه، هولو آنزیم

holotype	درست مونه، هولوتیپ
homeo box region 1	هومئوباکس منطقه 1
homeo box region 2	هومئوباکس منطقه 2
homeo box region 3	هومئوباکس منطقه 3
homeobox	هومئوباکس
homeostasis	اعتدال مزاج، تعادل، خود ایستایی/پایداری/تعادلی، هومئوواستازی
homeotic mutation	جهش هومئوتیک
hometic genes	ژنهای هومتیکی
homing receptor	گیرنده جهت یاب (ی)، گیرنده حسی بازگشت به خانه، هومینگ ریسپتور
hommopolymer	پلیمر یکنواخت، هموپلیمر
homocysteine	هوموسیستئین
homocystinuria	هوموسیستینوری
homogametic sex	جنسیت جورزامه ای/یک جور/هوموگامتی
homogeneous	جور، متجانس، همانند، همجنس، همگن، یک دست، یکسان
homogeneously staining regions	مناطق رنگ آمیزی همگن
homogenetic induction	انگیختگی/اینداکشن همگن
homogenization	همسان سازی، هموژنیزه کردن
homograft	جور پیوند، هموگرافت
homokaryon	جور/هم هسته هوموکاریون
homologous	مشابه، نظیر، همانند، همتا، همساخت، همسان همگون، هومولوگ
homologous chromosome	کروموزوم جور/همساخت
homologous compounds	ترکیبات هم رده
homologous protein	پروتئین جور/مشابه/همگون
homologous recombination	باز ترکیبی مشابه، رکومبیناسیون هومولوگ، نوترکیبی همگون
homologue gene	ژن مشابه/همانند/همسان
homology	برابری، تجانس، تشابه، همانندی، همتایی، هم

	ردیفی، همسانی
homology modeling	مدل سازی همسانی
homopolymer tailing	اتصال هموپلیمری/هم پلیمری
homothalism	جور ریسگی، هموتالیسم
homotropic enzyme	آنزیم همگرا/هوموتروپیک
homozygote	جور تخم، خالص
homozygous	جور تخم/یوغ، خالص
horizontal disease transmission	انتقال افقی بیماری
horizontal inheritance	وراثت اتوزومی مغلوب
hormone	اورمون، گیزن، هورمون
hormone response element	عامل واکنش هورمون
horseradish peroxidase	پر اکسیداز ترب کوهی
hose	شلنگ، لوله (خرطومی/لاستیکی)
hospital information system	سیستم اطلاعات بیمارستان
host	انگل پذیر/دار، میزبان
host cell	سلول میزبان
host factor	عامل میزبانی
host range	برد میزبانی/انگل پذیری
host restriction	محدودیت های میزبان
host vector system	ترکیب/سیستم ناقل-میزبان
host-cell reactivation	باز فعالی سلول میزبان
host-controlled restriction	محدودیت مهار شده توسط میزبان
host-vector	ناقل-میزبان
hot spot	منطقه حساس، هات اسپات
hour	ساعت
housekeeping genes	ژنهای اداره کننده/خانه داری/هاوس کیپینگ
HOX1 (= homeo box region 1)	اچ.او.ایکس.1
HOX2 (= homeo box region 2)	اچ.او.ایکس.2
HOX3 (= homeo box region 3)	اچ.او.ایکس.3
HPFH (= hereditary persistence of fetal hemoglobin)	اچ.پی.اف.اچ.
HPLC (= high-pressure/high	اچ.پی.ال.سی.

performance liquid chromatography)	
HPP (= high profilerative potential)	اچ.اچ.پی.
HPRH (= hypo plastic right heart)	اچ.پی.آر.اچ.
HPRT (= hypoxanthine phospho ribosyl transferase)	اچ.پی.آر.تی.
HRE (= hormone response element)	اچ.آر.ئی.
HS (= hereditary spherocytosis)	اچ.اس.
HSC (= hemopoietic stem cells)	اچ.اس.سی.
HSR's (= homogeneously staining regions)	اچ.اس.آر. (ها)
HSV (= herpes simplex virus)	اچ.اس.وی.
HTgL (= hepatic triglyceride-lipase)	
human artificial chromosome	کروموزوم مصنوعی انسان
human artificial episomal chromosome	کروموزوم اپیزومی انسانی مصنوعی، کروموزوم اپیزومی مصنوعی انسان
human chronic gonadotropin	گونادوتروپین انسانی کرونیک، گونادوتروپین تدریجی انسانی
human embryonic stem cell	سلول پایه جنینی انسان، یاخته دودمانی جنینی انسان
human equivalent concentration	تجمع/تراکم هم ارز انسانی
Human Genome Project	پروژه ژنوم انسان
human growth factor	عامل رشد انسانی
human growth hormone	هورمون رشد انسان
Human Immunodeficiency Virus	ویروس افت ایمنی انسان
human leucocyte antigen system	سیستم آنتی ژن لوکوسیتی انسانی

human leucocyte antigens	آنتی ژن های یاخته های سفید انسان، پادگن های گلبول سفید انسانی
human superoxide dismutase	سوپر اکسید دیسموتاز انسانی
human thyroid-stimulating hormone	هورمون محرک تیروئید انسان
humanized antibody	پادتن انسانی شده
humidifier	دستگاه بخور/رطوبت ساز/مرطوب کننده
humoral immune response	پاسخ ایمنی خونی، واکنش ایمنی مزاجی
humoral immunity	ایمنی خونی
humoral response	پاسخ خونی، واکنش مزاجی
humoral-mediated immunity	ایمنی متعادل خونی، مصونیت متعادل مزاجی
Hunter's syndrome	سندرم هانتر
Huntington's disease	بیماری هانتینگتون
hut operon	اپرون هات
HVS (= herpes virus saimiri)	اچ.وی.اس.
Hx (= hypoxanthine)	اچ. اکس.
hyaluronic acid	اسید هیالورونیک
hybrid	آمیخته، آمیزه، حیوان دورگه، دورگه، سه رگه، گیاه پیوندی، هیبرید
hybrid vigor (=heterosis)	قدرت دو رگه
hybrid zone	منطقه دورگه، نواره آمیخته
hybrid-arrested translation	نسخه برداری هیبرید متوقف شده
hybridization	آمیختگی، دورگه شدن، هم تیرگی، هیبریداسیون، هیبریدشدگی
hybridization probe	پرب/پروب/ردیاب/کاوشگر/زشانگر هیبریداسیون
hybridization probing	کاوشگری با هیبریدسازی، نشانگذاری با دورگه سازی
hybridization surface	سطح هیبریدیزاسیون/هم تیرگی
hybridized	هیبرید شده

224

hybridoma	هیبریدوما
hybrid-release translation (HRT)	نسخه برداری هیبرید رها شده
hydrated	آبپوشیده، آبدار
hydration	آبپوشی، آبدار کردن، آبدهی، آبش، آبگیری، آب وندی
hydrazine	هیدرازین
hydrazinolysis	هیدرازینولیز
hydrofluoric acid cleavage	شکافتگی اسید هیدروفلئوریک.
hydrogen	هیدروژن
hydrogen bond	اتصال/پیوند هیدروژنی
hydrogen peroxide	پراکسید هیدروژن
hydrogen sulfide	سولفید هیدروژن
hydrogenation	هیدروژن دار سازی، هیدروژن گیری، هیدروژنه کردن
hydrolysis	آب کافت (ی)، آب گسستگی، تجزیه آبی، هیدرولیز
hydrolytic	آبکافتی
hydrolytic cleavage	شکافتگی آبکافتی
hydrolyzable	آبکافت شدنی
hydrolyzate	هیدرولیزات
hydrolyze	آبکافت/هیدرولیز کردن
hydrophilic	آب پذیر، آب خواه، آب دوست، آبگرا، هیدروفیلیک
hydrophobic	آب گریز/ناپذیر/هراس، ضد آب، هیدروفوب (ی)
hydroponics	آب کشت
hydroxy praline	هیدروکسی پرالین
hydroxylation reaction	واکنش هیدروکسیله کردن
hydroxylysine	هیدروکسیلایزین
Hyl (= hydroxylysine)	
hymenium	هایمنیوم
Hyp (= hydroxy praline)	
hyperacute rejection	دفع حاد عضو پیوندی
hypercholesterolemia	بالا بودن کلسترول خون، هایپرکلسترولمی
hyperchromicity	هایپرکرومیسیته
hyperfiltration	هایپر فیلتراسیون
hyperimmune	فوق ایمن

hypernatremia	بالا بودن سدیم خون
hyperoxaluria	هایپراوکسالوری
hyperplasia	هایپرپلازی
hyperpolarization	بیش قطبیت، هایپر پولاریزاسیون
hyperprolinemia	هایپرپرولینمی
hypersensitive response	واکنش خیلی حساس
hypersensitivity	بیش/پر حساسیتی
hyperthermophilic	بیش گرما خواه/دوست
hypertrophic cardiomyopathy	کاردیومیوپاتی هایپرتروفیک
hypertrophy	هایپرتروفی
hyphae	ریشه قارچها، نخینه ها، هایفا، هیف ها
hypo plastic right heart	قلب هیپوپلاستیک سمت راست
hypoderm	زیرپوست
hypodermal	زیرپوستی
hypodermic	زیرپوستی
hypoglycemia	کم قند خونی
hypophosphataemic bone disease	بیماری استخوانی هیپوفسفاتائمیک
hypophosphataemic ricket	راشیتیسم مقاوم به ویتامین دی.، ریکت هایپوفسفاتمیکی
hypophysis	زیرمغزی
hypostasis	رسوب، زیر ایستایی، هیپوستازی
hypothalamus	زیر نهنج، هیپوتالاموس
hypotrichosis	هیپوتریکوزی
hypoxanthine	هیپوزانتین
hypoxanthine phospho ribosyl transferase	ترانسفراز هیپوزانتین فسفو ریبوزیل، هیپوگزانتین فسفوریبوزیل ترانسفراز
hypoxanthine–aminopterin-t hymidine	هیپوزانتین-آمینوپترین-تیمید ین
hypoxantine-guanine phospho-ribosyl transferase	ترانسفراز هیپوزانتین-گوانین فسفو-ریبوسیل، هیپوزانتین-گوانین فسفو-ریبوسیل ترانسفراز
hypoxia	کاهش اکسیژن بدن، کم

225

	اکسیژنی
hysteresis	تعادل رطوبتی دوگانه، هیسترزیس

I

I (= isochromosome)	آی.
IB (= immuno blotting)	آی.بی.
IBC	آی.بی.سی.
icosahedral virion	ویریون آیکوزاهدرال
IddM (= insulin-dependent diabetes mellitus)	آی.دی.دی.ام.
IDDM (= insulin-dependent diabetes mellitus)	آی.دی.دی.ام.
ideal protein concept	مفهوم پروتئین ایده آل
Idem (denotes the stemline karyotype in subclones)	آیدم
ideogram	ایدئو گرام
Ider (= isoderivative chromosome)	آیدر
idioblast	ایدیو بلاست
idiochromatin	ایدیو کروماتین، گزین تنده
idiochromosome	ایدیو کروموزوم، گزین فام تن
idiogram	ایدیوگرام، گزین نگاره
idiophase	ایدیوفاز
idiotope	ایدیوتوپ
idiotroph	ایدیو تروف
idiotrophic mutant	جهش یافته ایدیو تروفیک/گزین پرور
idiotype	به جور، تیپ ایده آل، گزین مونه
idiovariation	ایدیو واریاسیون، تغییر ارثی، گزین وردایی
IDL (= intermediate-density lipoprotein)	آی.دی.ال.
idling reaction	واکنش درجا
I-DNA	دی.ان.ای.نوع آی.
IF (= initiation factor)	آی.اف.
IF2 (= initiation factor 2)	آی.اف.2
IGF (= insulin growth factor)	آی.جی.اف.

IgG (= immunoglobin G)	آی جی.جی.
IGHD (= isolated human growth hormone deficiency)	آی.جی.اچ.دی.
IgM (= Immunoglobin M)	آی جی. ام.
Ile (= iso-leucine)	
illegal traffic	آمد و شد/ترافیک/داد و ستد غیر قانونی
imidazole	ایمیدازول، هسته اصلی هیستیدین
immediate early gene	ژن اولیه فوری
immobilization	بی جنبش سازی، بی حرکت سازی، تثبیت
immortal	نامیرا
immortalization	نامیرا سازی
immortalizing gene	ژن نامیرا سازی
immortalizing oncogene	ژن توموری پایدار
immune escape	فرار مصون
immune function	عملکرد/کارکرد ایمنی
immune globulin	گلوبولین ایمنی
immune interferon	انترفرون ایمنی
immune modulator	مدولاتور ایمنی
immune percipitate	رسوب ایمنی
immune reaction	واکنش ایمنی
immune response	پاسخ/عکس العمل/واکنش ایمنی
immune sera	پیماب های مصونی، سرم های ایمنی
immune system	دستگاه ایمنی/دفاعی بدن، نظام دفاعی بدن
immune-stimulating complexes	کمپلکس های محرک ایمنی
immunity	ایمنی، مصونیت
immunization	ایمن/مصون کردن
immuno blotting	ایمونو بلاتینگ
immunoadhesin	ایمونو ادهسین
immunoadsorbent	ایمونو ادسوربانت
immunoaffinity	ایمونو افینیتی
immunoassay	ایمنی سنجی
immunobiology	ایمونوبیولوژی، زیست شناسی مصونیت

immunoblotting	ایمونوبلاتینگ	implant radiation	تشعشع درون کاشت
immunocompetent	ایمنی قابل قبول	import	اهمیت، دلالت داشتن بر،
immunocompromised	به خطر افتاده از لحاظ ایمنی		مفهوم، وارد کردن
immunocompromised host	میزبان به خطر افتاده از لحاظ	importer	وارد کننده
	ایمنی	imprinting	نقش بندی/پذیری
immunoconjugate	ایمونو کانجوگیت	improved mass selection	انتخاب انبوه بهینه
immunocontraception	جلوگیری از بارداری از لحاظ	improved variety	تنوع بهینه
	ایمنی	in situ hybridization	هم تیرگی در محل
immunocyte	ایمونوسیت	in vitro fertilization	تلقیح این ویترو
immunodeficiency	کمبود مصونیت، نارسایی	in vitro marker	نشانگر این ویترو
	ایمنی	in vitro marker	مکمل سازی نشانگر این
immunodiffusion	انتشار ایمنی	complementation	ویترو
immunodominant	غالب ایمن، مصون غالب	inactivation	بی فعالیتی، عدم فعالیت،
immuno-enhancing	افزایش/بهبود/تقویت ایمنی		نافعالی
immunofluorescence	مصون پرتوافشانی	inapparent infection	آلودگی غیر واضح
immunogen	ایمنی زا	inborn error	اختلال توارثی
immunogenetics	ژنتیک ایمنی	inbred line	رگه خویش آمیز
immunogenic	ایمنی زایی	inbred strain	جوره/رقم/سویه/واریته درون
immunogenic protein	پروتئین ایمنی زا		آمیخته
immunoglobin G	ایمونوگلوبین جی.	inbreeding	تخم کشی بسته، تولید مثل
immunoglobin M	ایمونوگلوبین ام.		بین خودی، خویش آمیزی،
immunoglobolin	ایمونوگلوبولین		درون آمیزی/زاد
(immunoglobulin)			گیری/همسری
immunoglobulin	ایمونوگلوبولین، گلوبولین	inbreeding coefficient	معیار هم خونی
	ایمنی	inbreeding depression	پس روی خویش آمیزی،
immunologic	ایمنی شناختی		کاهش درون آمیزی
immunological homeostasis	هومئواستازی ایمونولوژیکی	INC (= incomplete	آی.ان.سی.
immunological rejection	پس زدن ایمونولوژیکی	karyotype)	
immunological screening	غربال ایمونولوژیکی	incapacitate	از توان انداختن، سلب
immunological tolerance	ظرفیت ایمونولوژیکی		صلاحیت کردن،
immunology	ایمونولوژی		عاجز/ناتوان کردن
immunomagnetic	ایمنی-مغناطیسی	incapacitating agent	عامل ناتوان کننده
immunopotentiator	ایمونوپتانسیاتور	incapacitation	از توان/کار انداری، ناتوان
immunosensor	حسگر ایمنی		سازی
immunosuppressive	ایمنی کاه/کوب، بازدارنده	incidence	بروز، رویداد، مورد، وقوع
	ایمنی	incidence rate	نرخ شیوع، میزان اتفاق/بروز
immunosuppressive therapy	درمان بازدارنده ایمنی	incipient species	گونه های در حال تظاهر
immunotherapy	ایمن درمانی، درمان ایمنی	incision enzyme	آنزیم برش
immunotoxin	ضد سم	inclusion body	جسم درون بسته/گنجیده
IMP (= inosine	آی.ام.پی.	income	درآمد، ورودی
mono-phosphate)		incompatibility group	گروه ناهمخوان/ناخوانا

English	Persian
incomplete dominance	غالبیت ناقص، غلبه جزئی
incomplete karyotype	کاریوتیپ/هسته مونه ناقص
incontinentia pigmenti	اینکونتیننتیا پیگمانتی
incremental feeding culture	کشت ازدیاد تغذیه ای
incubation	برتخم نشینی، جوجه کشی، خواباندن، خوابیدن، درون کمون، دوره شکل گیری/کمون
incubation period	درون نهفتگی، دوره جوجه کشی/کمون
independent assortment	تفکیک مستقل
independent cutting method	روش انقطاع مستقل
indicator	اندیکاتور، شاخص، شناساگر، شناسه، نشانه
indicator gene	ژن شاخص
indicator species	گونه شاخص/شناساگر/نشانه
indigenous	بومی
indirect contact	تماس غیر مستقیم
indirect interaction	تداخل غیر مستقیم
indirect source	منبع غیر مستقیم
indirect transmission	انتقال/سرایت غیر مستقیم
individual risk	ریسک فردی
induced fit	جفت شدن القایی
inducer	القا کننده/القاء گر
inducible enzyme	آنزیم القاء/سازش پذیر، آنزیم قابل القاء
inducible promoter	پروموتور/پیش محرک القاء پذیر
induction	استقرا، القاء، القایش، انگیختگی
induration	اندیوریشن، سخت شدن همراه با تورم، مستحکم سازی
infarction	انفارکتوس، سکته قلبی
infectibility	ابتلاء پذیری، قابلیت سرایت/عفونت
infection	آلودگی، ابتلاء، بیماری عفونی، سرایت، عفونت، مرض
infectious	واگیر دار، عفونی، مسری
infectious aerosol	افشانه آلوده کننده
infectious agent	عامل عفونت زا/آلوده کننده
infectious disease	بیماری عفونی
infectious material	ماده عفونی
infectiousness	آلودگی، سرایت، عفونت
infective	عفونت زا، عفونی، گند زا
infectivity	عفونت زایی، گند زایی
infestation	آلودگی، آلودگی به انگل/حشرات، سرایت، عفونت انگلی
infiltrate	از صافی رد کردن، تراویدن، رسوخ/نفوذ دادن/کردن
infiltration	تراوش، راه یابی، رخنه، رسوخ، نفوذ
information exchange	تبادل اطلاعات
information RNA	آر.ان.ای. اطلاعاتی
informational molecule	ملکول خبر رسان
informofer	اینفورمرفر
informosome	اینفورموزوم
in-frame start codon	کدون آغازین در چارچوب
in-frame stop codon	کدون پایانی در چارچوب
infrared	مادون قرمز
ingestion	بلع، فروبری، قورت دادن
inhalation	استنشاق، بخور، تنفس
inhalation exposure	قرارگیری در معرض استنشاق
inhaler	افشانه تنفسی، اینهیلر
inheritable	ارثی
inheritance	ارثیه، صفات ارثی
inheritance of acquired characteristic	به ارث بردن مشخصه اکتسابی
inherited	به ارث برده
inherited characteristic	مشخصه اکتسابی
inhibition	بازدارندگی، بازداری، تعدیل، سرکوب، وقفه، منع، مهار کنندگی
inhibitor	بازدارنده، سرکوبگر
inhibitor of sister chromatid separation	سرکوبگرجداسازی کروماتیدهای خواهر
inhibitory zone	منطقه بازداری

228

English	Persian
initial cell	سلول اولیه
initiation	آغاز، آغازگری، پیدایش، تحریک، شروع
initiation codon	رمز آغازین، کدون آغاز
initiation factor (IF)	عامل آغازین/اولیه
initiation factor 2	عامل آغازین 2
initiator	آغازگر
initiator codon	اینیشیتور کدون، کدون شروع کننده
injector	انژکتور، تزریق کننده، درون پاش، سوخت پاش، فشاننده، وسیله تزریق
innate immune response	واکنش ایمنی درون ذاتی
innate immune system	سیستم ایمنی سرشتی، نظام ایمنی مادر زادی
innocuousness	بی آسیبی/زیانی/ضرری
Ino (= inosine)	
inoculum	مایه آبله کوبی/تلقیح
inorganic phosphate	فسفات غیر آلی
inosine	اینوزین
inosine mono-phosphate	اینوزین مونو-فسفات
inosine tri-phosphate	اینوزین تری فسفات
inositol	اینوزیتول
inositol 1,4,5-tri-phosphate	تری فسفات-اینوزیتول،1،5،4
inositol lipid	اینوزیتول لیپید
Inr (= initiator)	
inroganic	غیر آلی، کانی، معدنی
INS (= insertion)	آی.ان.اس.
insect	حشره
insect cell culture	کشت سلول حشره
insecticide	حشره کش، دافع حشرات، ماده حشره کش
insertion	دخول، درون جای دهی، رخنه
insertion inactivation	غیر فعال سازی درون جای دهی
insertion mutation	جهش دخولی/رخنه ای
insertion or duplication	درون جای دهی یا نسخه برداری
insertion sequence	توالی درون جای دهی
insertion vector	ناقل درون جای دهی
insertional knockout system	سیستم ناک اوت دخولی
insidious	تدریجی، غافلگیر کننده، نهانی
in-silico	این سیلیکو
in-silico biology	این سیلیکو بیولوژی
in-silico screening	این سیلیکو اسکرینینگ، غربالگری کامپیوتری
in-situ	به موضع، در جا، درجای خود/طبیعی، در محل، در محل طبیعی
insitu (= in-situ)	
in-situ condition	وضعیت در جا
in-situ conservation	حفاظت در محل
in-situ conservation of farm animal genetic diversity	حفاظت تنوع ژنتیکی دام و طیور در محیط طبیعی
in-situ gene bank	مخزن ژن در محل/در جای خود/درجای طبیعی/در محل طبیعی/در جا/به موضع
Insp₃ (= inositol 1,4,5-tri-phosphate)	
insulin	انسولین
insulin growth factor	عامل رشد انسولین
insulin-dependent diabetes mellitus	ملیتوس دیابتی وابسته به انسولین
insurance value	ارزش بیمه گذاری
intake	درون جذب، مدخل، مکش، ورودی
intake rate	میزان درون جذب/ورودی، نرخ مکش/ورودی
intasome	اینتازوم
integral membrane protein	پروتئین غشای اصلی
integrated	تلفیقی، منسجم
integrated crop management	مدیریت تلفیقی گیاهان زراعی
integrated disease management	مدیریت تلفیقی بیماری
integrated pest management (IPM)	مدیریت تلفیقی آفات
integration	انتگراسیون
integration factor	عامل تلفیق

English	Persian
integrin	انتگرین
intein	اینتئین
intended release	آزادسازی مورد نظر، انتشار دلخواه/مطلوب، رهاسازی دلخواه
interallelic complementation	مکمل سازی بین آللی
interamolecular association	جفت شدن بین مولکولی
interband	اینترباند، میان نوار
intercalary deletion	حذف اینترکالاری/تداخلی/میان بافتی
intercalating agent	عامل تداخلی/موتاسیون زا
intercellular adhesion molecule	ملکول چسبندگی بین سلولی
intercellular transport	ناقل بین سلولی
intercistronic region	ناحیه بین سیسترونی
intercropping	کشت درهم ردیفی
interference	تداخل عمل
interferon	انترفرون، اینترفرون
interferon inducer	القاگر اینترفرون
interferon-beta	اینترفرون بتا
intergenerational equity	بی طرفی بین نسلی
intergenic region	منطقه بین نژادی، ناحیه بین گونه ای
interkinesis	اینترکینزیس
interleukin	اینترلوکین
intermediary filament	رشته حد وسط، رشته سیتواسکلتی واسطه
intermediary metabolism	دگر گوهرش میانجی، سوخت و ساز واسطه ای، متابولیسم واسطه ای
intermediate-density lipoprotein	لیپوپروتئین با غلظت متوسط
internal radiation	تابش داخلی/درونی
internal ribosome entry site	محل ورود ریبوزوم داخلی
internalin	اینترنالین
interphase	اینتر فاز
interphase cycle	چرخه اینترفاز
interspecific hybrid	دورگه بین گونه ای/دوگونه
interspecific variation	گوناگونی
interstitial cell-simulating	هورمون اینترستیشیال

English	Persian
hormone (ICSH)	محرک سلول
intervening sequences	توالی های مداخله گر
intoxication	مستی، مسمومیت
intracellular	داخل/درون سلولی/یاختگی
intracellular membrane	غشاء درون سلولی
intracellular transport	انتقال درون سلولی
intradermal	درون پوستی
intravenous antibiotic	آنتی بیوتیک داخل وریدی، پادزیست درون سیاهرگی
intravenous therapy	درمان تزریقی/داخل وریدی
intrinsic protein	پروتئین اصلی/ذاتی/درونی
introduced species	گونه وارد شده
introduction	بومی سازی، معرفی، وارد سازی
introgression	دخول، وارد کردن، ورود
introgressive	هم تیرگی درون رونده
intron	انترون
inulin	اینولین
INV (= inversion)	آی.ان.وی.
INV (= inverted)	آی.ان.وی.
invagination	توی خود برگشتی، توی هم رفتگی، درون نیامی، غلاف شدگی
invasin	اینوازین، پروتئین ترا شامه ای
invasive	تهاجمی
invasive species	گونه مهاجم
invasiveness	قدرت تهاجم
inventorying	فهرست برداری
inversion	برگشتگی، تقلیب، معکوس شدگی، وارونش، وارونگی، واژگونی
inverted	معکوس
inverted micelle	گروه ملکولی وارونه
inverted terminal repeat	تکرار پایانه معکوس
investigational new drug	داروی جدید تحقیقی
in-vitro	خارج بدنی، در شیشه/محیط کشت/لوله آزمایش
in-vitro culture	کشت در شیشه
in-vitro evolution	تکامل/فرگشت در محیط کشت
in-vitro selection	گزینش در محیط کشت

in-vivo	در جاندار/زنده/موجود زنده/نسج زنده، محیط زنده
ion	یون
ion channel	مجرای/منفذ یون
ion exchanger	مبادله کننده یونی
ion trap	تله یونی
ion-channel-binding toxin	سم اتصال مجرای یون
ion-exchange chromatography	رنگ نگاری تبادل یونی
ionization	تجزیه یونی، یونش، یونیزاسیون، یونیزه شدن/کردن
ionized atom	اتم یونیده
ionizing	یوننده، یونیزه کننده
ionotropic	گرایش یونی
IR (= infrared)	آی.آر.
IRES (= internal ribosome entry site)	آی.آر.ئی.اس.
iron bacteria	باکتری آهن
iron deficiency anemia	آنمی کمبود آهن، کم خونی ناشی از کمبود آهن
iron response protein	پروتئین واکنش آهن
IRP (= iron response protein)	آی.آر.پی.
irradiation	پرتو افشانی/تابی/درمانی /دهی، تشعشع
irritability	برانگیختگی، تاثیر/تحریک پذیری
ISCOM (= immune-stimulating complexes)	آی.اس.سی.او.ام.
isoacceptor tRNA	تی.آر.ان.ای. ایزو آکسپتور
isochromosome	ایزوکروموزوم
isoderivative chromosome	کروموزوم هم اشتقاقی
isodisomy	ایزودیزومی
isoflavin	ایزوفلاوین
isoflavone	ایزوفلاون
isoflavonoid	ایزوفلاوونوئید
isogenic lines	سلسله های ایزوژنیکی
isogenic stocks	نژادهای ایزوژنیک
isograft	ایزو-پیوند، ایزوگرافت

isolated human growth hormone deficiency	کمبودمجزای هورمون رشد انسانی
isolation	انزوا، تنها/جدا سازی، جدا (یش)/(یی) عایق بندی
isoleucine	ایزولوسین
iso-leucine	ایزولوسین
isomer	ایزومر، هم پار/فرمول/هسته
isomerase	ایزومراز
isoprene	ایزوپرن
isosemantic substitution	جایگزینی ایزوسمانتیکی
isotachophoresis	ایزو تاکو فورز
isothiocyanates	ایزوتیوسیانات ها
isotope	ایزوتوپ، یکسانگرد، هم پروتون/سان
isozyme	ایزو آنزیم، ایزوزیم، جور زیما/زیمایه
ISS (= inhibitor of sister chromatid separation)	آی.اس.اس.
itaconic acid	اسیدایتاکونیک
itching	خارش (ی)/دار، خاریدن
ITF (= integration factor)	آی.تی.اف.
ITP (= inosine tri-phosphate)	آی.تی.پی.
ITR (= inverted terminal repeat)	آی.تی.آر.
IVF (= in vitro fertilization)	آی.وی.اف.
IVSs (= intervening sequences)	آی.وی.اس. ها

J

J chain	زنجیره جی.
Jak (= janus kinase)	
janus kinase	کیناز جانوس
jasmonate cascade	آبشیب ژاسمونات
jasmonic acid	اسید ژاسمونیک
JGS (= juvenile galactosia lidosis)	جی.جی.اس.
Jimson weed	علف جیمسون
joining (J) segment	بند/قطعه اتصال
jugular vein	سیاهرگ گردن

jumping gene	ژن پرشی/جا به جا شدنی/متحرک
juncea	خردل صحرایی/وحشی
junctional sliding	جانکشنال اسلایدینگ
junk DNA (= selfish DNA)	دی.ان.ای.آشغال
juvenile galactosia lidosis	گالاکتوزیا لیدوز نابالغ
juvenile hormone	هورمون نابالغ

K

kainic acid	اسید کاینیک
kairomone	کایرومون
kallidin	کالیدین
kallikerin	کالیکرین
kanamycin	کانامیسین
kanR	کن آر
Kaposi's sarcoma	سارکومای کاپوزی
Kappa particles	قطعات کاپا
karnal bunt	کارنال بونت
karyogamy	کاریوگامی، هسته زامی
karyogram	کاریوگرام، هسته نگاره
karyokinesis	کاریوکینزیس، هسته جنبی
karyon	کاریون، هسته یاخته
karyosome	کاریوزوم
karyotype	کاریوتیپ، نماد کروموزومی، هسته مونه
karyotyper	هسته مونه گر
kasugamycin	کازوگامیسین
katabolism	کاتابولیسم
katal (Kat)	کاتال
katanin	کاتانین
kb (= kilobase)	کیلو باز
kDa (= kilodalton)	
kefauver rule	قانون کفاور
Kennedy disease	بیماری دیستروفی میوتونیک، مرض کندی
kephalin	کفالین
keratan sulfate	سولفات کراتان
keratin	شاخینه، کراتین، ماده شاخی
keratinization	کراتینیزه سازی

keratinocyte	کراتینوسیت
keratocyte	کراتوسیت
keratosulfate	کراتوسولفات
keto acid	اسید کیتو
ketogenesis	کیتوژنز
ketone	کیتون
ketone body	پیکره کیتون
ketonemia	کیتونمی
ketonuria	کیتونوری
ketoprofen	کیتوپروفن
ketose	کیتوز
ketosis	کیتوزیس
keystone species	گونه مبنا
killed vaccine	واکسن کشته
killer cell	سلول کشنده
killer T cell (=cytotoxic cell)	سلول تی. کشنده
kilobase (Kb)	کیلوباز
kilobase pair	جفت کیلو باز
kilodalton	کیلودالتون
kinase	کیناز
kindred	خاندان، خویشاوندان
kinesin	کینزین
kinetic	جنبشی
kinetic proofreading	ویرایش کینتیک
kinetin	کینتین
kinetochore	کینتوکور
kinetosome	کینتوزوم
kinin	کینین
kininase	کینیناز
kininogen	کینینوژن
kinome	کینوم
kinomere	کینومر
kirromycin	کایرومایسین
kistrin	کیسترین
Klinefelter syndrome	سندروم/نشانگان کلاین فلتر
knapsack	نپساک
knockdown	ناک داون
knockin	ناک این
knockout	از پا در آوردن، از کارانداختن، انداختن، خارج/خراب/کله پا کردن

knockout mouse	موش ناک اوت
knottins	ملکولهای گره ای
kojic acid	اسید کوجیک
Konzo (= lathrism)	کونزو
koseisho	کوسیشو
kozak sequence	توالی کوزاک
kpnl	کی.پی.ان.ال.
kratein	کراتئین
krebs cycle	چرخه/دوره/سیکل کربس
krüppel protein	پروتئین کروپل
Kunitz trypsin inhibitor	بازدارنده تریپسین کونیتز

L

LA (= left atrium)	ال.ای.
LA (= linkage analysis)	ال.ای.
label	اتیکت، برچسب، علامت
labile	بی ثبات، تاثیر/تغییر پذیر، سست، فعال، گذرنده، لغزنده، ناپایدار
labile toxin	توکسین/سم بی ثبات/سست/لابایل
labling	برچسب، برچسب زدن، علامت گذاشتن، نشاندار کردن
lac operon	اپرون لاک/لاکتوز
lac selection	گزینش لاک
laccase	لاکاز
lachrymal fluid	مایع اشکی/لاکریمال
lactalbumin	لاکتآلبومین
lactase deficiency	کمبود لاکتاز
lactate dehydrogenase	دهیدروژناز لاکتاز
lactate de-hydrogenase	دهیدروژناز لاکتات
lactic dehydrogenase	لاکتیک دهیدروژناز
Lactobacillus	لاکتوباسیل
lactoferricin	لاکتوفریسین
lactoferrin	لاکتوفرین
lactoferritin	لاکتوفریتین
lactogen	لاکتوژن
lactoglobulin	لاکتوگلوبولین
lactonase	لاکتوناز

lactoperoxidase	لاکتوپروکسیداز
lactose	لاکتوز
lactose repressor	سرکوبگر لاکتوز
lag phase	دوره/مرحله تاخیر/تاخر/لنگی
lagging strand	رشته پسرو
LAK cell (= lymphokin activated killer cell)	سلول ال.ای.کی.
lambda bacteriophage	باکتری خوار لاندا
lambda phage	باکتری خوار/فاژ لاندا
laminin	لامینین
lampbrush chromosome	کروموزوم لمپ براش
landfill	خاکچال
landfill gas	گاز زباله دان
landrace	رقم محلی، نژاد بومی
lane	خط، راه، مسیر
langerhans cell	سلول لانگرهانس
lanolin	لانولین
lantibiotic	لانتیبیوتیک
lariat	کمند
laser capture microdissection	ریزشکافی لیزری
laser inactivation	غیرفعال سازی لیزری
LAT (= latency-associated transcript)	ال.ای.تی.
late effect	تاثیر تازه
late gene	ژن های تازه
late protein	پروتئین تازه
latency	کمون، نهان بودگی، نهانی، نهفتگی
latency period	دوره نهفتگی، زمان درنگ
latency-associated transcript	نسخه برداری مربوط به نهفتگی
latent virus	ویروس خفته/نهفته
lateral transfer	انتقال جانبی
lathrism	لاتریسم
lathyrism	لاتیریسم
laurate	لوریت (یک اسید چرب)
lauric acid	اسید لوریک
lavendustin	لاوندوستین
law of dominance	قانون غالبیت
law of independent	قانون تفکیک مستقل (ژنها)

assortment	
law of segregation	قانون جداشدن یا انفصال
lawn	چمن، لایه باکتری (روی محیط کشت)
lazaroid	لازاروئید ها
LCAT (= lecithin-cholesterol-acyl-transferase)	ال.سی.ای.تی.
LCR (= locus control region)	ال.سی.آر.
LD 50	میزان کشندگی 50 درصدی
LDH (= lactate de-hydrogenase)	ال.دی.اچ.
LDL (= low-density lipoprotein)	ال.دی.ال.
LE (= lupus erythematosus)	ال.ئی.
leader	پیشرو، راهنما، نوساقه
leader peptide	پپتید پیشرو
leader sequence	توالی راهنما
leading strand	رشته پیشرو
leaky mutant	جهش یافته ناقص
leaky mutation	جهش بی صدا/خفیف/ساکن
lear	لیر
lecithin	لسیتین
lecithin-cholesterol-acyl-transferase	ترانسفراز لیستئین-کلسترول-آسیل
lectin	لکتین
left atrium	دهلیز چپ
leghemoglobin	لگوگلوبین، لگ هموگلوبین
legume	بنشن، سبزی، غلاف، گیاه خوردنی، نیام، نیامک، لوبیا
legumin	لگومین
leishmaniasis	لیشمانیاز
lentinan	لنتینان
LEP (= leptotene)	ال.ئی.پی.
leptin	لپتین
leptin receptor	گیرنده لپتین
leptonema (= leptotene)	لپتونما
leptotene	لپتوتین
Lesch Nyhan syndrome	سندروم/نشانگان لش نایهان

lethal	کشنده، مهلک
lethal concentration	غلظت کشنده
lethal factor	عامل کشنده
lethal gene	ژن کشنده
lethal mutant	جهش یافته کشنده
lethal mutation	جهش کشنده
lethal toxin	زهرابه مهلک، سم کشنده
Leu (= leucine)	
leucine	لوسین
leucine zipper	زیپ لوسین
leukemia inhibitory factor	عامل سرکوب لوکمی
leukocyte	سفید گویچه/یاخته، گلبول سفید، لوکوسیت
leukocytosis	سفید گویچه تبسی، لوکوسیتوز
leukopenia	سفید گویچه کاستی، لوکوپنی
leukotriene	لوکوترین
leupeptin	لوپپتین
level	تراز، حد، سطح، صاف، مسطح
levorotary	چپ بر/گرد/گردان/گردانی/گرد ش
levulose	لو لوولوز
Leydig cell	سلول لایدیگ
LGMD2A (= limb-girdle muscular dystrophy type 2A)	ال.جی.ام.دی.2.آ.
LH (= luteinizing hormone)	ال.اچ.
LI repeat sequence	توالی تکراری ال.آی
liability	مسئولیت
liaison	ارتباط، رابط، رابطه، ربط
library	مخزن
LIF (= leukemia inhibitory factor)	ال.آی.اف.
life cycle	چرخه زندگی
life history	تاریخچه زندگی (از لقاح تا مرگ)
lifetime	دوره، طول/مدت زندگی/عمر، عمر

English	Persian
ligand	لیگاند
ligand-activated transcription factor	عامل نسخه برداری فعال شده توسط لیگاند
ligandin	لیگاندین
ligase (DNA)	سنتتاز، لیگاز
ligate	اتصال، لیگیت
ligation	انعقاد، بستن رگ، بند لیگاسیون
light chain	زنجیره سبک
light-chain variable	متغیر زنجیره سبک
light-harvesting complex	کمپلکس برداشت سبک
light-scattering	پخش سبک
lignan	لیگنان
lignin	چوب مایه، چوبینه لیگنین، ماده چوب (ی)
lignocellulose	لیگنوسلولز
limb-girdle muscular dystrophy type 2A	دیستروفی عضلانی کمر وپائین تنه نوع 2آ.
limit dexterin	دکسترین معیار
limiting factor	عامل محدودکننده
limonene	لیمونن
lincomycin	لینکومایسین
line	رگه
LINE (= long interspersed nuclear element)	ال.آی.ان.ئی.
lineage	اجداد، اصل و نسب، اعقاب، تبار، خاندان، دودمان
linear energy transfer	انتقال خطی انرژی
line-source	چشمه خطی
line-source delivery system	سیستم انتقال چشمه خطی
linewidth	عرض خط
linkage	اتصال، ارتباط، بست، بند، پیوستگی، پیوند، پیوند یافتگی، وابستگی
linkage analysis	تحلیل اتصال
linkage group	گروه پیوسته
linkage map	نقشه وابستگی/پیوستگی
linked enzyme assay	سنجش آنزیم وابسته
linked gene/marker	ژن/نشانگر پیوسته
linker	اتصال دهنده
linker DNA	دی.ان.ای.رابط
linking	اتصال
linoleic acid	اسید لینولئیک
linolenic acid	اسید لینولنیک
lipase	لیپاز
lipemia	لیپمی
lipid (= fat)	چربی، لیپید
lipid bilayer	دولایه چربی
lipid micelle	میسل لیپید
lipid monolayer	تک لایه لیپید
lipid raft	شناور چربی
lipid sensor	حسگر چربی/لیپید
lipid vesicle	ریزکیسه چربی
lipidomics	لیپیدومیکز
lipoamide	لیپوآمید
lipocaic	لیپوکائیک
lipofection	لیپوفکشن
lipofuscin	لیپوفوسین
lipogenesis	چربی زایی، لیپوژنز
lipolysis	چربی کافت (ی)، لیپولایز
lipolytic enzyme	آنزیم تجزیه کننده/کاهنده چربی
lipophilic	چربی/لیپید دوست/خواه
lipo-poly saccharide	لیپو پلی ساکارید
lipopolysaccharide (LPS)	لیپوپلی ساکارید
lipoprotein	لیپو پروتئین
lipo-protein lipase	لیپو پروتئین لیپاز
lipoprotein-associated coagulation	انعقاد ناشی از لیپوپروتئین
liposome	چربی تن، لیپوزوم
lipotropin	لیپو تروپین
lipovitellin	لیپوویتلین
lipoxidase	لیپوکسیداز
lipoxygenase	لیپوکسی ژناز
lipoxygenase null	لیپوکسی ژناز صفر
liquid	آب، آب گونه، بی ثبات، مایع
liquid crystal	کریستال مایع
liquid-crystalline phase	فاز کریستالین مایع
liquid-scintillation counting	شمارش درخشش-مایع
listeria monocytogene	لیستریا مونوسیتوژن
lithotroph	لیتوتروف

live cell array	آرایه سلول زنده
live vaccine strain	جوره، رقم، سویه، واریته واکسن زنده
liver	بیماری کبد، جگر، جگر سیاه، کبد
living modified organism	موجود زنده تغییر یافته
lobe	لپ
localized	محلی، منطقه ای، موضعی
localized melting	ذوب منطقه ای
loci	لوکای، موقعیت
lock-and-key model	مدل قفل و کلیدی
lock-washer structure	ساختار لاک-واشر
locomotion	انتقال، تغییر مکان، جا به جایی، حرکت
locus	جایگاه، لوکوس، مکان، مکان هندسی
locus control region	منطقه/ناحیه کنترل لوکوس
log growth phase	مرحله رشد لگاریتمی
log phase (= logarithmic phase)	
logarithmic phase	مرحله لگاریتمی/نمایی
London Fog foggers	سمپاش/فاگر (های) لندن فاگ
long arm of a human chromosome	شاخه بلند کروموزوم انسانی
long interspersed nuclear element	عنصر پراکنده هسته ای طولانی
long terminal repeat	تکرار پایانه بلند
long-acting thyroid stimulator	محرک تیروئیدی دراز مدت
long-range biological standoff detection system	سیستم شناسائی ابر بیولوژیکی دوربرد
long-term culture-initiating cells	سلولهای آغاز گر کشت بلند مدت
loricrin	لوریکرین
loss-of-function mutation	جهش از بین برنده کارآئی
low calcium response plasmid	پلاسمید پاسخگو به کلسیم کم
low level waste	پسماند کم پرتو
low-density lipoprotein	لیپو پروتئین کم غلظت/با غلظت پائین

LPL (= lipo-protein lipase)	ال.پی.ال.
LPS (= lipo-poly saccharide)	ال.پی.اس.
LT (= labile toxin)	ال.تی.
LTC-IC (= long-term culture-initiating cells)	ال.تی.سی.-آی.سی.
LTR (= long terminal repeat)	ال.تی.آر.
LUC (= luciferase)	ال.یو.سی.
luciferase	لوسیفراز
luciferin	لوسیفرین
luliberin	لولیبرین
lumen	حفره، روزن، کاواک، لومن، مجرا، محفظه مرکزی غده
lumichrome	لومیکروم
lumiflavin	لومیفلاوین
luminesce	نور افشان/زا
luminescence	پرتو، روشنایی، لیانندگی، لیانی، نور، نور افشانی/زایی
luminescent assay	سنجش نورافشانی/نورزائی
luminophore	لومینوفور
lupine	باقلا مصری
lupus	سل جلدی، قرحه آکله، لوپوس
lupus erythematosus	لوپوس اریتماتوز
lutein	زرده، لوتئین
luteinizing hormone	هورمون محرک جسم زرد، هورمون محرک یاخته بینابینی پرولان بی
luteolin	لوتئولین
luteolysin	لوتئولایزین
luteolysis	لوتئولایز
lux gene	ژن لاکس/نور
lux protein	پروتئین لاکس/نور
lyase	لیاز
lyate ion	یون لیات
lycopene	لیکوپن
lymph	تنابه، لنف
lymphadenitis	التهاب گره های لنفی
lymphatic	تنابه ای، لنفی
lymphatic endothelium	درون پوشه لنفاوی
lymphocyte	لنفوسیت، لنف یاخته

lymphocytic choriomeningitis virus	ویروس کوریومننژیت لنفاوی
lymphocytosis	ازدیاد لنفوسیت ها در خون، لنفوسیتوز
lymphogranuloma	بزرگی غدد لنفاوی، بیماری هوچکین، لنفوگرانولوما
lymphokin activated killer cell	سلول کشنده فعال شده توسط لنفوکین
lymphokine	لنفوکاین
lyochrome	لیو کروم
lyonization	لیونیزاسیون
lyophilization	خشکانیدن انجمادی، یخ خشکانی
Lys (= lysine)	
lysate	لیزات
lyse	از بین بردن، کافتن، کافته، شدن، کشتن، لیز
lysine	لیزین
lysis	انحلال، انهدام، پاره گی، تجزیه، تلاشی، کافتندگی، لیز
lysogen	لیزوژن
lysogenesis	لیزوژنز
lysogenic	کافتی زا، لیزوژنیک
lysogenic infection cycle	چرخه عفونت لیزوژنیک
lysogeny	کافتی زایی، لیزوژنی
lysophosphatidylethanolamine	لیزوفسفاتیدیل اتانولامین
lysosomal storage disease	بیماری تجمع لیزوزومی
lysosome	کافنده تن، کافینه تن، لیزوزوم
lysostaphin	لیزواستافین
lysozyme	کاف زیما، کافتی زیما، لیزوزیم
lytic	تجزیه ای، تحلیلی، کافتی، کافنده، مربوط به تجزیه
lytic cycle	چرخه کافتی
lytic infection	آلودگی کافنده
lytic infection cycle	چرخه عفونت کافتی
lytic pathway	مسیر کافتی

M

M (= male)	ام.
M (= mitosis)	ام.
M I (= first meiotic metaphase)	ام.1
M II (= second meiotic metaphase)	ام.2
M phase	مرحله ام.
M13	ام. 13
MAbs (= monoclonal antibodies)	
MAC (= mammalian artificial chromosome)	ام.ای.سی
MACE (= membrane anchor cleaving enzyme)	ام.ای.سی.ئی.
machine	چارا، چرخ کردن، دستگاه، ماشین (ی)،میونگ
macrolide	ماکرولاید
macromolecule	بزرگ مادیزه، ماکرومولکول، ملکول درشت
macromutation	درشت جهش، ماکروموتاسیون
macronutrient	درشت خوراک، ماکرونوترینت
macrophage	بیگانه خوار، درشت خوار، کلان خوار، ماکروفاژ
macrophage activiation	فعال سازی درشت خوار
macrorestriction map	نقشه ماکرو ریستریکشن
macule	لکه کوچک پوستی/جلدی، ماکول
maculopapular	ماکولاپاپولار
Mad Cow Disease	بیماری جنون گاوی
magainin	ماگائینین
magic bullet	گلوله جادویی
magnesium sulfate	سولفات منیزیم
magnetic antibody	پادتن مغناطیسی
magnetic bead	دانه های جادویی
magnetic cell sorting	جدا سازی سلول مغناطیسی
magnetic labeling	علامت گذاری مغناطیسی
magnetic particle	ذره مغناطیسی

Maillard reaction	واکنش میلارد
major groove	فرورفتگی بزرگ
major histocompatibility complex	عارضه عمده بافت سازگاری، کمپلکس بافت سازگاری اصلی، کمپلکس هیستوکمپتیبیلیته اساسی
maker gene	ژن سازنده
maker rescue	نجات سازنده
malate dehydrogenase	دهیدروژناز مالات
male	مذکر، نر
maleness	نرینگی
male-sterile	نرعقیم
malformation	بد ریختی/شکلی، ناقص الخلقگی، ناهنجاری، نقص عضو
malic acid	اسید مالیک
malignant	بدخیم، خطرناک، زهرآلود، سمی، کشنده، مهلک
malignant edema	آماس بدخیم، خیز مهلک
malnutrition	بد غذائی، بدی تغذیه، تغذیه ناقص، سوء تغذیه
MALT (= mucosa-associated lymphoid tissue)	ام.ای.ال.تی.
mammalian artificial chromosome	کروموزوم مصنوعی پستانداران
mammalian cell culture	کشت سلول پستاندار
mammary tumor virus	ویروس تومور پستان
mandrake root	ریشه تمیس/مهر گیاه
mannan	مانان
mannan oligosaccharide	مانان الیگوساکارید
mannanoligosacchariddes	مانانولیگو ساکارید ها
mannitol	مانیتول
mannogalactan	مانوگالاکتان
mannose	مانوز
mannosidostreptomycin	مانوسیدواسترپتومایسین
mannuronic acid sidostreptomycin	سیدواسترپتومایسین اسید مانورونیک
manometry	مانومتری
MAP (= mitogen activated	ام.ای.پی.

protein)	
map distance	فاصله نقشه
map unit	واحد نقشه
maple syrup urine disease (MSUD)	بیماری ادرار میپل سیروپی/شیره افرائی
mapping	بازنمایی، نقشه کشی، نگاشت
MAR (= matrix-associated region marker chromosome)	ام.ای.آر.
Marfan syndrome	سندروم مارفان
marine bacteria	باکتری دریائی
marine microorganism	میکروارگانیسم دریائی
marine toxin	توکسین/زهرابه/سم دریایی
marker	علامت، نشان، نشانگر، نشانه
marker enzyme	آنزیم نشانگر
marker sequence	توالی نشانگر
marker-assisted breeding	تخم کشی به کمک نشانه
marker-assisted breeding	اصلاح نژاد به کمک نشانه
marker-assisted selection	انتخاب به کمک نشانه
Marple Aerosol Generator	مولد افشانه مارپل
marrow repopulation ability	استعداد بازسازی مغز استخوات
marsh gas microorganisms	میکروارگانیسم های گاز مرداب
mash	پوره، له
mask	پرده، پوشانه، پوشش، ماسک (زدن)، نقاب
mass screening	جداسازی توده ای، غربال کردن گسترده
mass selection	انتخاب انبوه
mass spectrometer	طیف سنج جرمی
mass-applied genomics	ژن شناسی کاربرد توده ای
mass-casualty biological weapon	جنگ افزار بیولوژیکی کشتار انبوه
massively parallel signature sequencing	توالی یابی حجیم موازی امضائی
mast cell	سلول ماست، یاخته بافت هم بند، یاخته ماستوسیت

maternal inheritance	توارث مادری
maternal serum AFP	سروم مادری ای.اف.پی.
maternal serum alfa fetoprotein (MSAFP)	آلفافیتوپروتئین سروم مادری
mating	جفت گیری
mating type	نوع جفتگیری
matrix metalloproteinase	ماتریکس متالو پروتئیناز
matrix-associated region marker chromosome	کروموزوم نشانگر منطقه وابسته به ماتریکس
matroclinal inheritance	توارث متمایل به مادر
maturase	ماتوراز
maxicell	ماکسی سل
maximal medium	محیط کشت ماکسیمال
maximum contaminant level	حد اکثر سطح آلاینده
maximum enzyme velocity	حداکثر سرعت آنزیم
maximum permissible concentration	حداکثر تجمع/غلظت مجاز
maximum residue level	حداکثر پس مانده مجاز
maximum sustainable yield	حداکثر برداشت پایدار/معقول
maysin	مایسین
Mb (= megabase pairs)	ام.بی.
Mb (= myoglobin)	ام.بی.
MBP (= myelin basic protein)	ام.بی.پی
MCG (= multiple cloning group)	ام.سی.جی.
MCH (= mean corpuscular hemaglobin)	ام.سی.اچ.
MCHC (= mean corpuscular hemaglobin concentration)	ام.سی.اچ.سی.
MCK (= muscle creatine kinase)	ام.سی.کی.
MCV (= mean corpuscular volume)	ام.سی.وی.
MDP (= muramyl di-peptide)	ام.دی.پی.
MDS (= myelo-dysplastic syndrome)	ام.دی.اس.
mean corpuscular hemaglobin	میانگین هموگلوبین کورپوسکولار
mean corpuscular hemaglobin concentration	غلظت میانگین هموگلوبین کورپوسکولار
mean corpuscular volume	میانگین حجم کورپوسکولار
mean lifetime	زمان میانگین عمر، میانگین طول عمر
mean residue weight	میانگین وزن رسوبی
measurement	اندازه گیری، پیمایش، سنجش
mechanical cleaning	تمیز کردن مکانیکی
MeCP (= methyl-CpG-binding protein)	
median lethal dose	متوسط دوز کشنده واسطه
mediator	
medical control	کنترل پزشکی، مهار پزشکی
medical informatics	انفورماتیک/داده شناسی پزشکی
medifoods	غذاهای طبی
Mediterranean fruit fly	مگس میوه/سرکه مدیترانه ای
medium	محیط، محیط بدن، محیط عمل، محیط کشت، واسطه، وسیله
medium chain saturated fats	چربی های اشباع شده زنجیره محیط
medium chain triacyglyceride	تریاسیگلیسرید های زنجیره محیط
medium chain triglyceride	تری گلیسرید دارای 8-10 اتم کربن.
medium intake rate	متوسط نرخ ورودی، میزان متوسط جذب درونی
mega base pair	درشت جفت باز، مگا جفت باز
megabase	کلان باز، مگاباز
megabase cloning	دودمان سازی کلان باز
megabase pairs	جفت مگاباز
megaDalton	مگا دالتون
megakaryocyte stimulating factor	عامل تحریک یاخته کلان هسته ای
megaplasmid	مگاپلاسمید
mega-yeast artificial	کروموزوم مصنوعی

chromosome	کلان-مخمر
meiosis	تقسیم کاهشی، کاستمان، میوز
melanin	ملانین
melanocyte	ملانوسیت
melanoidin	ملانوئیدین
melanoma	ملانوما
MELAS (= mitochondrial myopathy, encephalopathy, lactic acidosis and stroke-like)	ام.ئی.ال.اس.
melatonin	ملاتونین
melibiose	ملیبیوز
melitriose	ملیتریوز
melting	ذوب سازی/شدن
melting temperature	دمای ذوب
membrane	پرده، پوسته، شامه، غشاء
membrane affinity seperation	جداسازی تمایل غشائی
membrane anchor cleaving enzyme	آنزیم شکافنده لنگر غشائی، آنزیم شکندی (شکننده)
membrane channel	کانال/مجرا غشاء
membrane digestion	هضم غشائی
membrane filter bioreactor	بیوراکتور فیلتر غشائی
membrane filtration	تصفیه غشایی
membrane transport	انتقال غشایی
membrane transporter protein	پروتئین انتقالگر غشایی
MEN (= multiple endocrine neoplasia)	ام.ئی.ان.
Mendelian inheritance	توارث مندلی
mendelian inheritance in man	توارث مندلی در انسان
Mendelian transmission	انتشار/انتقال مندلی
meninge	شامه گان
meningitis	شامه آماس، مننژیت
menopause	یائسگی
mental retardation	عقب ماندگی ذهنی
mentation	تقسیم دی.ان.ای.، تکه سازی
mercaptoethanol	مرکاپتواتانول

mercapturic acid	اسید مرکاپتوریک
mercury	جیوه، زیبق، سیماب، مرکور
Mercury Knapsack mistblower	اسپری مرکوری ناپساک
meromixis	مرومیکزی
merozygote	مروزیگوت
MERRF (= myoclonic epilepsy with ragged red fibers)	ام.ئی.آر.آر.اف.
mesenchymal adult stem cell	سلول/یاخته بنیادی/دودمانی بالغ میان آگنه ای
mesenchymal stem cell	یاخته بنیادی/دودمانی میان آگنه ای
meso carbon atom	اتم کربن مزو/میانی
meso compound	ترکیب مزو (میانی)
mesoderm	میان پوست
mesodermal	میان پوستی
mesodermal adult stem cell	یاخته بنیادی/دودمانی بالغ میان پوستی
mesodermic	میان پوستی
mesophile	میان دما
mesoscale	مزواسکیل، مقیاس متوسط
mesosome	مزوزوم
mesotocin	مزوتوسین
messenger RNA (mRNA)	آر.ان.ای. اطلاعاتی، آر.ان.ای. پیک، اسید ریبونوکلئیک پیامبر
Messenger™	مسنجر
Met (= methionine)	مت
meta chromatic leano dystrophy	دیستروفی متاکروماتیک لینو
metabolic	دگرگشتی
metabolic disturbance	به هم خوردگی متابولیکی، به هم ریختگی سوخت و سازی
metabolic engineering	طراحی متابولیکی، مهندسی سوخت وساز
metabolic flux analysis	تحلیل جریان سوخت و ساز، تحلیل سیلان متابولیکی
metabolic pathway	گذرگاه

240

English	Persian
	دگرگوهرشی/سوخت و سازی
metabolic poison	سم متابولیکی
metabolic pool	حوضچه متابولیکی
metabolic product	فرآورده دگر گوهرشی، فرآورده سوخت وساز، دگرگوهره
metabolism	دگرگشت، دگرگونی، سوخت و ساز، متابولیسم
metabolite	فرآورده سوخت وساز، دگرگشته، دگر گوهره، متابولیت، محصول متابولیسم
metabolite profiling	نمای متابولیتی
metabolome	متابولوم
metabolomics	متابولومیک
metabolon	متابولون
metabonomic signature	امضای متابونومیکی
metabonomics	متابونومیک
metacentric chromosome	کروموزوم متاسانتریکی، فام تن پس میانپار
metachromatic granule	ریزدانه دگر رنگ، گرانول متاکروماتیکی
metafemale	فرا/فوق/متا ماده
metalloenzyme	آنزیم فلزی
metalloflavoprotein	متالوفلاووپروتئین
metalloprotein	پروتئین فلزی
metallothionein	متالوتیونئین
metamodel	متا مدل
metanomics	متانومیکز
metaphase	پس چهره، متافاز
metaphase chromosome	کروموزوم متافازی
metaplasia	تغییر بافت، دگر دشتاری، متاپلازی
metastasis	دگر دیسی، فراگستری، متاستاز
meteorology	اقلیم شناسی، جو شناسی، علم کاینات هوا، هواشناسی
meter	اندازه گیر، -سنج، سنجه، -شمار، شمارگر، کنتور،

English	Persian
	متر
metered	با کنتور اندازه گیری شده، کنتور دار
metering	اندازه گیری
methanol	الکل تقلیبی، متانول
methionine	متیوناین، متیونین
methotrexate	متوترکسات
methyl jasmonate	متیل ژاسمونات
methyl salicylate	متیل سالیسیلات
methyl violet	بنفش متیلی
methylated	متیل دار شده
methylation	متیلاسیون، متیل دار کردن
methyl-CpG-binding protein	پروتئین اتصال متیل-سی.پی.جی.
methyl-directed mismatch repair	ترمیم ناهماهنگ توسط متیل
methylene blue	آبی متیلنی
methyl-guanine-DNA methyl-transferase	متیل ترانسفراز متیل-گوانین-دی.ان.ای.
methylophilus methylotrophus	متیلوتوفوس متیلوفیلوس
mevalonic acid	اسید موالونیک
MGMT (= methyl-guanine-DNA methyl-transferase)	ام.جی.ام.تی.
MHC (= major histocompatibilty complex)	ام.اچ.سی.
Mi (= microphtalmic)	ام.آی.
micelle	میسل
micro sensor	ریز حسگر
micro total analysis system	سیستم تحلیل تظاهر ژن
micro total analytical system	سیستم تحلیلی تظاهر ژن
microaerophile	اکسیژن خواه
microarray	ریز آرایه
microbalance	ترازوی حساس/میکرو، میکرو بالانس
microbe	زیوه، میکروب
microbial fermentation	تخمیر میکروبی
microbial filaments	فیلمانهای میکروبی
microbial film process	پروسه فیلم میکروبی

English	Persian
microbial floc	فلاک میکروبی
microbial mat	تشکچه میکروبی
microbial physiology	تنکار شناسی میکروبی، فیزیولوژی میکروبی
microbial polysaccharide	پلی ساکارید میکروبی
microbial sensor	حسگر میکروبی
microbial source tracking	ردیابی منشاء میکروبی
microbial technology	فناوری میکروبی
microbicide	میکروب کش
microbiological cell	سلول میکروبیولوژیکی
microbiology	زیوه شناسی، میکروبیولوژی
microbodymicrocarrier	میکروکاریر میکروبادی
microbology	میکروب شناسی
microchannel fluidic devices (= microfluidics)	دستگاههای ریز لوله سیالاتی
micro-electromechanical system	سیستم ریز-الکترومکانیکی، سیستم میکرو الکترومکانیکی
microenvironment	خرد محیط
microfilament	ریز میله
microfluidic chip	تراشه ریز سیال
microfluidics	ریز سیال شناسی
microgram	میکروگرم
microinjection	ریز تزریق
micromachining	ریز ماشینکاری
micrometer	ریز سنج، میکرومتر
micro-meter	میکرون، میکرومتر
micromodification	ریز تغییر
micron	میکرون
MicronAir	مایکرون ایر
Micronair AU4000	مایکرون ایر ای.یو. 4000
Micronair spray nozzle	پستانک/دهانه اسپری مایکرون ایر
microorganism	جاندار میکروسکوپی، موجود ذره بینی، میکروارگانیسم
micro-organism (=microorganism)	میکروارگانیسم
microparticle	ذرات بسیار ریز
microperoxisome	میکروپراکسی زوم
microphage	خرد/ریز خوار
microphtalmic	میکروفتالمیک

English	Persian
micropipette	ریز پیپت
microplasts	میکروپلاست ها
micropropagation	ریز ازدیادی
micro-RNAs	میکرو آر.ان ای. ها
microsatellite DNA	دی.ان.ای. ریز قمر
microscopy	ریزبینی، میکروسکوپی
microsequencing	ریز توالی گری
microsphere	ریز سپهر
microsystems technology	فناوری ریز سیستمها
microtome	ریز بر، میکروتوم
microtubule	ریز لوله (چه)
microwave bombardment	بمباران ریزموجی/میکرو ویو
mid-oleic sunflowers	آفتاب گردان های میان-اولئیکی
mid-oleic vegetable oils	روغن گیاهی میان-اولئیکی
migration inhibition factor	عامل سرکوب مهاجرت
mild effect	تاثیر خفیف/ملایم
milled	آسیا شده، فرزکاری شده، کنگره دار
milling	آجیده/آسیا/فرز کاری/کردن، کنگره دار کردن
millipore filteration	تصفیه میلیپور
Millon reaction	واکنش میلون
Millon's reagent	معرف میلون
MIM code number (= mendelian inheritance in man)	شماره کد ام.آی.ام.
mimetics	تقلیدی
mimetics	میمتیک
mineralocorticoid	مینرالوکورتیکوئید
mini	مینی
mini	کوچک
mini cell	ریز سلول
mini preparation	ریز آماده سازی
minimal medium	محیط کشت حداقل/مینیمم
minimal risk level	حداقل سطح ریسک
minimized domain	محدوده مینیم شده/مینیمایز شده
minimized protein	پروتئین حداقل/به حداقل رسیده/کوچک شده/مینیم شده/کمینه

minimum tillage	حداقل کشت و زرع	mixed cropping	کشت درهم
miniprotein	مینی پروتئین	mixed culture	کشت مخلوط
miniprotein domain	محدوده یک مینی پروتئین	mixed-function oxygenase	اکسیژناز کارکرد آمیخته
minisatellite	ریز قمر	mixing	آمیختن، اختلاط، امتزاج،
minor base	باز فرعی		مخلوط کردن
minor groove	فرورفتگی کوچک	MLD (= meta chromatic	ام.ال.دی.
minus strand	رشته منفی	leano dystrophy)	
minute volume	حجم در دقیقه	MLV (= murine leukemia	ام.ال.وی.
mischarging	در رفتن اشتباهی،	virus)	
	میسشارژ کردن	MMR (= methyl-directed	ام.ام.آر.
miscoding	رمزگذاری غلط	mismatch repair)	
misdivision haploid	هاپلوئید میسدیویژن	MMTV (= mouse mammary	ام.ام.تی.وی.
mismatch repair	اصلاح نامناسب، ترمیم	tumor virus)	
	ناهماهنگ	model organism	جاندار نمونه، موجود زنده
missense mutation	جهش بی شعور/نادرست		مدل
mist	مه، بخار، بخار گرفتن	modelling	الگو/اندام/نمونه/مدل سازی،
Mistblower	میست بلوئر		طراحی
MIT (= monoiodo-tyrosine)	ام.آی.تی.	modification enzyme	آنزیم مدیفیکاسیون
mithramycin	میترا مایسین	modifier	تغییر دهنده
mithridatism	میتریداتیزم	modifying factor	عامل تغییر دهنده
mitigation	تخفیف، تسکین، تعدیل	modifying gene	ژن تغییر دهنده
mitochondria	دشته تن ها، راکیزه ها،	modon	مودون
	میتوکندری ها	modulation	مدولاسیون
mitochondrial ATPase	آ.ت.پآز میتوکندریائی	moiety	سهم، نیم، نیمه
mitochondrial DNA	دی.ان.ای. میتوکندریایی	mold	الگو، بوزک، ریخت، قالب،
(mtDNA)			قالب گرفتن، کپک، کپک
mitochondrial myopathy,	میوپاتی، انسفالوپاتی،		زدن
encephalopathy, lactic	لاکتیک اسیدوز و شبه	mold deterioration	تخریب کپک
acidosis and stroke-like	سکته میتوکندریائی	mole	توده (درون رحم)، خال، خال
mitochondrion	دشته تن، راکیزه، میتوکندری		گوشتی، مل، موش کور
mitogen	میتوز زا، میتوژن		مول مولکول گرم
mitogen activated protein	پروتئین فعال شده توسط	molecular beacon	شعاع نور ملکولی
	میتوژن	molecular biology	زیست شناسی ملکولی
mitogen-activated protein	آبشار/آبشیب/پیشار پروتئین	molecular breeding	نژادگیری مولکولی
kinase cascade	کیناز فعال شده توسط	Molecular Breeding ™	مولکولار بریدینگ
	میتوژن	molecular bridge	پل ملکولی
mitogen-activated protein	پروتئین کیناز فعال شده	molecular chaperone	مراقب ملکولی
kinases (MAPK)	توسط میتوژن	molecular cloning	دودمان سازی ملکولی
mitomycin	میتومایسین	molecular diversity	تنوع/گوناگونی ملکولی
mitosis	تقسیم رشتمانی، رشتمان،	molecular evolution	تکامل/فرگشت مولکولی
	میتوز، هسته جنبی	molecular fingerprinting	انگشت نگاری ملکولی

243

molecular genetics	ژنتیک ملکولی
molecular knife	کارد مولکولی
molecular lithography	لیتوگرافی ملکولی
molecular machine	دستگاه/ماشین ملکولی
molecular mass	جرم/وزن ملکولی
Molecular Pharming ™	مولکولار فارمینگ
molecular profiling	نمای ملکولی
molecular sieve	الک/غربال ملکولی
molecular vehicle	حامل/ناقل ملکولی
molecular weight	وزن ملکولی
molecule	مولکول
Moloney murine leukemia virus	ویروس لوکمی موشی مولونی
MO-MLY (= Moloney murine leukemia virus)	ام.او-ام.ال.وای.
monarch butterfly	پروانه مونارک
monensin	موننزین
monkeypox	آبله میمون
mono zygote	تک تخمکی، مونوزیگوت
monoamine oxidase	اکسیداز مونوآمین
monocistron	مونوسیسترون
monoclonal	تک بنیانی/دودمانی
monoclonal antibodies	پادتن های تک دودمانی
monocotyledon	تک لپه ای، مونوکوتیلدون
monoculture	تک کشت
monocyte	تک هسته، گویچه تک هسته
monoecious	تک پایه، نر-ماده
monogenic	تک زا، تک زاد، تک ژن، تک منشاء
monogenic disorder/trait	اختلال تک ژنی، بی نظمی تک زایی
monoiodo-tyrosine	مونوئیدو تیروزین
monolepsis	مونولپسی
monomer	تک پار، زیر واحد، مولکول تک واحدی، مونومر
mononuclear	تک/یک هسته ای
mononuclear cell	سلول تک هسته ای
mononucleotide	مونونوکلئوتید
monoploid	تک لاد، مونوپلوئید، نیم دانه، نیمگان، نیمه گان،

	هاپلوئید
monosaccharide	تک/یک قندی، مونوساکارید
monosome	مونوزوم
monosomic	مونوزومی
mono-thio-glycerol	مونو-تیو-گلیسرول
monounsaturated fat	چربی تک اشباع نشده
monounsaturated fatty acid	اسید چرب تک اشباع نشده
monozygotic twins	دوقلوهای یکسان/مونوزیگوت/تک تخمکی
morbidity	ابتلاء، بیماری، شیوع مرض، فساد، ناخوشی
moribund	دم مرگ، رو به مرگ/نزع، محتضر
morphogenetic	ریخت زا
morphology	اندام/ترکیب/ریخت شناسی، ریخت شکل شناسی، مورفولوژی
MOS (= mannanoligo-saccariddes)	ام.او.اس.
mosaic	موزائیک(ی)
mosquito	پشه
motor protein	پروتئین حرکتی
mounted	کار گذاشته شده، نصب شده
mouse mammary tumor virus	ویروس تومور پستان موش
mouse-ear cress	کرس گوش موشی
movable genetic element	عامل متحرک ژنتیکی
MPOD (= myelo-peroxidose deficiency)	ام.پی.او.دی.
MPS (= muco-poly-saccharidosis)	ام.پی.اس.
MR (= mental retardation)	ام.آر.
MRA (= marrow repopulation ability)	ام.آر.ای.
mRNA (= messenger RNA)	ام.آر.ان.ای.
MRV (= mucosa rota virus)	ام.آر.وی.
MS (= multiple sclerosis)	ام.اس.
MSAFP (= maternal serum AFP)	ام.اس.ای.اف.پی.

multiple sclerosis	چند سختینگی، مالتیپل اسکلروز
multiple sulfatase deficiency	کمبود سولفاتاز چندگانه
multiplex	چند بخشی/تایی/شکلی/گانه /عضوی، مرکب
multiplex assay	سنجش/عیارگیری مرکب
multiplex PCR	پی.سی.آر. مرکب
multiplication cycle	چرخه تکثیر
multipotent	چند خاصیتی
multipotent adult stem cell	یاخته دودمانی بالغ چند خاصیتی (قابل تمایز)
multivalent	چند تایی/ظرفیتی
munition	جنگ افزار/مهمات
muramyl di-peptide	دی پپتید مورامیل
murein	مورئین
murine	مربوط به جوندگان، موراین
murine leukemia virus	ویروس لوکمی موشی
muscle creatine kinase	کیناز کریاتین عضلانی
Mut (= mutator gene)	
Mut (= mutator protein)	
mutable gene	ژن قابل جهش
mutagen	جهش زا، موتاژن
mutagen dipoxybutane	دیپوکسی بوتان جهش زا/موتاژن
mutagenesis	جهش زایی
mutagenic compound	ترکیب جهش زا
mutagenicity	جهش زایی، قابلیت ایجاد جهش
mutant	جهش یافته، موتان
mutase	موتاز
mutate	جهش دادن/کردن/یافتن
mutation	تغییر، جهش، دگرگونی، موتاسیون
mutation breeding	انتخاب از طریق جهش، به نژادی جهشی
mutator gene	ژن جهشگر
mutator protein	پروتئین جهنده
mutein	موتئین
muton	موتون
mutualism	تعاون، همیاری

MSD (= multiple sulfatase deficiency)	ام.اس.دی.
MSUD (= maple syrup urine disease)	ام.اس.یو.دی.
mtDNA	دی.ان.ای. ام.تی.
MTG (= mono-thio-glycerol)	ام.تی.جی.
MTV (= mammary tumor virus)	ام.تی.وی.
MTX (= Mrthotrexate)	ام.تی.اکس.
mucin	موسین
mucoid	بلغمی/مخاط ماند/مخاطی
mucoitin sulphate	سولفات موکوئیتین
muco-poly-saccharidosis	موکو-پلی-ساکاریدوز
Mucor	موکور
mucosa rota virus	ویروس موکوسا روتا
mucosa-associated lymphoid tissue	بافت لنفوئیدی وابسته به موکوس
mucous membrane	غشاء مخاطی
multi-agent munition	جنگ افزار/مهمات چند عاملی
multi-copy plasmid	پلاسمید/دشتیزه چند نسخه ای
multi-drug resistance	مقاومت داروئی چند گانه
multienzyme system	سازگان چند آنزیمی
multifactorial	چند عاملی
multifactorial disease	بیماری چند عاملی
multigene family	خانواده چند ژنی
multigenic	چند ژنی
multi-layered high-efficiency particulate air mask	ماسک چند لایه کاربرد بالای ذرات معلق در هوا
multi-locus probe	پروب/پرب/ردیاب/کاوشگر/ز شانگر چند نقطه ای
multiple aleurone layer (MAL) gene	ژن چند لایه آلورونی
multiple allele	آلل چند گانه
multiple cloning group	گروه همسانه ساز چندگانه
multiple cropping	کشت چندگانه
multiple endocrine neoplasia	نئوپلازی اندوکرین چندگانه
multiple myeloma	میلومای چند گانه

MYC (= MYC gene/ oncogene)	ام.وای.سی.
MYC gene/oncogene	ژن/انکوژن ام.وای.سی.
mycelium	مایسلیوم
mycoagglutinin	مایکوآگلوتینین
mycobacterium tuberculosis	مایکوباکتری سل
mycobiont	مایکوبیونت
mycorrhizae	قارچ-ریشه، میکوریز
mycotoxin	زهرابه، قارچ زهر، میکوتوکسین
myelin	میلین
myelin basic protein	پروتئین بازی مایلین
myelitis	التهاب طناب نخاعی/نخاع
myeloblastin	میلوبلاستین
myelo-dysplastic syndrome	سندروم مایلو-دیسپلاستیک
myeloma	تومور مغز استخوان، میلوما
myelo-peroxidose deficiency	کمبود مایلو-پروکسی دوز
myocard(ium)	عضله/ماهیچه دل/قلب، میوکارد
myoclonic epilepsy with ragged red fibers	اپیلپسی میوکلونیکی با فیبرهای قرمز آشفته
myoelectric signal	پیام الکتریکی ماهیچه
myofibril	میوفیبریل
myoglobin	میوگلوبین
myograph	ماهیچه نگار
myography	ماهیچه نگاری
myosin	میوزین
myotonic dystrophy	دیستروفی میوتونیک
myristoylation	مایریستولاسیون
myrothecium verrucaria	مایروتسیوم وروکاریا
myxomycetes	قارچهای مخاطی، مایزومایست
myxovirus	مایزوویروس، ویروس مخاطی
MZ (= mono zygote)	ام.زد.

N

n (= nano =10^{-9})	ان.
NAD, reduced form	فرم کاهش یافته ان.ای.دی.
NAD$^+$ (= nicotinamide	ان.ای.دی. مثبت

adenine dinucleotide)	
NADH (= NAD, reduced form)	ان.ای.دی.اچ.
nadir	حضیض، سمت القدم، نهایت افت
NADP, reduced form	فرم کاهش یافته ان.ای.دی.پی.
NADP$^+$ (= nicotinamide adenine dinucleotide phosphate)	ان.ای.دی.پی.مثبت
NADPH (= NADP, reduced form)	ان.ای.دی.پی.اچ.
nafcillin	نافسیلین
naive T cells	سلولهای تی. ساده
naked DNA	دی.ان.ای برهنه
naked gene	ژن برهنه
naloxone	نالوکسون
nanism	کوتولگی
nanobiology	ریز بیو لوژی
nanobot	ریز روبات
nanocochleate	نانوکوکلیات
nanocomposites	ریز ترکیبها
nanocrystal molecules	مولکولهای ریز کریستال
nanocrystals	ریز بلور
nanofluidics	ریز سیالات
nanogram	نانوگرم
nanolithography	ریز لیتوگرافی
nanometer	نانومتر
nanoparticle	ریز ذره
nanopore	ریز منفذ
nanopore detection	تشخیص ریز منفذ
nanoscience	ریز دانش
nanoshells	ریز گلوله
nanotechnology	ریز تکنولوژی
nanotube	ریز لوله
nanotube membrane	غشاء ریز لوله
nanowire	ریز سیم
napole gene	ژن ناپول
narcosis	بی هوشی، تخدیر، خواب رفتگی، هوش بری
naringen	نارینژن

246

NARK gene	ژن ان.ای.آر.کی.	nematodes	کرم های حلقوی/رشته ای/گرد/نخی شکل، نخسانه ها، نماتود ها
nascent cleavage	تقسیم در حال شکل گیری	neoantigen	پادگن جدید/نوساخته/نوظهور
nascent DNA	دی.ان.ای. در حال شکل گیری		
nascent RNA	آر.ان.ای. در حال شکل گیری	neocarzinostatin	نئوکارزینواستاتین
nasopharynx	حلق-بینی، حلق و بینی، گلو و بینی	neoendemics	نئواندمیکز
native conformation	انطباق/سازش طبیعی	neolignan	نئولیگنان
native species	گونه بومی	neomycin	نئومایسین
native structure	ساختمان بومی	neoplasia	نئوپلازی
naturaceuticals (=nutraceuticals)	نوتراسوتیکال	neoplasm	بافت نوساخته، تومور، نودشتینه، نئوپلاسم
natural forest	جنگل طبیعی	neoplastic growth	رشد مربوط به نوسازی/نئوپلاستی
natural immunity	ایمنی ذاتی/طبیعی		
natural killer	قاتل طبیعی	nephelometer	نفلومتر
natural killer cell (NK cell)	یاخته قاتل طبیعی	nephrocalcin	نفروکالسین
natural selection	انتخاب طبیعی	nephron	گردیزه
natural source	منبع طبیعی	nephrotis syndrome	سندروم نفروتیس
NDV (= newcastle disease virus)	ان.دی.وی.	NER (= nucleotide excision repair)	ان.ئی.آر.
nearest neighbor sequence analysis	تحلیل توالی نزدیکترین همسایه	nerve growth factor	عامل رشد عصب
near-infrared spectroscopy	طیف نمایی نزدیک به مادون قرمز	nested PCR	پی.سی.آر. جایگزین
		net present value	ارزش خالص کنونی
near-infrared transmission	انتقال مادون قرمز	netilmicin	نتیلمایسین
nebulin	نبولین	neural	عصبی، نورال
nebulizer	نبولایزر	neural tube defects	نقص لوله عصبی
necrobiosis	مرده زیستی، نکروبیوز	neuraminidase	نورامینیداز
necrosis	بافت مردگی، مرگ بافت/سلولها/نسج، فساد نسج	neurobiologist	پی زیست شناس، زیست شناس اعصاب
		neurobiology	پی زیست شناسی، زیست شناسی اعصاب
necrotic ulcer	زخم بافت مرده/نکروزی	neuro-fibromatosis	نورو-فیبروماتوز
necrotizing lymphadenitis	التهاب گره های لنفی کشنده نسج/نکروز آور	neuroglia	پی بان
		neurologic sequelae	ضایعه عصب شناختی
needs assessment	برآورد نیاز، تخمین مایحتاج	neuron	پی یاخته، نورون، یاخته عصبی
neem tree	درخت زیتون تلخ		
negative control	کنترل منفی	neuropsychiatric	عصبی-روانی، نوروسایکیاتریک
negative strand (= antisense strand)	رشته مکمل مثبت/منفی		
		neurotoxin	پی زهرابه، سم عصب (ی)
negative supercoiling	ابرپیچش منفی، نگاتیو سوپر کویلینگ	neurotransmitter	تراگسیلنده عصبی، ناقل عصبی

neutraceuticals	نوتراسوتیکال(ها)
neutral protease	پروتئاز خنثی
neutriceuticals	نوتریسوتیکال(ها)
neutropenic	نوتروپنیک
neutrophil	خنثی خواه، نوتروفیل
new animal drug application	کاربرد داروی جدید حیوانی
new drug application	کاربرد داروی جدید
new species	گونه های جدید
newcastle disease virus	ویروس بیماری نیوکاسل
NF (= neuro-fibromatosis)	ان.اف.
NGF (= nerve growth factor)	ان.اف.جی.
NH₂ (= amino group)	ان.اچ.2.
NHEJ (= non-homologus end joining)	ان.اچ.ئی.جی
niche	آشیانه، تورفتگی، زیستخوان، شکاف، کنام، موضع بوم شناختی، موقعیت مناسب
nick	پریدگی، شکاف، شکستگی
nick translation	ترجمه شکاف
nicked circle	دایره شکافته/شکسته
nicotinamide adenine dinucleotide	آدنین دی نوکلئوتید نیکوتین امید
nicotinamide adenine dinucleotide phosphate	آدنین دی نوکلئوتید فسفات نیکوتین امید
nif gene	ژن نیف
ninhydrin reaction	واکنش نینهیدرین
nisin	نیسین
nitrate	نمک اسید نیتریک، نیترات
nitrate bacteria	باکتری نیترات
nitrate reductase	نیترات ردآکتاز
nitrate reduction	احیاء/کاهش نیترات
nitric oxide	اکسید نیتروژن، نیتریک اکسید
nitric oxide synthase (NOS)	نیتریک اکسید سینتاز
nitrification	تبدیل نیتروژنی، شوره سازی/گذاری، نیتراته شدن
nitrifying bacteria	باکتری های شوره ساز
nitrilase	نیتریلاز

nitrite	نمک اسید نیترو، نیتریت
nitrocellulose	نیترات سلولز، نیتروسلولز
nitrogen base	باز نیتروژن دار، باز نیتروژنه
nitrogen cycle	چرخه/گردش ازت/نیتروژن
nitrogen fixation	تثبیت ازت
nitrogen metabolism	دگر گوهرش نیتروژن، سوخت و ساز ازت، متابولیسم ازت
nitrogenase	نیتروژناز
nitrogenase system	سیستم نیتروژناز
nitrogenous base	باز ازت/نیتروژن ساز
nitrogenous fertillizer	کود نیتروژنی
NK (= natural killer)	ان.کی.
nm (= nanometer)	ان.ام.
NMR (= nuclear magnetic resonance)	ان.ام.آر.
nod gene	ژن ناد
node	برآمدگی، بند، عقده، گره، ورم
nodulation	ندولاسیون
nodule	برآمدگی مخچه، گره ریز، گرهک، ندول
nogalamycin	نوگالامایسین
nojirimycin	نوجیریمایسین
nonchromosomal protein	پروتئین غیر کروموزومی
non-coding DNA	دی.ان.ای. غیر رمزگذار
non-coding parts of a gene	اجزاء غیر رمز ساز ژن
non-coding strand	رشته غیر رمز گذار
non-consumptive value	ارزش غیر مصرفی
nondisjunction	عدم/نا گسستگی
nonenteric	غیر روده ای
non-equilibrium theory	نظریه غیر تعادلی
non-equilibrium theory	تئوری غیر تعادلی
nonessential amino acid	اسید آمینه غیر ضروری
non-exclusive goods	کالاهای غیر انحصاری
nonheme-iron protein	پروتئین آهن غیر همی
nonhiston protein	پروتئین اسیدی/غیر هیستونی
non-homologous recombination	نوترکیبی غیر همتا/ناهمتا
non-homologus end joining	پیوند انتهای غیر همسان
non-indigenous species	گونه های غیر بومی

noninvasive	غیر تعارضی
nonmass casualty agent	عامل تلفات/کشتار غیر انبوه
non-Mendelian inheritance	وراثت غیر مندلی
non-point source	منبع نان-پوینت/عمومی
non-polar	غیر قطبی
nonpolar group	گروه غیر قطبی
nonproliferation	عدم تکثیر/افزونی/گسترش
nonsecretor	غیر مترشحه
nonsense codon	رمز/کودون بی معنی
nonsense mutation	جهش بی معنی
nonspecific symptom	عوارض غیر مشخص
non-starch polysaccharide	پلی ساکارید/چندقندی بدون نشاسته/فاقد نشاسته
non-target	غیر هدف
nontarget organism	جاندار غیر هدف
nontranscribed spacer	حد فاصل نسخه برداری نشده
nontraumatic	غیر تروماتیک/ضربه ای
non-use value	ارزش غیر کاربردی
non-viral retroelement	عنصر پسگرای غیر ویروسی
nonvolatile	نا/غیر فرار
no-observed adverse effects level	سطح عدم مشاهده اثر سوء
no-observed effects level	سطح عدم مشاهده اثر
normalizing selection	انتخاب متعادل کننده، گزینش بهنجار کننده
northern blot/transfer	نورترن بلات/ترانسفر
northern blotting	نورترن بلاتینگ
northern corn rootworm	کرم ریشه ذرت شمالی
northern hybridization	هم تیرگی شمالی، هیبریداسیون نورترن
Norum's disease	بیماری نوروم
NOS terminator	پایانه ان.او.اس.
nosocomial spread	اشاعه/شیوع بیمارستانی
notification	اخطاریه، اطلاع، اطلاعیه، گزارش
no-tillage crop production	تولید محصول کشاورزی بدون زراعت
novel trait	خصیصه جدید
novobiocin	نووبیوسین
nozzle	افشانک، پاشنده، پستانک،

	روزنه، دهانه، شیپوره، مایع پاش، ناودانک
NS (= nephrotis syndrome)	ان.اس.
N-segment	جزء/قطعه ان.
NTDS (= neural tube defects)	ان.تی.دی.اس.
N-terminal	پایانه ان.
nuclear DNA	دی.ان.ای.هسته ای
nuclear envelope	پوشش هسته ای
nuclear hormone receptor	گیرنده هورمون هسته ای
nuclear magnetic resonance	انعکاس مغناطیسی هسته ای، رزونانس مغناطیسی هسته
nuclear matrix	ماتریکس هسته ای
nuclear matrix protein	پروتئین ماتریکس هسته ای
nuclear receptor	گیرنده هسته ای
nuclear transfer	انتقال هسته ای
nuclear transplantation	پیوند هسته ای
nuclease	نوکلئاز
nuclease-free reagent	معرف بدون نوکلئاز
nucleic acid	اسید نوکلئیک
nucleic acid hybridization	هم تیرگی اسید نوکلئیک
nucleic acid probe	پرب/پروب/ردیاب/کاووشگر/از شانگر اسید نوکلئیک
nucleic base	باز هسته ای
nuclein	نوکلئین
nucleocapsid	هسته پوشه
nucleoid	نوکلئوئید هسته سا/واره
nucleolus	هستک
nucleophilic group	گروه هسته دوست/خواه
nucleoplasm	پروتوپلاسم هسته سلول، دشته هسته، نوکلئوپلاسم
nucleoprotein	پروتئین هسته، نوکلئو پروتئین
nucleosidase	نوکلئوزیداز
nucleoside	نوکلئوزید
nucleoside analog	آنالوگ نوکلئوزید
nucleoside diphosphate sugar	قند دفسفات نوکلئوزید
nucleosome	نوکلئوزوم، هسته تن

nucleosome phasing	مرحله ای کردن نوکلئوزوم
nucleotid	نوکلئوتید
nucleotide	نوکلئوتید
nucleotide excision repair	ترمیم برش نوکلئوتیدی
nucleus	کانون هسته سلول، مرکز، مغز، هستک، هسته
null hypothesis	فرضیه صفر
null model	مدل صفر
nullisomy	نولیزومی
nutraceuticals	مواد غذائی طبی
nutriceuticals (=nutraceuticals)	نوتریسوتیکال
nutricine	نوتریسین
nutrient	مغذی
nutrient limitation	محدودیت های مواد مغذی
nutrigenomics	نوتریژنومیکس
nutritional epigenetics	اپی ژنتیک/پس زاد شناسی تغذیه ای
nystatin	نایستاتین

O

OAT (= ornithine amino-transferase)	او.ای.تی.
obligate	اجباری کردن
obligate aerobic	هوازی اجباری
observational study	بررسی/مطالعه مشاهده ای
occupation theory of agonist action	تئوری اشغال فعالیت آگونیست
occurrence	اتفاق، پیدایش، تشکیل، رویداد، فراوانی، وقوع
ocher codon	کدون آشر
ochratoxin	اکراتوکسین
oculo-dental-digital syndrome	سندروم چشم-دهان-انگشت
ODD (= oculo-dental-digital syndrome)	او.دی.دی.
odorant binding protein	پروتئین بو
ODP (= ozone-depleting potential)	او.دی.پی.
OFAGE (= orthogonal field	او.اف.ای.جی.ئی.

alternation gel)	
OFD (= oral-facial-digital)	او.اف.دی.
offspring	زاده ها
OI (= osteogenesis imperfecta)	او.آی.
oil-free	بدون روغن
Okazaki fragment	تکه اوکازاکی
oleate	اولئات، نمک اسید اولئیک
oleic acid	اولئیک اسید، ترشای روغن
oleosome	اولئوزوم
oligo (dT) cellulose	الیگو دی.تی. سلولز
oligofructan	اولیگوفروکتان
oligofructose	اولیگو فروکتوز
oligomer	اولیگومر، کم/چند پار
oligomycin	اولیگومایسین
oligonucleotide	اولیگونوکلئوتید
oligonucleotide probe	پرب/پروب/ردیاب/کاووشگر/ز شانگر اولیگونوکلئوتید
oligopeptide	اولیگو پپتید
oligos	اولیگوز
oligosaccharide	اولیگو ساکارید
oligotrophic	اولیگوتروفیک، کم پرور
olivomycin	اولیوومایسین
omega-3 fatty acid	اسید چرب امگا-3
oncogene	تومورزا، ژن توموری غده زا، سرطان زا
oncogene FOS	انکوژن اف.او.اس.
oncogenesis	سرطان زایی
oncology	انکولوژی، تومور/سرطان/غده شناسی
oncomodulin	انکومدولین
onconavirus	انکوناویروس
oncostatin M	انکواستاتین ام.
onset	آغاز بیماری، حمله، هجوم
oocyte	تخم/تخمک نابالغ/نارس/نرسیده/یاخته ، سلول تخم
oogenesis	تخمک زایی، تشکیل تخمک یا اووم
opal codon	کدون اوپال
open circular	چرخه ای باز

open environment	محیط باز
open pollination	دگر گشن، گرده افشانی باز
open promoter complex	کمپلکس مبلغ باز، مجموعه پروموتر باز، مجموعه پیش محرک باز
open reading frame	چار چوب باز خوانی باز
operator	اپراتور، اجرا کننده کارگردان، گرداننده، عامل، عملگر، مجری
operon	اوپرون
opines	اوپاین ها
opportunistic infection	آلودگی/عفونت فرصت طلب
opportunity cost	هزینه فرصت
opsonin	اوپسونین
opsonization	اوپسونین سازی
optical activity	فعالیت نوری، نورورزی
optimum food	غذای بهینه/مطلوب
optimum pH	پ.هاش بهینه/مطلوب
optimum temperature	حرارت بهینه/مطلوب
option value	ارزش انتخاب
optrode	نورکاو
oral-facial-digital	دهانی-صورتی-انگشتی
orbital	اوربیتال
ORC (= origin recognition complex)	او.آر.سـی.
ordered clone map	نقشه همانندسازی منظم
ORF (= open reading frame)	او.آر.اف.
organ	آلت، ابزار، اندام، بخش، عضو
organ culture	کشت اندام
organelle	ارگانل، اندام سلولی، اندامک، اندام کوچک
organic compounds	ترکیبات آلی
organism	ارگانیسم، اندامگان، جاندار، سازواره، سامانه، موجود زنده، دستگاه
organisms with novel traits	جاندار دارای نشانویژه های جدید
organogenesis	اندام ریخت گیری، اندام زایی، تشکیل عضو
organotroph	آلی پرور، ارگانوتروف
ORI (= origin of DNA replication)	او.آر.آی.
origin	اصل، خاستگاه، سرچشمه، مبداء، منشاء
origin of DNA replication	منشاء همانند سازی دی.ان.ای.
origin of replication	مبداء همانند سازی
origin recognition complex	کمپلکس شناسائی منشاء
ornithine	اورنیتین
ornithine amino-transferase	آمینوترانسفراز اورنیتین
ornithine transcarbamylase deficiency	کمبود اورنیتین ترانسکارب‌آمیلاز
oropharyngeal	اوروفارینژیال، دهانی-حلقی
orosomucoid	اوروسموکوئید
orphan drug	داروی غیر اقتصادی
orphan genes	ژن های بیکاره
orphan receptor	گیرنده های جفت نشده
orthogonal field alternation gel	ژل تغییر قائمی زمینه
ortholog	ارتولوگ
orthologous genes	ژنهای ارتولوگ
orthophosphate cleavage	شکافتگی ارتوفسفات
oscillin	اوسیلین
osmosis	اسمز، تراوش، تراوندگی، گذرندگی، نفوذ
osmotic pressure	فشار اسمزی
osmotins	اسموتین ها
ossein	اوسئین
osteoarthritis	آرتروز استخوان
osteocalcin	استئوکالسین
osteogenesis imperfecta	استئوژنز ایمپرفکتا
OTC (= ornithine transcarbamylase deficiency)	او.تی.سی.
otocephaly	اوتوسفالی، گوش سری
outbreak	اپیدمی، شیوع
outbreeding	برون نژادگیری
outcrossing	برون چلیپائی، زادگیری چلیپائی
outcrossing	دگر آمیزی
OVA (= Ov-albumin)	او.وی.ای.
ovalbumin	آلبومین تخم، اووالبومین

overatropinization	آتروپین سازی بیش از حد، فرا آتروپین سازی
overlap genes	ژن های هم پوشان
overlapping genes	ژن های هم پوشان
overlapping reading frame	چهارچوب بازخوانی همپوش
oversight	سرپرستی، سهو، غفلت، نظارت
overt	آشکار، علنی، عیان
overt release	آزاد سازی علنی
overwinding	فراپیچی
ovum	تخمک
oxalate	اکسالات
oxalate oxidase	اکسالات اکسیداز
oxalic acid	اسید اگسالیک
oxcycephaly (= acrocephalia)	
oxidant	اکساینده، اکسنده، اکسیدان، اکسید کننده، عامل اکسایش
oxidase	اکسیداز
oxidation	اکسایش، اکسیداسیون، اکسیده شدن/کردن
oxidation-reduction reaction	واکنش اکسایش-کاهش
oxidative phosphorylation	فسفوریلاسیون اکسایشی
oxidative stress	تنش اکسایشی
oxidizing agent	عامل اکسندگی/اکسید کنندگی
oxycephalic (=acrocephalic)	
oxygen consumption	مصرف اکسیژن
oxygen deficiency	کمبود اکسیژن
oxygen deficient atmosphere	جو کم اکسیژن
oxygen free radical	بنیان بی اکسیژن
oxygen supplementation	مکمل سازی اکسیژنی
oxygen transfer	انتقال اکسیژن
oxygenase	اکسیژناز
oxyhemoglobin	اکسی هموگلوبین
oxymyoglobin	اکسی میوگلوبین
oxytetracycline	اکسی ترا سایکلین
oxytocin	اکسی توسین
ozone depletion	کاهش ازن
ozone hole	حفره ازن

ozone layer	لایه ازن
ozone-depleting potential (ODP)	ازن کاهی
ozonolysis	ازونولایزیس

P

P (= phosphate)	پی.
p (= pico =10^{-12})	پی.
P (= short arm of chromosome)	پی.
p arm (= short arm of a human chromosome)	شاخه پی.
P1 (= first meiotic prophase)	پی.1
P1 (= phosphatidyl-inositol)	پی.1
p53 (= Protein 53)	پی.53
PAC (= P1-derived artificial chromosome)	پی.ای.سی.
pachymeningitis	التهاب سخت شامه
pachynema (= pachytene)	پاکینما
pachytene	پاکیتن، ستبر نوار
packaging	بسته بندی
packed bed reactor	رآکتور بستر فشرده
packed column	ستون فشرده
paclitaxel	پاکلیتاکسل
PADP (= poly-adenylate binding protein)	پی.ای.دی.پی.
PAGE (= poly-aclamide gel electrophoresis)	پی.ای.جی.ئی.
paint	رنگ، رنگ زدن/کردن، مواد رنگ کاری
paleontology	دیرین شناسی
palindrome	پالیندروم، مساوی الطرفین
palindromic sequence	توالی جناس قلب/مساوی الطرفین
palladium	پالادیوم
palliative	تسکین دهنده/مسکن
palmitate	پالمیتات
palmitic acid	اسید پالمیتیک، جوهر نخل
palytoxin	پالی توکسین
pancreas	پانکراس

pancreastatin	پانکرااستاتین
pancreatic hormone	هورمون پانکراتیکی
pancreatic juice	عصاره پانکراتیکی
pancreatic lipase	لیپازپانکراتیکی
pancreatic thread protein	پروتئین رشته ای پانکراتیکی
pancreatic trypsin inhibitor	سرکوبگر تریپسین پانکراتیکی
pancreatin	پانکراتین
pandemic	بیماری جهانگیر، همه گیر، همه گیری گسترده
panmixia	جفت گیری تصادفی، پانمیکسیا، هر آمیزی
panmixis	پانمیکسی
panose	پانوز
pantetheine	پانتتئین
pantothemic acid	اسید پانتوتمیک
papain	پاپائین
papaverine	پاپاورین
papillomavirus	ویروس پاپیلوما
papovavirus	ویروس پاپوا
parabiosis	پارابیوز
paracentric inversion	واژگونی پاراسنتریکی
paraclinical	پیرابالینی
paracrine	پاراکرین
paraffin-embedded tissue	بافت جایگزین در پارافین
parafusin	پارافوزین
paralytic cobra toxin	سم فلج کننده کبرا
paralytic shellfish toxin	سم فلج کننده صدف(ها)
paramyosin	پارامیوزین
paramyxoviridae	پارامایوکسوویروس(ها)
paramyxovirus	پارامایوکسوویروس
parapatric speciation	پاراپاتریک اسپشیشن
paraprotein	پاراپروتئین
paraquat	پاراکوات
pararetrovirus	پارارتروویروس، شبه رتروویروس
parasite	انگل، جانور/قارچ انگلی، طفیلی، گیاه طفیلی
parasitic	انگلی/طفیلی
parasitic disease	بیماری انگلی
parasitism	انگل بودن، انگلی، زندگی

	انگلی/طفیلی، حالت
	انگلی
parasitoid	انگل ماند
parataxonomist	پاراتاکسونومیست، پرا/فرا آرایه شناس
parathormone	پاراتورمون، پاراتیرین، هورمون پاراتیروئیدی
paravertebral	اطراف/جنب/نزدیک مهره ای
parental organism	ارگانیزم والد
parents	پدر و مادر، والدین
paresis	پارزی، فلج خفیف/ناقص
parthenocarpic	بکربارده
parthenocarpy	بکرباردهی، پارتنوکارپی
parthenogenesis	بکر زایی
parthenogenetic	بکرزا
partial digestion	هضم ناقص
partial linkage	پیوستگی جزئی/ناقص
partial monosomy	حذف نسبی
partial trisomy	تریزومی ناقص
particle	پاریزه، خرد،خردک، خرده، دانه، ذره، ریزک، ریزه
particle cannon	توپ ذره ای
particle gun	توپ ذره ای
partition coefficient	ضریب تجزیه/جدا سازی
partitioning agent	عامل جداسازی
parturition	تولد، زادآوری زایمان، وضع حمل، ولادت
party concerned	طرف درگیر/مربوطه
party of export	طرف صادر کننده
party of import	طرف وارد کننده
party of origin	طرف مبدا
party of transit	طرف ترانزیت
parvovirus	پارو ویروس
PAS (= periodic acid)	پی.ای.اس.
passive immunity	ایمنی زودگذر/غیر فعال/ناکنش ور
passive use value	ارزش کاربرد غیر فعال
PAT gene	ژن پی.ای.تی.
patent	آشکار، ثبت شده، حق انحصاری اختراع، روشن، واضح

patent ductus arteriosus	آرتریوسوس مجرای باز، پاتنت دوکتوس آرتریوسوس
paternal	وابسته به پدر
paternity index	اندیکس/شاخص پدری
pathogen	عامل بیماری زا
pathogen toxin	زهرابه/سم بیماری زا
pathogenesis	بیماری زایی، پاتوژنز، تکوین بیماری
pathogenesis related proteins	پروتئین های وابسته به عوامل بیماری زا
pathogenic	بیماری زا، پاتوژنیک، مولد بیماری
pathogenic agent	عامل بیماری زا
pathogenic organism	ارگانیزم/جاندار بیماری زا
pathognomonic	شناسه بیماری
pathologic	آسیب شناختی، بیماری زایی، پاتولوژیک
pathologic process	فرآیند آسیب شناختی/بیماری زایی
pathophysiology	فیزیولوژی بیماری شناسی
pathway	راه، گذرگاه، مجرا، مسیر، معبر
pathway feedback mechanisms	سازوکار های پس خور مسیر (شیمیائی)، مکانیزم فید بک مسیر
pattern biomarker	نشانه زیستی الگو
patulin	پاتولین
PBG (= porpho-bilino-gen)	پی.بی.جی.
PBL (= peripheral blood lymphocyte)	پی.بی.ال.
PBMNC (= peripheral blood mono-nuclear cells)	پی.بی.ام.ان.سی.
pBR322	پی.بی.آر.322
PBS (= phosphate-buffered saline)	پی.بی.اس.
PC (= phosphatidyl-choline)	پی.سی.
PCC (= premature chromosome condensation)	پی.سی.سی.
PCd (= premature centromere division)	پی.سی.دی.

PCNA (= proliferating cell nuclear antigen)	پی.سی.ان.ای.
PCOD (= poly-cystic ovarian disease)	پی.سی.او.دی.
PCR (= polymerase chain reaction)	پی.سی.آر.
PDA (= patent ductus arteriosus)	پی.دی.ای.
PDGF (= platelet-derived growth factor)	پی.دی.جی.اف.
PDHC (= pyruvate de-hydrogenase complex)	پی.دی.اچ.سی.
PE (= phosphatidyl-ethanolamine)	پی.ئی.
pedigree	پیشینه، تبار، دودمان، دودمانه، شجره، شجره نامه، نژاد
pedigree analysis	تجزیه شجره ای
PEG (= poly-ethylene glycol)	پی.ئی.جی.
pellagra	پلاگرا
penecillium citrinum	پنیسیلیوم سیترینوم
penetrance	انفاذ، پدیدار شوندگی
penicillium patulum	پنیسیلیوم پاتولوم
pentose	پنتوز
PEP (= phospho-enol-pyruvate)	پی.ئی.پی.
PEPCK (= phospho enol-pyruvate carboxylase)	پی.ئی.پی.سی.کی.
pepsin	پپسین
peptid	پپتید
peptidase	پپتیداز
peptide	پپتید
peptide bond	پیوند پپتیدی
peptide elongation factor	عامل دراز شدگی پپتید
peptide linkage	اتصال پپتیدی
peptide mapping	بازنمایی پپتید
peptide nanotube	ریز لوله پپتیدی
peptide T	پپتید تی.

peptidergic	پپتیدرژیک
peptidoglycan	پپتیدوگلیکان
peptido-mimetic	پپتیدو-میمتیک
peptidyl transferase	پپتیدیل ترانسفراز
peptidyle site (P-site)	جایگاه پپتیل
peptone	پپتون
per capita intake rate	نرخ جذب سطحی سرانه، نرخ ورودی سرانه
percolating filter	صافی پرکولاتوری
percutaneous	پرکوتانوس، پوستی، تراپوستی
percutaneous umbilical blood sampling	نمونه برداری زیر جلدی از خون بند ناف
perforin	پرفورین
pericardium	برون شامه قلب
pericentric inversion	واژگونی پریسنتریک
periderm	پیراپوست
perikaryon	پریکاریون، پیراهسته، جسم سلولی
perinuclear space	فضای پری نوکلئار
periodic acid	اسید تناوبی
periodic acid-Schiff reaction	واکنش تناوبی اسید-شیف
periodic transfer	انتقال دوره ای
periodicity	تناوب، چرخه تولید مثل، خصلت تناوبی، دوران، دوره ای بودن، نوبه ای
periodontium	بافتهای پوشاننده دندان، پیرا دندان
peripheral blood lpymphocyte	لنفوسیت خون جانبی
peripheral blood mono-nuclear cells	سلولهای تک هسته ای فرعی خون
periplasm	پریلازم، پیرادشته
peritoneal	پریتونیال، صفاقی
peritoneal cavity	حفره صفاقی
peritoneal membrane	غشاء صفاقی
permeability barrier	سد تراوایی
pernicious anemia	آنمی پرنیشیاس
peromyscus species	گونه پرومایسکوس
peroxidase	پر اکسیداز
persistence	استقامت، استمرار، استواری، ایستادگی،

	پایداری، پایایی، مقاومت
person	خود، شخص، فرد، نفر
personal	خصوصی، شخصی، فردی
Pertussis toxin	سم پرتوسیس
pest	آفت
pest free area	منطقه بدون آفت
pest risk analysis	تحلیل احتمال خطر آفت
pesticide	آفت/حشره کش، سم دفع آفات
PET (= paraffin-embedded tissue)	پی.ئی.تی.
PET-PCR (= RT-PCR out of paraffin-embedded tissue)	پی.ئی.تی.-پی.سی.آر.
petri dish	ظرف پتری
petrochemicals	فرآورده های پتروشیمی
petrochemistry	پتروشیمی
PFGE (= pulse field gel electrophoresis)	پی.اف.جی.ئی.
PG (= prostaglandin)	پی.جی.
pH	پ. هاش
Ph (= Philadelphia chromosome)	پی.اچ.
phage	باکتری/ترکیزه خوار، فاژ
phage display	نمایش باکتری خوار
phagocyte	بیگانه خوار، سلول بیگانه/میکروب خوار، فاگوسیت
phagocytic cell	یاخته بیگانه خوار
phagocytize	بیگانه خواری کردن، بیگانه خوردن، فاگوسیتایز
phagocytosis	بیگانه/ریزه خواری، فاگوسیتوز، یاخته خواری
phagosome	فاگوزوم
PHA-LCM (= phyto-hem-agglutinin-sti mulated lymphocyte conditioned medium)	پی.اچ.ای.-ال.سی.ام.
pharmacodynamics	دارو دینامیک شناختی
pharmacoenvirogenetics	ژنتیک داروشناختی محیطی
pharmacogenetics	دارو زاد شناسی
pharmacogenomics	دارو ژن شناسی

pharmacognosy	دانش داروبابی، مبحث داروشناسی، مفردات پزشکی
pharmacokinetics	دارو جنبش شناسی
pharmacology	دارو شناسی
pharmacophore	دارو زا
pharmacovigilance	دارو هشیاری
pharming	فارمینگ
phase	فاز
phasmis	فازمیز
Phe (= phenylalanine)	
phenazine methosulphate	متاسولفات فنازین
phenocopy	فنوکپی
phenolic hormone	هورمون فنولی
phenomics	فنومیکز
phenotype	پدیدگان، رخ مانه/مون، ریخته/شکل ظاهری، فنوتیپ
phentolamine	فنتولامین
phenylalanine	فنیل آلانین
phenylketonuria	فنیل کتونوری
phenylmethylsulfonyl fluoride (PMSF)	فلوئورید فنیلمتیلسولفونیل
phenylmethylsulfonylfluoride	فنیل متیل سولفونیل فلوئورید
phenyl-thio-carbomide	فنیل تیوکاربومید
phenyl-thio-hydration	فنیل تیوهیدراسیون
pheromone	واگیزن، فرومون
philadelphia chromosome	کروموزوم فیلادلفیا
phlebovirus	فلبو ویروس
phloem	فلوئم
phloretin	فلورتین
phlorizin	فلوریزین
phloroglucinol	فلوروگلوسینول
phocomelia	فوکوملی
phorbol ester	استر فوربول
phosgene	فسژن، فوسژن
phosphatase	فسفاتاز
phosphate	فسفات
phosphate group	گروه فسفات
phosphate transporter gene	ژن حامل فسفات

phosphate-buffered saline	محلول نمکی بافری فسفاتی
phosphate-group energy	انرژی گروه فسفات
phosphatidyl choline	فسفاتیدیل کولین
phosphatidyl serine	فسفاتیدین سرین
phosphatidyl-choline	فسفاتیدیل کولین
phosphatidylethanolamine	فسفاتیدیلاتانولآمین
phosphatidyl-ethanolamine	فسفاتیدیل-اتانولآمین
phosphatidylinositol cycle	چرخه فسفاتیدیلینوزیتول
phosphatidyl-serine	فسفاتیدیل سرین
phosphinothricin	فسفینو تریسین
phosphinothricin acetyltransferase	فسفینو تریسین استیل ترانسفراز
phosphinotricine	فسفینو تریسین
phospho enol-pyruvate carboxylase	فسفو-انول-پیروویت کاربوکسیلآز
phosphodiester	فسفو دی استر
phosphodiester bond	پیوند فسفو دی استر
phosphodiesterase	فسفو دی استراز
phospho-enol-pyruvate	فسفو-انول-پیروویت
phosphofructokinase	فسفو فروکتوکیناز
phosphoglucomutase	فسفوگلوکوموتاز
phospholamban	فسفولامبان
phospholipase	فسفو لیپاز
phospholipid	فسفو لیپید
phosphonomycin	فسفونومایسین
phosphoramidite	فسفورآمیدایت
phosphorus	تابنده، رخشا، فسفر_)ی(
phosphorylation	فسفات افزایی، فسفریلاسیون، فسفریله کردن
phosphorylation potential	استعداد/توان فسفات افزایی
phosphoserine	فسفوسرین
phosphotransferase	فسفوترانسفراز
phosphotyrosine	فسفوتیروزین
phosphovitin	فسفو ویتین
phosphramidon	فسفورآمیدون
photoautotroph	فوتواتوتروف، نورخودپرور
photochemical reaction	واکنش نورشیمیائی
photolithotrophic bacteria	باکتری فوتولیتوتروف، باکتری های نور کانی پرور

photoluminescence	فوتولومینسنس
photolysis	فوتولیز
photon	شیدپار، فوتون
photoorganotroph	نورآلی پرور، فوتوارگانوتروف
photoperiod	نور دوره، فتو پریود
photophore	اندام نوری
photophosphorylation	فتو فسفریزه شدن، فسفری شدن نوری
photoreactivation	فعال سازی نوری
photoreceptor	گیرنده نوری
photorhabdus luminescens	فوتورابدوس لومینسانس
photosensitivity	حساسیت به نور
photosynthesis	سنتز نوری، فتو سنتز، فروغ آمایی
photosynthetic phosphorylation	فسفریزه شدن فتو سنتزی/نور ساختی
phototroph	فوتوتروف
phototropic	نورگرا
phototropism	نورگرایی
PHP (= pseudo-hypo-parathyroidism)	پی.اچ.پی.
phragmoplast	فراگموپلاست
phylasalaemin	فیلاسالائمین
phyletic evolution	تکامل نژادی، فرگشت تباری
phylogenetic	تبار زایشی دودمانی
phylogenetic constraint	محدودیت دودمانی
phylogenetic profiling	طرح دودمانی
phylum	سلسله، شاخه
physical containment level	سطح حصر فیزیکی و ایمنی
physical map (of genome)	نقشه فیزیکی (ژنوم)
physiologic	کار اندام شناختی
physiological	کار اندام شناختی
physiologically active compound	ترکیب فعال فیزیولوژیکی
physiologist	کار اندام شناس
physiology	تنکار شناسی، کار اندام، کار اندام شناسی، فیزیولوژی
physostigmine	فیزوستیگمین
phytase	فایتاز
phytate	فیتات

phytic acid	اسید فیتیک
phytoalexin	فیتو آلکسین
phytochemical	گیاه شیمیائی
phytochrome	رنگ گیاهی، فیتو کروم، گیارنگ
phytoene	فیتوئین
phytoestrogen	استروژن گیاهی
phyto-hem-agglutinin-stimulated lymphocyte conditioned medium	محیط کشت تنظیم شده با لنفوسیت تحریک شده توسط فیتو-هم-آگلوتینین
phytohormone	هورمون گیاهی
phyto-manufacturing	تولید فیتو
phytonutrient	ماده غذایی گیاهی
phytopharmaceutical	گیاه دارویی، گیاه-داروئی
phytophthora	فیتوفتورا
phytoplankton	پلانکتون گیاهی، شناوران گیاهی، گیاهان شناور، گیاه پلانکتونی
phytoremediation	گیاه پالائی
phyto-sterol	فیتو استرول
phytosterols	استرول های گیاهی
phytotoxin	سم گیاهی
P$_i$ (= inorganic phosphate)	پی.اندیس آی.
PI (= paternity index)	پی.آی.
pica	آشغال/خاک خوری، ویار
pico	پیکو-
picogram	پیکوگرم
picorna	پیکورنا
picornaviridae	پیکورناویروس(ها)
picornavirus	پیکورناویروس
piericidin	پیریسیدین
piezoelectric	بارابرقی، پیزوالکتریک
pigmentation	پیگمنتاسیون، رنگیزه دارشدن، رنگی شدن
pilus	پیلوس، کرک
pink bollworm	کرم غوزه
pinocytosis	پینوسیتوز، قطره خواری
pipette	پیپت
Pitt-3	پیت-3
pituitary gland	غده هیپوفیز
PK (= pyruvate kinase)	پی.کی.

PKU (= phenylketonuria)	پی.کی.یو.
placebo	دارونما، داروی بی اثر/کاذب، دل خوشکنک، گول زنک، ماده بی اثر
placenta	جفت
plague	آفت، طاعون، گرفتار/مبتلا کردن، مصیبت
plague meningitis	مننژیت همه گیر
plakoglobin	پلاکوگلوبین
plankton	پلانکتون، شناوران
planned release	رها سازی عمدی
plant	رستنی، کارخانه، کاشتن، گیاه (ی)، نبات
plant functional attributes	صفات عملکردی گیاه
plant hormone	هورمون گیاهی
plant sterol	استرول گیاهی
plant toxin	سم گیاهی
Plantibodies™	پلانتیبادیز
plantigens	ژن های گیاهی
plaque	پلاک، پولک
plasma	پلاسما، پیش مایه، خونابه، دشتینه
plasma cell	یاخته پلاسما/دشتینه
plasma membrane	پوسته خارجی سلول، غشاء دشتینه ای/سیتوپلاسمی
plasma protein	پروتئین پلاسما (یی)
plasma protein binding	اتصال پروتئین پلاسما
plasmacyte	پلاسماسیت
plasmalemma	پلاسمالما، غشاء خارجی پروتوپلاسم
plasmalogen	پلاسمالوژن
plasmapheresis	پلاسما فرسیز
plasmic membrane	غشای یاخته
plasmid	پلاسمید، دشتیزه
plasmid amplification	تکثیر پلاسمیدی
plasmin	پلاسمین
plasminogen	پلاسمینوژن
plasmocyte	دشته یاخته
plasmodesma	پلاسمودسما
plasmodium	پلاسمودیوم
plasmogamy	پلاسموگامی

plasmolysis	پلاسمولیز، دشته کافتی
plasmon	پلاسمون
plasmoptysis	پلاسموپتیز
plastid	پلاست، پلاستید، دشتاره، دیسه
plastidome	پلاستیدوم، دشتاره مونه
plastocyanin	پلاستوسیانین
plastogene	پلاستوژن
plastoquinone	پلاستوکینون
plasto-quinone	پلاستوکینون
platelet	پلاک، گرده، گرده خون، یاخته لخته شده
platelet activating factor	عامل فعال کننده پلاک
platelet aggregation	تجمع/تراکم پلاک
platelet-derived growth factor	عامل رشد مشتق از پلاک
platelet-derived wound growth factor	عامل رشد زخم مشتق از پلاک
platelet-derived wound healing factor	عامل التیام زخم مشتق از پلاک
plating	آبکاری، روکش دادن
plecksterin	پلک استرین
plectonemic coiling	چنبره پیچیده رشته
pleiotropic	چند اثر/جانبه/رخ/گرا
pleiotropism	پلئیتروپیزم
pleiotropy	پلئیتروپی، چند رخی/گرایی
pleistocene	پلئیستوسن، پیشین پدید
pleocytosis	پلئوسیتوز
PLP (= pyridoxal-5-phosphate)	پی.ال.پی.
plug flow digester	پلاگ فلو دایجستر
pluripotent	چند قوه زا
pluripotent stem cell	سلول پایه/دودمانی چند قوه زا
plus and minus screening	غربالگری مثبت و منفی
plus strand	رشته مثبت
plus ultra-violet light of the A wavelength	نور مثبت ماورای بنفش با طول موج ای.
Plys (= poly-L-lysine)	
PMFS (= phenylmethylsulfonylfluo	پی.ام.اف.اس.

258

English	Persian
ride)	
PNP (= purine nucleoside phosphorylase)	پی.ان.پی.
podophyllotoxin	پودوفیلوتوکسین
point mutation	جهش نقطه ای
point-source	منبع نقطه ای
poisoning	مسموم کننده (گی)، مسمومیت
Pol (= polymerase)	
polar	قطبی
polar body	گویچه قطبی
polar group	گروه قطبی
polar molecule	ملکول قطبی
polar mutation	جهش قطبی
polarimeter	قطب سنج، قطبیت سنج
polarity	پلاریته، قطبیت، قطبی شدن/بودن
polarity mutation	جهش قطبیتی
polarization	قطبش
polarize	قطبیدن
polarized	قطبیده
pollutant	آلاینده
polluter	آلوده گر
pollution	آلودگی
poly(A) polymerase	پلیمراز پلی آدنین
poly-aclamide gel electrophoresis	الکتروفورز ژلی پلی آکلامیدی
polyacrylamide gel	پلی آکریل آمید ژل
polyacrylamide gel electrophoresis (PAGE)	الکتروفورز ژلی پلی آکریل آمید
poly-adenylate binding protein	پروتئین اتصال پلی آدنیلات
polyadenylated	پلی آدنیله شده
polyadenylation	پلی آدنیله سازی
polycation conjugate	پلیکیشن کانجوگیت
polycistronic	چند سیسترونی
polycistronic mRNA	ام.آر.ان.ای. چند سیسترونی
polyclinic	مجتمع درمانی
polyclonal antibody	پادتن چند کلونی
polyclonal antibody response	واکنش پادتن چند کلونی
polyclonal response	واکنش چند کلونی
poly-cystic ovarian disease	بیماری تخمدان پلی-سیستیک
polydactyl	پلی داکتیل
polyethylene glycol (PEG)	پلی اتیلن گلیکول
polygalacturonase	پلی گالاکتوروناز
polygene	پلی ژن، ژن فرعی
polygenic	پر/چند زاد، پر/چند ژنی، پلی ژنیک
polygenic inheritance	وراثت چند ژنی/فاکتوری
polyhydroxyalkanoate	پلی هیدروکسی آلکانوآت
polyhydroxyalkanoic acid	پلی هیدروکسی آلکانوئیک اسید
polyhydroxylbutylate	پلی هیدروکسیل بوتیلات
polylinker	اتصال دهنده چند تائی، پلی لینکر
poly-L-lysine	پلی ال.لیزین
polymer	بسپار، پر پار، پلیمر، پلی مر
polymerase	پلیمراز
polymerase chain reaction (PCR)	واکنش زنجیره ای پلیمراز
polymerase I	پلیمراز 1
polymerase I and transcript release factor	عامل رها سازی پلیمراز 1 و رو نوشت
polymerization	پلیمریزاسیون
polymorphism	پر ریختی، چند ریختی/شکلی/گونگی، دگر دیسی
polymorphonuclear granulocyte	گویچه سفید دانه ای چند هسته ای
polymorphonuclear leukocyte	گویچه سفید چند هسته ای
polynucleotide	پر/پلی نوکلئوتید
polynucleotide phosphorylase	پلی نوکلئوتید فسفریلاز
polypeptide	پلی پپتید
polyphenol	پلی فنل
polyploid	پلیپلوئید پر/چندگان/لاد
polyploidy	پلی پلوئیدی
polyprotein	پلی پروتئین
polyribosome	پر/چند رناتن، پلی ریبوزوم

polysaccharide	پلی ساکارید، پر قند، چند قنده، چند قندی
polysome (= polyribosome)	پلی زوم
polytene chromosome	کروموزوم پلی تن
polyunsaturated fatty acids	اسید های چرب پر اشباع نشده
polyvalent	پلی والنت، چند ارزشی
polyvalent vaccine	واکسن چند بنیانی
population	تعداد جمعیت، سکنه، نفوس
population genetics	ژنتیک جمعیت
population viability analysis	تحلیل قابلیت زیست/زیستایی جمعیت
porin	پورین
porpho-bilino-gen	پورفوبیلینوژن
porphyrins	پورفیرین
port	بندر، دریچه، روزنه چشم، سوراخ، لنگر گاه، مرز
portable spray	اسپری دستی، افشانه قابل حمل
port-a-cath	پورتاکات
portal	باب، دروازه ای، مربوط به مدخل باب کبدی
portal vein	سیاهرگ باب
position effect	اثر موقعیت/وضعیت
positional cloning	دودمان/همسانه سازی موقعیتی
positive control	کنترل مثبت
positive supercoiling	فوق مارپیچ مضاعف
postaglandin	پستاگلاندین
postaglandis A	پستاگلاندیس ای.
postexposure	پس از تماس
postnatal selection	انتخاب پس از تولد
post-transcriptional gene silencing	فرونشانی پسا نسخه برداری ژنی
post-transcriptional processing	فرآیند پسا نسخه برداری
post-translational modification	تغییر پسا ترجمه ای
post-translational modification of protein	تغییر پروتئین پساترجمه
potassium	پتاسیم

potassium iodide	یدید پتاسیم
potential receiving environment	محیط تحویل گیرنده بالقوه
potentiates	تقویت اثر، توانمند سازی
powder	پودر، پودر کردن/زدن، ساییدن، گرد
powdered	پودر/خشک شده، پودر زده، گرد شده، ساییده
powered	برقی، به توان رسیده، قدرت گرفته
poxvirus	ویروس آبله/پوکس
PP (= protein purification)	پی.پی.
PPHP (= pseudo-pseudo-hypo-para thyroidism)	پی.پی.اچ.پی.
PP$_i$ (= pyro-phosphate ion)	پی.پی.اندیس آی.
PQ (= plasto-quinone)	پی.کیو.
Prader-Willi syndrome	سندروم پرادر-ویلی
pRb (= retinoblastoma protein)	
Pre Con vector (= promoter conversion vector)	ناقل پری کان
prebiotics	پریبایوتیکز
precautionary approach	رویکرد اخطارانه، رهیافت اخطارانه ای، نگرش احتیاطی
precautionary principle	اصل احتیاطی
precautionary zone	منطقه احتیاطی/اخطاری
preclinical	پیش درمانگاهی/کلینیکی
precursor	پیش تاز/ساز/شرط/ماده/مرحله/نیاز، صورت ابتدایی، متقدم
predator	شکارچی، شکارگر (حیوان)، طعمه گیر
prednisolone	پردنیسولون
pre-initiation complex	مجموعه پیش آغاز
premature centromere division	تقسیم سانترومر نابهنگام/زود هنگام
premature chromosome condensation	فشردگی کروموزوم نابهنگام/زود هنگام

pre-messenger RNA	آر.ان.ای.پیش پیغامبر
pre-mRNA	پیش ام.آر.ان.ای.
prenatal diagnosis	تشخیص پیش از تولد/زایشی
preparative chromatography	کروماتوگرافی آماده سازی
preparative ultracentrifuge	فوق سانتریفوژ آماده سازی
prepotency	پیش قوت
preprohormone	پری پروهورمون
preproinsulin	پریپرو انسولین
preprophase band	نوار پریپروفاز
Pre-RC (= pre-replication complex)	پری-آر.سی.
pre-replication complex	کمپلکس پیش از نسخه برداری
pressure	تحت فشار قرار دادن، فشار
pre-tRNA	پیش تی.آر.ان.ای.
prevalence	استیلا، رواج، شیوع، عمومیت، غلبه
prevalence rate	میزان شیوع
prevalence survey	بررسی شیوع
prevention	بازداری، پیش گیری، جلوگیری، ممانعت
previously-vaccinated individual	فرد پیش واکسینه شده
Pribnow box	جعبه پریبناو
primary cell	سلول اولیه
primary constriction	فرورفتگی ابتدائی
primary forest	جنگل اولیه
primary immune response	پاسخ ایمنی اولیه
primary productivity	محصول دهی/فرآوری اولیه
primary structure	ساختار نخستین
primary transcript	نسخه برداری اولیه
primary value	ارزش اولیه
primase (= DNA primase)	پریماز
primaverose	پریماوروز
primer	آماده ساز، پرایمر، چاشنی، خرج، ماده اولیه
primidine dimmer	دیمر پریمیدین
primordium	پریم اوردیوم
primosome	پریموزوم
prion	پریون

prior informed consent	اعلام موافقت قبلی، موافقت از قبل اعلام شده
private opportunity cost	هزینه فرصت اختصاصی
private value	ارزش اختصاصی
Pro (= proline)	
proanthocyanidin	پروآنتوسیانیدین
proband	پروباند، پیش گدازه، سر تبار/دودمان
probe (= DNA probe)	پرب، پروب، تحقیق، جستجو، ردیاب، کاوشگر، کندو کاو، معاینه، نشانگر
probiotics	پروبایوتیکز
procaryotes	شبه هسته داران،پیش هسته ای ها
process validation	تضمین فرآیند
processing	پروسس کردن
processivity	پروسسیتیوته
procollagen	پرو/پیش کولاژن
prodrome	پیش در آمد/زمینه، علامت/نشانه اولیه
product	تولید، دستاورد، فرآورده، محصول
production environment	محیط تولید
production function	کارکرد/عملکرد تولید
production trait	نشانویژه/ویژگی تولید
proenzyme	پیش آنزیم/زیما
profilin	پروفیلین
proflavin	پروفلاوین
progenitor	جد
progeria	پیری زود رس، پیش پیری
progesterone	پروژسترون
programmed cell death	مرگ سلولی برنامه ای/برنامه ریزی شده
prohormone	پیش هورمون
proinsulin	پیش انسولین
project leader	راهبر پروژه
prokaryotes	پیش هسته ای ها، شبه هسته داران
prokephalin	پروکفالین
prolactin	پرولاکتین
prolamine	پرولامین

proliferating cell nuclear antigen	آنتی ژن هسته سلولی تکثیر شونده
proline	پرولین
promiscous DNA	دی.ان.ای.ولگرد
promoter	تقویت/فعال/شروع کننده، پیش برنده، راه انداز
promoter conversion vector	ناقل تبدیل پروموتور
pronucleus	پیش هسته
proofreading	ویرایش
proopiomelanocortin	پرواوپیوملانوکورتین
propeptide	پیش پپتید
prophage	پروفاژ
prophase	پروفاز، پیش چهر/هنگام
propionic acid	اسید پروپیونیک
propioyl-CoA carboxylase	پروپیول-کوآ کاربوکسیلاز
proplast	پیش دیسه
proplastid	پروپلاستید
propolypeptide	پروپلی پپتید
proponent	طرفدار
proprietary	اختصاصی، تجارتی، تصاحب گرانه، خصوصی، مالکانه
propylene glycol	پروپیلین گلیکول
prospective study	بررسی قریب الوقوع، مطالعه آتی
prostaglandin	پروستا گلاندین
prostaglandin endoperoxide synthase	پروستا گلاندین اندو پر اکسید سنتاز
prostate	غده پروستات
prostatitis	التهاب غده پروستات
prosthetic group	گروه افزایشی/پروستتیک
protamine	پروتامین
protease	پروتاز
proteasome	پروتئازوم
proteasome inhibitor	بازدارنده پروتئازوم
protectant	محافظ، نگهدارنده
protected area	منطقه حفاظت شده
protection of human health and environment	حفاظت از بهداشت و محیط زیست انسان
protective action zone	منطقه عمل حفاظتی
protective clothing	پوشاک محافظ
protein	پروتئین

Protein 53	پروتئین 53
protein array	آرایه پروتئین
protein biochip	زیست تراشه پروتئین
protein bioreceptor	زیست گیرنده پروتئین
protein C	پروتئین سی.
protein chip	تراشه پروتئین
protein conformation	انطباق پروتئین(ی)
protein encoding	رمزگذاری پروتئین
protein engineering	طراحی/مهندسی پروتئین
protein exotoxin	برون زهر پروتئین
protein expression	تظاهر پروتئین
protein folding	تاشدن/چین خوردگی پروتئین
protein inclusion body	جسم میانبار پروتئینی
protein interaction analysis	تحلیل کنش متقابل پروتئین
protein kinase	پروتئین کیناز
protein kinase C	پروتئین کیناز سی.
protein microarray	ریز آرایه پروتئینی
protein purification	پاکسازی پروتئین
protein quality	کیفیت پروتئین
protein S	پروتئین اس.
protein sequencer	توالی یابی پروتئین
protein signaling	مخابره پروتئینی
protein splicing	اتصال پروتئینی
protein structure	ساختار پروتئین
protein toxin	سم پروتئینی
protein tyrosine kinase	تیروزین کیناز پروتئینی
protein tyrosine kinase inhibitor	بازدارنده تیروزین کیناز پروتئینی
proteinaceous infectious particle	ذره آلوده پروتئیناسیوز
proteinase	پروتئیناز
protein-based lithography	لیتوگرافی مبتنی بر پروتئین
protein-coupled receptor	گیرنده جفتی پروتئین
protein-protein interaction	بر هم کنش پروتئین-پروتئین، واکنش متقابل پروتئین-پروتئین
proteolysis	پروتئولیز
proteolytic	پروتئولیتیک، پروتئین کافت
proteolytic enzyme	آنزیم پروتئولیتیک، آنزیم پروتئین کافت، پروتئاز

proteome	پروتئوم، محتوای پروتئینی سلول
proteome chip	تراشه پروتئوم/محتوای پروتئین سلولی
proteomics	پروتئومیکز، پروتئین شناسی
prothrombin	پروترومبین
protista	پروتیستا
protocol	پروتکل، تشریفات، تفاهم نامه، توافق نامه
protoderm	پیش پوست
protodermal	پیش پوستی
protomer	پروتومر
proto-oncogene	ژنهای سلولی سرطان زا
proto-oncogene	پروتوانکوژن
protoplasm	پروتوپلاسم، پیش دشته/مایه، درون یاخته
protoplast	پروتوپلاست، پیش دش/یاخته، تک یاخته، جرم زنده، ماده حیاتی سلولی
protoplast fusion	امتزاج پروتوپلاست
prototroph	پروتوتروف، پیش پرور، خود پرور، خود غذا
protoxin	پروتوکسین
protozoa	پروتوزوا، پیش زیان
provirus	پرو/پیش ویروس
provitamin	پیش ویتامین
PRPP (= 5-phospho-ribosyl-1-pyro phosphate)	پی.آر.پی.پی.
prune belly syndrome	سندروم شکم آلوئی
pruritic	خارش دار، خارشی
PRV (= pseudo-rabies virus)	پی.آر.وی.
PS (= phosphatidyl-serine)	پی.اس.
PS (= pulmonary stenosis)	پی.اس.
pseudo autosomal region	منطقه شبه اتوزومی
pseudoallele	شبه آلل
pseudodominance	شبه غلبه
pseudogene	شبه ژن، ژن کاذب/واره
pseudohemophilia	شبه هموفیلی
pseudo-hypo-parathyroidism	شبه هیپو پاراتیروئیدیزم

pseudomonas	شبه موناس
pseudo-pseudo-hypo-parath yroidism	شبه شبه هیپو پاراتیروئیدیسم
pseudo-rabies virus	ویروس شبه هاری
pseudo-xanthoma elasticum	شبه زانتوما الاستیکوم
psi	پوند بر اینچ مربع، پی.اس.ای.
psoralen	پسورالن، داروی داءالصدف
psoralene	پسورالن
Pst1 (= Pst1 endonuclease)	پی.اس.تی.1.
Pst1 endonuclease	اندونوکلئاز پی.اس.تی.1
psychrophile	سرما دوست/خواه
psychrophilic enzyme	آنزیم سرما دوست/خواه
PTC (= phenyl-thio-carbomide)	پی.تی.سی.
pter	انتهای بازوی کوتاه
pterostilbenes	پتروستیل بن (ها)
PTH (= parathormone)	پی.تی.اچ.
PTH (= phenyl-thio-hydration)	پی.تی.اچ.
PTRF (= polymerase I and transcript release factor)	پی.تی.آر.اف.
public good	مصلحت همگانی، نفع عمومی
PUBS (= percutaneous umbilical blood sampling)	پی.یو.بی.اس.
puffer	باد بزن، دودکن
pulmonary stenosis	استنوز ریوی
pulse field gel electrophoresis	الکتروفورز ژلی پالس فیلد (میدان ضربه ای)
pulse shape discrimination	تبعیض/تشخیص شکل ضربه
pulse-chase experiment	آزمایش پالس-چیز
pulse-field gradient gel electrophoresis	الکتروفورز ژلی شیب میدان ضربه ای
pungi stick	چوب خاردار/پانجی
pure culture	کشت خالص
purine	پورین
purine nucleoside phosphorylase	فسفوریلاز نوکلئوزید پیورین
puromycin	پورومایسین
pus	چرک

push package	مواد دفع شدنی
putrefaction	تعفن
PUVA (= plus ultra-violet light of the A wavelength)	پی.یو.وی.ای.
PWS (= Prader-Willi syndrome)	پی.دابلیو.اس.
PXE (= pseudo-xanthoma elasticum)	پی.اکس.ئی.
pyralis	پیرالیس، حشره ذرت اروپایی
pyranose	پیرانوز
pyrenoid	پیرنوئید، دانک
pyrethrins	پایرترین
pyrexia	پایرکسیا
pyridoxal	پیریدوکسال
pyridoxal-5-phosphate	فسفات پیریدوکسال-5
pyridoxamine	پیریدوکسامین
pyridoxine	پیریدوکسین
pyrimidine	پیریمیدین
pyrogen	آتشزا، تب آور، تب زا، عامل مولد تب، عامل تب آور
pyrogenic toxin	سم تب آور/زا
pyronin Y	پیرونین وای.
pyrophosphate cleavage	شکافتگی پیروفسفات
pyro-phosphate ion	یون پیروفسفات
pyrrolizidine alkaloid	پیرولیزیدین آلکالوئید
pyruvate de-hydrogenase complex	کمپلکس دهیدروژناز پایروویت
pyruvate kinase	پیروویت کیناز
pyruvic acid	اسید پیروویک
pytalin	پیتالین
pyuria	ادرار چرک دار، چرک شاشی

Q

q arm (= long arm of a human chromosome)	شاخه کیو
QB	کیو.بی.
Q-band	کیو. باند
QH2 (= reduced coenzyme Q)	کیو.اچ.2.
Q-RT-PCR (= quantitative reverse transcriptase PCR)	کیو.-آر.تی.-پی.سی.آر.

quadruple cropping	کشت چهار گانه
quadrupole ion trap	تله یونی چهارقطبی/کوادروپول
qualitative spectrometric analysis	تحلیل کیفی اسپکترومتری/به کمک طیف سنج
quantitative character	شاخصه کمی
quantitative inheritance	توارث صفات کمی
quantitative phenotype	فنوتیپ کمی
quantitative reverse transcriptase PCR	نسخه برداری معکوس کمی پی.سی.آر.
quantom speciation	تخصیص کمی
quantum dot	نقطه کوانتومی
quantum tag	برچسب کوانتومی
quantum wire	سیم کوانتومی
quantum yield	واگذاری کوانتمی
quarantine	در قرنطینه نگهداشتن/بودن، قرنطینه، قرنطینه کردن
quarantine pest	آفت قرنطینه
quasi-option value	ارزش شبه گزینه
quaternary period	دوره چهارم/کواترنری
quaternary structure	ساختار چهارم
quelling	سرکوب، فرونشانی
quencher dye	رنگ سیر/اشباع کننده
quenching	اشباع/سیر کردن
quercetin	کو ارستین
quick-stop	ایست سریع، توقف فوری/کوتاه
quinolone	کینولون
quorum sensing	حس حد نصاب

R

R point (= restriction point in the cell cycle)	نقطه آر.
R5P (= ribose-5-phosphate)	آر.5. پی.
RACE (= rapid amplification of cDNA ends)	آر.ای.سی.ئی.
racemase	راسماز

264

racemate	راسمات
racemic	راسمیک
radiation	اشعه، انرژی تابشی، تابش، پرتو افشانی، تشعشع
radiation genetics	ژنتیک تشعشعی
radicular	ریشه ای
radioactive	پرتوزا، پرتو کنش ور، رادیواکتیو
radioactive isotope	ایزوتوپ پرتوزا/رادیواکتیو، همسان پرتوزا
radioactive marker	نشانگر رادیواکتیو
radioactivity	پرتوزایی
radiobiology	رادیوبیولوژی
radioimmunoassay	ایمنی سنجی پرتویی
radioimmunotechnique	تکنیک ایمنی پرتویی
radioisotope	ایزوتوپ پرتوزا، رادیو ایزوتوپ
radiolabeled	برچسب رادیواکتیو
radiotherapy	اشعه/پرتو درمانی، رادیوتراپی
raffinose	رافینوز
raft	شناور، کلک
raman optical activity spectroscopy	طیف نمائی فعالیت نوری رامان
random amplified polymorphic DNA	دی.ان.ای. چند ریخت تکثیر یافته تصادفی
random coil	چنبره تصادفی
random fragments	تکه های تصادفی
random genetic drift	جریان ژنتیکی تصادفی
randomized	انتخاب شده تصادفی، بختی سازی شده، بختین، بر خورده
rapamycin	راپامایسین
rapid amplification of cDNA ends	تقویت سریع پایانه های سی.دی.ان.ای.
rapid cycle DNA amplification (RCDA)	تقویت چرخه سریع دی.ان.ای
rapid microbial detection	تشخیص/شناسائی سریع میکروب
rare bases	بازهای کمیاب
rare cutter	برنده کمیاب، ریرکاتر
ras (= ras oncogene)	راس

ras gene	ژن راس
ras oncogene	ژن سرطان زای موشی
ras protein	پروتئین راس
rash	بثورات جلدی، جوش، دانه های پوستی، کهیر
rational drug design	طراحی معقول دارو
rational expectation	انتظار منطقی، توقع عقلانی
RB (= retino blastoma)	آر.بی.
Rb (= retinoblastoma)	آر.بی.
r-banding	آر.-بندینگ، نوار بندی آر.
RBC (= red blood cell)	آر.بی.سی.
RBS (= RNA binding site)	آر.بی.اس.
RC Replication Complex	کمپلکس همانند سازی آر.سی.
rcp (= rearrangement)	آر.سی.پی.
RCV (= replication-competent virus)	آر.سی.وی.
RDB (= reverse dot blot)	آر.دی.بی.
rDNA small laboratory animals	حیوانات کوچک آزمایشگاهی آر.دی.ان.ای. ئی
rDNA techniques	تکنیکهای آر.دی.ان.ای. ئی
RE (= response element)	آر.ئی.
RE (= RNA extraction)	آر.ئی.
reactivate	بازفعال سازی
reactive oxygen species	گونه باز فعال اکسیژن
reading frame	چهارچوب خواندن/قرائت
reading frame shift	جابجائی چارچوب قرائت
reagent	معرف، واکنشگر
reagin	رآژین
reannealing	باز آبکاری، واتاباندن
rearrangement	باز آرائی
reassociation	پیوند مجدد
ReC (= recombinant chromosome)	
recalcitrant seed	دانه مقاوم
receiving party	طرف دریافت کننده
receptor	پذیرنده، گیرنده
receptor fitting (RF)	ریسپتور فیتینگ
receptor mapping	بازنمایی گیرنده
receptor population	جمعیت گیرنده
receptor tyrosine kinase	تیروزین کیناز گیرنده

English	Persian
receptor-mediated	با واسطه گیرنده
receptor-mediated endocytosis	درون یاختگی به واسطه گیرنده
recessive	مغلوب، نهفته
recessive allele	آلل مغلوب
recessive gene	ژن نهفته
recessive oncogene	ژن توموری نهفته
recipient cell	سلول گیرنده
reciprocal crosses	تقاطع دوطرفه
reciprocal externality	اکسترنالیتی متقابل
reciprocal recombination	نوترکیبی دوطرفه
reciprocal translation	ترجمه دوطرفه
recirculation	باز چرخش/چرخانی
recognition sequence (site)	توالی (جایگاه) شناسائی
recognition site	جایگاه شناسائی
recombinant	بازترکیب، رکمبینان، نوترکیب
recombinant chromosome	کروموزوم نوترکیب
recombinant DNA	دی.ان.ای. نوترکیب
recombinant DNA microorganism	میکروارگانیسم های نوترکیب
recombinant DNA molecule	مولکول دی.ان.ای. نوترکیب
recombinant DNA technology	تکنولوژی دی.ان.ای. نوترکیب
recombinant protein	پروتئین رکمبینان/نوترکیب
recombinase	رکومبیناز
recombination	اتصال/ترکیب مجدد، بازترکیبی، رکمبیناسیون، نوترکیبی
recombination frequency	وفور نوترکیبی
recon	شناسائی
recruitment	بکار گماری
recurrent risk	ریسک عود کننده/برگشتی
recycling	باز یافت/یابی
red blood cell	گلبول قرمز خون، ر
redement napole gene	ژن ردمنت ناپول
redness	سرخی، قرمزی
redox	اکسایش و کاهش اکسیداسیون و احیاء، ردوکس
reduced coenzyme Q	کوآنزیم کیو. کاهش یافته
reduced flavin adenine	دینوکلئوتید فلاوین آدنین
dinucleotide	تضعیف شده
reduction	احیاء، اکسیده شدن، کاهش، کاهیدگی
reduction division	تقسیم کاهشی
redundancy	افزونگی، افزونی، زائد بودن، زیادی، مازاد بر احتیاج بودن، مازادی
reference concentration	تراکم/غلظت مرجع
reference laboratory	آزمایشگاه مرجع
refillable	قابل پر شدن/کردن مجدد
refractile body	بدنه/پیکره/تنه منکسر کننده/شکننده نور، ریفرکتایل بادی
refraction	انکسار/شکست (نور)
regeneration	باززایی، بازسازی، تجدید حیات، ترمیم اندام، نوپدیدی، نوزائی
regioselective	رژیوسلکتیو
regiospecific	رژیواسپسیفیک
regulated article	ماده تنظیمی، شیئی تحت کنترل
regulatory DNA sequence	توالی دی.ان.ای. تنظیمی
regulatory element	عامل تنظیم کننده، عنصر تنظیمی
regulatory enzyme	انزیم تنظیم کننده/تنظیمی
regulatory gene	ژن تنظیم کننده، ژن نظم دهنده
regulatory sequence	توالی تنظیمی
regulon	رگولون
rehydration	باز آبدهی
reiterated genes	ژنهای برگشتی
relapse	بازگشت، برگشت، عود
relaxed	آزاد، بدون پیچ، راحت
relaxed circle plasmid	پلاسمید دایره ای/مدور سست
relaxed plasmid	پلاسمید سست
relaxin	ریلاکسین
release factor	عامل رها سازی
releasing hormone	هورمون رها سازی
remediation	درمان
renaturation	بازگشت طبع، برگشت به

English	Persian
	حالت طبیعی
renature	به حالت طبیعی باز گرداندن
renewable resource	منبع تجدید پذیر
renin	پنیر مایه، رنین
renin inhibitor	بازدارنده رنین
rennin	رنین
reoviridae	رئوویروس (ها)
reovirus	رئوویروس
reperfusion	باز ریزش، باز وامیختگی
repetetive DNA	دی.ان.ای.تکراری
repiration	ریپایریشن
replacement vector	ناقل جایگزینی
replica plating	آبکاری همانند
replicase	رپلیکاز
replication	تکثیر، تکرار، تولید مجدد، همانند سازی، همتا سازی
replication factor C	عامل همانندسازی سی.
replication fork	دوشاخه همانند سازی
replication licensing factor	عامل مجوز همانند سازی
replication origin	سرچشمه/منشاء همانند سازی
replication protein A	پروتئین همانند سازی ای.
replication-competent virus	ویروس مستعد همانند سازی
replicative form of M13	شکل همانند ساز ام.13
replicon	رپلیکون
replisome	رپلیزوم
reporter	گزارشگر
reporter gene	ژن گزارشگر
repressible enzyme	آنزیم قابل سرکوب
repression	از کارافتادگی تظاهر ژنی، بازدارندگی، سرکوب، واپس رانی، فرونشانی
repressor	بازدارنده، رپرسور، مانع شونده، مهار کننده
RER (= rough endoplasmic reticulum)	آر.ئی.آر.
residual risk	خطر پسماندی
residue	پس مانده، تفاله، ته نشین، رسوب

English	Persian
residue weight	جرم رسوبات
resistance	مقاومت
resistant	مقاوم
resolving power	قدرت رفع
resource	امکان (ات)، ذخائر، منابع
respirable	قابل تنفس
respiration	تنفس، دم زنی، دمش
respiratory distress	ناراحتی تنفسی
respiratory mucosa	لایه مخاطی دستگاه تنفسی
respiratory quotient	ضریب تنفسی
respiratory syncytial virus	ویروس تنفسی سینسیتیال
respiratory tract	راه/مجرای تنفسی
respirometer	تنفس سنج
response	پاسخ، پاسخ به محرک، و اکنش
response element	عامل پاسخ
restoration	آبادسازی، احیاء، اصلاح، ترمیم، کشت مجدد، مرمت
restriction	محدودیت
restriction analysis	تحلیل محدودیت
restriction endonuclease	اندونوکلئاز محدود کننده
restriction enzymes	آنزیم های محدود کننده
restriction fragment length polymorphism (RFLP)	چند شکلی طولی قطعات محدود کننده
restriction map	نقشه محدود کننده
restriction point in the cell cycle	نقطه منع در چرخه سلولی
restriction sites	محلهای محدود کننده
restriction-modification system	سیستم تحول محدودیت
restrictive enzyme	آنزیم محدودکننده
resuscitation	احیاء، برانگیزش، تجدید حیات، زنده کردن
resveratrol	رزوراترول
reticulocyte	رتیکولوسیت
retinene	رتینن
retinitis pigmentosa	رتینایتیس پیگمنتوزا
retino blastoma (retinoblastoma)	رتینوبلاستوما

retinoblastoma protein	پروتئین رتینوبلاستوما
retinoic acid	اسید رتینوئیک
retinoid	رتینوئید، شبکیه مانند، شبه شبکیه
retinoid X receptor	گیرنده رتینوئید ایکس
retinol	رتینول
retroelements	عناصر پسرو
retrograde	برگشت دهنده/کننده، پسرو، عقب رونده، قهقرایی، نزولی
retropharyngeal	پشت حلقی
retrospective diagnosis	تشخیص پس نگر/بازنگرانه
retroviral vector	ناقل رترو ویروسی
retroviral-like element	عنصر شبه رتروویروس
retrovirus	رترو ویروس
reuse	بازمصرف
revealed preference approach	نگرش ترجیحی آشکار شده
reverse dot blot	دات بلات معکوس
reverse genetics	ژن شناسی معکوس
reverse osmosis	اسمز معکوس
reverse phase chromatography	کروماتوگرافی فاز معکوس
reverse transcriptase	ترانسکریپتاز معکوس
reverse transcriptase-PCR	ترانسکریپتاز-پی.سی.آر. معکوس
reverse transcription	نسخه برداری معکوس
reversed micelle	میسل معکوس
reversion	برگشت، ترجمه مجدد، رجوع، عودت
revolutions per minute	دور در دقیقه
RF (= release factor)	آر.اف.
RFC (= replication factor C)	آر.اف.سی.
RFLP (= restriction fragment length polymorphism)	آر.اف.ال.پی.
RFLP linkage analysis	آنالیز/تحلیل اتصال آر.اف.ال.پی.
rh	آر.اچ.
Rh factor	عامل آر.اچ.
rhabdoviridae	رابدوویروس (ها)
rhamnose	رامنوز
rhinovirus	راینو ویروس
rhizobia	ریزوبیوم ها
rhizobium	ریزوبیوم
rhizoremediation	باز اصلاحی ریشه ای
rhizosphere	ریزواسفر، ریشه سپهر، ریشه گاه
rhizospheric	ریشه گاهی
rho	رو
rho factor	عامل رو
rhodanese	رودانیز
rho-dependent terminator	پایان دهنده متکی به رو
rhodospin	رودواسپین
ri plasmid	پلاسمید ری
riboflavin	ریبوفلاوین
ribonuclease	ریبونوکلئاز
ribonucleic acid	اسید ریبونوکلئیک
Ribo-Nucleic Acid	اسید ریبونوکلئیک
ribonucleoprotein	ریبونوکلئوپروتئین
ribo-nucleo-protein	ریبونوکلئوپروتئین
ribonucleoside	ریبونوکلئوزید
ribonucleotide	ریبونوکلئوتید
ribose	ریبوز
ribose-5-phosphate	ریبوز-5-فسفات
ribosomal adaptor	وفق دهنده ریبوزومی
ribosomal DNA	دی.ان.ای. ریبوزومی
ribosomal ribonucleic acid	اسید ریبونوکلئیک ریبوزومی
ribosomal RNA (rRNA)	آر.ان.ای. ریبوزومی
ribosome	ریبوزوم، ریزدانه پالاد، رناتن
ribosome-binding site	جایگاه/محل پیوند ریبوزوم
riboswitch	ریبوسوئیچ
ribozyme	ریبوزیم
ribulose	ریبولوز
rice blast	رایس بلاست
ricin	رایسین
rifampicin	ریفآمپیسین
rifampin	ریفآمپین
Ringer's solution	محلول رینگر
risk	احتمال خطر، خطر، ریسک
risk assessment	ارزیابی احتمال خطر/ریسک
risk factor	عامل ریسک

risk management	مدیریت ریسک
risk management RNA	آر.ان.ای. گرداننده ریسک
risk patient	بیمار خطر پذیر
risk ratio	ضریب ریسک، نسبت خطر
risk reduction	کاهش ریسک
risk worker	کارگر خطر پذیر
ristocetin	ریستوستین
RLF (= replication licensing factor)	آر.ال.اف.
RMB (= Roche Molecular Biochemicals)	آر.ام.بی.
RNA (= Ribo-Nucleic Acid)	آر.ان.ای.
RNA binding site	محل اتصال آر.ان.ای.
RNA extraction	استخراج آر.ان.ای.
RNA interference	تداخل آر.ان.ای.
RNA polymerase	پلیمراز آر.ان.ای.
RNA probes	پرب/پروب/ردیاب/کاووشگر/ز شانگر های آر.ان.ای.
RNA processing	پردازش آر.ان.ای.
RNA splicing	بهم تابیدن آر.ان.ای.
RNA transcript	رونوشت آر.ان.ای.
RNA transcriptase	ترانسکریپتاز آر.ان.ای.
RNA vectors	ناقلین آر.ان.ای.
RNAse (= ribonuclease)	آر.ان.آز
RNP (= ribo-nucleo-protein)	آر.ان.پی.
ROb (= Robertsonian translocation)	
Robertsonian translocation	جابجاشدگی رابرتسونی
Roche Molecular Biochemicals	روش مولکولار بیوکمیکالز
rocket	راکت، موشک
rocket immunoelectrophoresis	ایمونو الکتروفورز موشکی
rod cell	سلول میله ای
roller bottle apparatus	دستگاه رولر باتل
rolling circle	حلقه رولینگ
rolling circle replication	همانند سازی حلقه رولینگ
room temprature	دمای اتاق
rootworm	کرم ریشه
rosemarinic acid	اسید رزمارینیک
rosetting	گل وبوته دادن

rotating biological contractor	روتیتینگ بیولوژیکال کنتراکتور
rotenone	روتنون
rotor fermentor	روتور فرمنتور
rough endoplasmic reticulum	رتیکولوم اندوپلاسمیک زبر، شبکه درون دشته ای زبر
roving gene	ژن جا به جا شدنی/سرگردان
RP (= retinitis pigmentosa)	آر.پی.
RPA (= replication protein A)	آر.پی.ای.
rpm (= revolutions per minute)	آر.پی.ام.
rRNA (= ribosomal RNA)	آر.آر.ان.ای.
RSV (= respiratory syncytial virus)	آر.اس.وی.
RT (= reverse transcription)	آر.تی.
rT (= room temprature)	آر.تی.
RT-PCR (= reverse transcriptase-PCR)	آر.تی.-پی.سی.آر.
RT-PCR out of paraffin-embedded tissue	بافت جایگزین آر.تی.-پی.سی.آر خارج از پارافین
RTVL (= retroviral-like element)	آر.تی.وی.ال.
rubitecan	روبیتکان
RuBP (= ribuose-1,5-bis-phosphate)	
rubratoxin	روبراتوکسین
rumen	شکمبه
rumenic acid	اسید رومنیک
runaway plasmid	پلاسمید فراری
rust	زنگ، زنگ آهن، زنگار، زنگ خوردگی/زدگی/زدن

S

S (= S phase)	اس.
S (= sedimentation coefficient)	اس.
S (= Svedberg unit)	اس.

English	Persian
s (second)	اس.
S phase (DNA replication phase)	فاز اس.
s1 nuclease	نوکلئاز اس.1
SA (= streptococal antigen)	اس.ای.
SA (= surface antigen)	اس.ای.
Sabin vaccine	واکسن سابین
saccharase	ساکاراز
saccharification	ساکاریفیکاسیون، قندی شدن
saccharomyces cereviciae (= baker's yeast)	ساکارومایسز سرویسیا
S-adeno homocysteine hydrolase	هیدرولاز اس.-آدنوهوموسیستئین
safe minimum standard	حداقل استاندارد ایمنی
safe safety	ایمنی/ضامن مطمئن
safe transfer	انتقال ایمن
safety factor	ضریب اطمینان، ضریب/عامل ایمنی
safety-pin morphology	ریخت شناسی سیفتی پین، مورفولوژی سنجاق قفلی ای
SAHH (= S-adeno homocysteine hydrolase)	اس.ای.اچ.اچ.
salicylic acid	اسید سالیسیلیک، ترشای بید
salinity tolerance	تحمل شوری
salmonella	سالمونلا
salmonella enteritidis	آنتریت سالمونلا، التهاب روده باریک ناشی از سالمونلا
salmonella tymphiumurium	تیمفیموریوم سالمونلا
salmonella typhimurium	تیمفیموریوم سالمونلا
salt tolerance	تحمل نمک
salting in	سالتینگ این
salting out	سالتینگ آوت
salvage pathways	مسیرهای بازیابی
salvarsan	سالوارسان
samesense mutation	جهش ساکت/هم معنی
sample applicator	اپلیکاتور نمونه گیری
Sanger's method	روش سانگر
Sanger-Coulson method	روش سانگر کولسون
sangivamycin	سانگیوا مایسین
SAP (= stress activated protein)	اس.ای.پی.
saponification	صابونی شدن/کردن
saponin	اشنان، چوبک، ساپونین، ماده کف آور
saponnin (=saponin)	ساپونین
saprophagy	ساپروفاژی
saprophyte	ساپروفیت
saprozoic	ساپروزوئیک
SAR (= scaffold attachment region)	اس.ای.آر.
sarafotoxin	سارافوتوکسین
sarcolemma	سارکولما
sarcoma	سارکوما
sarcosine	سارکوزین
sarkomycin	سارکومایسین
SARS	بیماری سارز
SASD (= sialic acid storage disease)	اس.ای.اس.دی.
satellite chromosome	کروموزوم اقماری
satellite DNA	دی.ان.ای.ماهواره ای
satellite RNA	آر.ان.ای. ماهواره ای
satellite virus	ویروس اقماری
satratoxin	ساتراتوکسین
saturated fatty acid	اسید چرب اشباع شده
saxitoxin	ساکسیتوکسین
SC (= secretory component)	اس.سی.
SCA (= sickle cell anemia)	اس.سی.ای.
scab	پوست (ه) زخم، دلمه
scaffold	اسکلت، بدنه پروتئینی، داربست، کالبد
scaffold attachment region	منطقه اتصال داربست
scaffolding protein	پروتئین داربستی
scale-up	افزایش مقیاس
scanning tunneling electron microscopy	ریزبینی الکترونی تونل زنی پویشی
scanning tunneling microscopy	ریزبینی تونل زنی پویشی
SCARMD (= severe, childhood, autosomal,	اس.سی.ای.آر.ام.دی.

recessive muscular dystrophy)	
SCE (= sister chromatid exchange)	اس.سی.ئی.
schizogone	شیزوگون
schizogony	شیزوگونی
SCID (= severe combined immuno-deficiency disease)	اس.سی.آی.دی.
sclerin	اسکلرین
sclerobasidium	اسکلروبازیدیوم
scleroprotein	اسکلروپروتئین، پروتئین سخت
sclerotium	اسکلروتیوم، سختینه
scotophobin	اسکوتوفوبین
scrapie	اسکراپی، فلج گوسفندی
screening	غربال کردن
scRNA (= small cytoplasmic RNA)	اس.سی.آر.ان.ای.
scrubbing	اسکراب کردن
SCT (= secondary constriction)	اس.سی.تی.
SDS (= sodium dodecylsulfate)	اس.دی.اس.
sebaceous gland	غده چربی
sebum	چربی مترشحه، سبیوم
second meiotic anaphase	دومین آنافاز میوتیک
second meiotic metaphase	دومین متافاز میوتی
second messenger	پیک ثانویه
secondary constriction	تنگای ثانویه
secondary forest	جنگل ثانویه
secondary immune response	پاسخ ایمنی ثانویه
secondary pneumonic plague	طاعون ریوی ثانوی
secondary septicemic plague	طاعون خونی ثانوی
secondary structure of protein	ساختار ثانوی پروتئین
secondary value	ارزش ثانویه
secretagogue	سکرتاگوگ
secretase	سکرتاز
secretin	سکرتین

secretogogue	سکرتوگوگ
secretor	تراونده، ترشح کننده، سکرتور
secretory component	جزء مترشح
secretory IgA	آی.جی.آی. مترشح
sector cell	سلول سکتوری/ترشحی
sedimentation coefficient	ضریب ته نشینی
seed bank	انبار/مخزن بذر
seedless fruit	میوه بدون بذر
seed-specific promoter	اسپسیفیک پروموتر دانه ای
segregation (= separation)	تفکیک
segregation distorter (SD)	برهم زننده تفکیک
seizure	تصرف، حمله صرعی/بیماری، قفل کردن
selectable marker	نشانه قابل انتخاب
selectable marker gene	ژن دارای نشانه قابل انتخاب
selectin	سلکتین
selection	انتخاب، گزینش
selection pressure	فشار انتخاب(ی)
selective estrogen effect	تاثیر استروژن گزینشی
selective estrogen receptor modulator (SERM)	مدولاتور گیرنده استروژن انتخابی
selector gene	ژن منتخب
selenocysteine	سلنوسیستئین
selenomethionine	سلنومتیونین
selenoprotein	سلنوپروتئین
self-assembling molecular machine	دستگاه مولکولی خود تجمع/برپا ساز
self-assembly	خود بر پائی
self-fertilization	خودباروری، خود گشنی
self-inactivating vector	ناقل خود خنثی گر
self-incompatibility	باخود ناسازگاری، ناخود سازگاری
selfish DNA	دی.ان.ای.آشغال/خودخواه
selfish DNA hypothesis	فرضیه دی.ان.ای. خودخواه
self-pollination	خود گرده افشانی، خودگشنی
semiconservative DNA	دی.ان.ای.نیمه محافظه کار
semiconservative replication	همانند سازی نیمه محافظه کارانه
semisynthetic catalytic	پادتن کاتالیزوری نیمه

antibody	مصنوعی	serotonin	سروتونین
senescence	پیری، پیرشدگی،	serotype	پیماب مونه، سروتیپ، نوع
	سالخوردگی، کهولت		سرمی
sense	احساس، حس (ی)،	serpin	سرپین
	سوهش (ی)،شعور، فهم،	serum	پیماب، خونابه، سرم
	مفهوم	serum albumin	آلبومین خون، سرم آلبومین
sense codon	کدون معنی دار	serum half life	نیمه عمر سرم
sense DNA	دی.ان.ای. معنی دار	serum immune response	واکنش ایمنی سرم
sense strand	رشته معنی دار	serum lifetime	طول عمر سرم
sensitivity	حساسیت	serum response factor	عامل پاسخ سروم
sensitization	حساس سازی	serum-trypsin-inhibitory	ظرفیت سرکوبگری سروم
sentinel surveillance	شناسائی	capacity	تریپسین
	سنتینلی/پاسداری	sessile	بدون پایک/دمگل، برجا، بی
separation	جدائی		پایه/دمبرگ/ساقه
sepsis	پلشتی، چرکی شدگی،	settle	ته نشین کردن/شدن،
	عفونت، گند		فروکش کردن، نشاندن
septic shock	شوک عفونی	settling	ته نشانی، ته نشستن، ته
sequela	بازمانده، جای زخم، دنباله،		نشین شدن، ته نشینی،
	عارضه باقیمانده، نشانه		رسوب، رسوب گذاری
	زخم	severe combined	بیماری نقص ایمنی مرکب
sequence	ترادف، ترتب، ترتیب،	immuno-deficiency	حاد
	تسلسل، توالی، رشته،	disease	
	زنجیره	severe, childhood,	پلاسیدگی عضلانی اتوزومی
sequence homology	همگونی در توالی	autosomal, recessive	مغلوب حاد کودکان،
sequence hypothesis	فرضیه توالی	muscular dystrophy	موسکولار دیستروفی
sequence map	نقشه توالی		مغلوب اتوزومی بچگانه
sequence tagged site	جایگاه نشاندار شده با توالی		حاد
sequence-tagged site (STS)	سایت نشاندار شده با توالی	sewage farm	کشتزار آبیاری شده با
sequencing	تعیین توالی		فاضلاب
sequencing gel	ژل تعیین توالی	sewage sludge	گل ولای فاضلابی
sequon	سکوئون	sewage treatment	پالایش فاضلاب
Ser (= serine)		sewer	پساب/فاضلاب رو
serial analysis of gene	تحلیل متوالی تظاهر ژن	sewerage	پساب/فاضلاب داری
expression		sewreage system	شبکه پساب/فاضلاب
serial passage	گذر متوالی	sex cell	سلول جنسی
sericin	سریسین	sex chromatin	کروماتین جنسی
serine	سرین	sex chromosome	کروموزوم جنسی
seroconversion	سروکانورژن	sex determination	تشخیص جنسیت
serologist	سرم شناس	sex factor	عامل جنسیت
serology	سرم شناسی	sex influenced dominance	برتری متاثر از جنسیت،
seronegative	سرم منفی		غالبیت تحت تاثیر جنس

sex limited	محدود به جنس
sex linkage	وابستگی جنسیتی
sex linked	وابسته به جنسیت
sex pilus	تاژک جنسی
sex-conditioned character	شخصیت شرطی شده توسط جنسیت
sexduction	القاء جنسی، سکسداکشن
sex-influenced character	شخصیت متاثر از جنسیت
sex-limited character	شخصیت محدود به جنس
sex-linked gene	ژن متصل به جنسیت
sexual conjugation	آمیختگی/هم یوغی جنسی
sexual reproduction	تولید مثل جنسی
sexually transmitted disease (STD)	بیماری مقاربتی/آمیزشی
SFC (= Spot-Formi cell)	اس.اف.سی.
shake culture	کشت ارتعاشی/شیک
sheep erythrocytes	اریتروسیت گوسفندی
shelter-in-place	پناهگاه-در-محل
shifting agriculture	کشت کوچان
shifting balance theory	فرضیه جابجائی توازن
shiga toxin	سم شیگا
Shigella	شیگلا
shikimate	شیکیمات
Shine-Dalgarno sequence	توالی شاین-دالگارنو
shock	برق گرفتگی، شوک، ضربه، ضربه روحی
shock fluid	مایع شوکی
short arm of a human chromosome	شاخه کوتاه کروموزوم انسانی
short arm of chromosome	شاخه کوتاه کروموزوم
short interfering RNA	آر.ان.ای. کوتاه تداخلی، توالی های مزاحم کوتاه آر.ان.ای.
short interspersed nuclear element	عنصر دارای فواصل هسته ای کوتاه
short limbed dwarfism	کوتولگی کوتاهی عضو
short tandem repeats	تکرارهای اتصال کوتاه
shotgun cloning method	روش همسانه سازی شات گانی
shotgun sequencing	توال یابی شات گانی
showdomycin	شودومایسین

showering	دوش گرفتن، زیر دوش شستن، شست وشو
shuttle vector	ناقل شاتلی
sialic acid	اسید سیالیک
sialic acid storage disease	بیماری تجمع اسید سیالیک
sialidase	سیالیداز
sialoadhesin	سیالو ادهسین
sib	فامیل، هم خون
sibling	فرزندان یک خانواده، هم نیا
sibling species	گونه هم نیا
sibmating (= crossing of sibling)	جفتگیری هم نیائی
sibriomotsin	سیبریوموتزین
sibship	هم نیائی
sickle cell anemia	آنمی سلول داسی شکل
side effect	اثرجانبی/اثرفرعی، عارضه جانبی، واکنش ثانوی
sieve element	عنصر غربالی
SIgA (= secretory IgA)	
sigma (= σ)	زیگما
sigma bond	پیوند سیگما
signal peptide (= leader peptide)	پتید علامتی
signal recognition particle	ذره شناسائی پیام
signal sequence	توالی علامتی
signal transduction	ترارسانی علامتی، فرا پیام انتقالی
signaling	علامت دادن، مخابره کردن
signaling molecule	ملکول پیام دهنده
signaling protein	پروتئین پیام دهنده
silencer	سرکوبگر
silencing	سرکوب
silent mutation	جهش خاموش/ساکت/ساکن
silica	سیلیس، سیلیکا
silk	ابریشم
silviculture	پرورش جنگل، جنگل پروری/کاری، کشت درختان
simian immunodeficiency virus	ویروس نقص ایمنی سیمیان/بوزینه ای

273

English	Persian
simian virus 40 (SV40)	ویروس سیمیان/بوزینه ای 40
simple protein	پروتئین ساده
simple sequence length polymorphism	پلی مورفیسم طولی توالی ساده، چندشکلی طولی توالی ساده
simple sequence repeat	تکرار توالی ساده
simulation	شبیه/مشابه/مدل سازی
sin nombre virus (SNV)	ویروس بی شماره
SIN vector (= self-inactivating vector)	ناقل اس.ای.ان.
sindbis virus	ویروس سیندبیس
SINE (= short interspersed nuclear element)	اس.ای.ان.ئی.
single burst experiment	آزمایش تک انفجاری
single cell culture	کشت تک سلولی
single cell isolation	جداسازی تک سلولی
single nucleotide polymorphism	چند ریختی تک نوکلئوتیدی
single strand	تک رشته
single-cell protein	پروتئین تک یاخته
single-copy DNA	دی.ان.ای.تک نسخه
single-nucleotide polymorphism	چند ریختی تک نوکلئوتید
single-site mutation	جهش تک جایگاهی
single-strand binding protein	پروتئین اتصال تک رشته
single-strand conformational polymorphism	پلی مورفیسم تک رشته ای تطبیقی
single-stranded DNA	دی.ان.ای. تک رشته ای
singlet oxygen	اکسیژن تک دانه
single-walled carbon nanotube	ریز لوله کربن تک دیواره
sirenin	سیرنین
sirtuin	سیرتوئین
sisomicin	سیزومیسین
sister chromatid	کروماتید خواهر
sister chromatid exchange	تبادل کروماتید خواهر
site-directed mutagenesis	استخلاف، جهش زایی جهت یافته، واردکردن/حذف جهت یافته
site-specific drug	داروی مختص به جایگاه
sitostanol	سیتوستانول
sitosterol	سیتوسترول
SIV (= simian immunodeficiency virus)	اس.ای.وی.
sizing	اندازه/دانه/شماره بندی، اندازه گیری، سایزینگ
sizing column	ستون اندازه گیری
sizing gel	ژل اندازه گیری
skeleton	استخوان بندی
skin	پوست، پوست کندن، پوسته، جلد
slant culture	کشت مایل
SLD (= short limbed dwarfism)	اس.ال.دی.
SLE (= systemic lupus erythematosus)	اس.ال.ئی.
slide	تیغه
slide cell culture	کشت اسلاید سل، کشت تیغه ای سلولی
slide cover	تیغک
slide culture	کشت اسلاید/تیغه ای
sliding filament model	مدل میلک لغزان
slime	گل، لجن، لعاب، ماده لزج
slime bacteria	باکتری گل ولای
slime fungi	قارچ گل و لای
slime layer	لایه گل و لای
slime molds	کپک گل و لای
slough	باتلاق، لجنزار، مرداب، نهر باتلاقی
slow component	جزء کند
sludge volume index (SVI)	ایندکس حجم گل و لای
small cytoplasmic RNA	آر.ان.ای. کوچک سیتوپلاسم
small interfering RNA	آر.ان.ای. کوچک تداخلی، توالی های مزاحم کوچک آر.ان.ای
small nuclear ribonucleoprotein (snRNP)	ریبونوکلئوپروتئین کوچک هسته
small nuclear RNA (snRNA)	ار.ان.ای. کوچک هسته ای
small nucleolar RNA	آر.ان.ای. کوچک نوکلئولار
small RNA	ار.ان.ای. کوچک

English	Persian
small ubiquitin-related modifier (SUMO)	تعدیل کننده کوچک وابسته به یوبیکوتین
SMD (= somatostatin)	اس.ام.دی.
SMG (= sub-mandibular glands)	اس.ام.جی.
smog	دودمه
smoke	بخار بیرون دادن، دود، دود دادن/کردن
smooth muscle	ماهیچه صاف
smut	آلوده به سیاهک، زنگ سیاه، زنگ گیاهی، سیاهک، لکه
Sno RNA (= small nucleolar RNA)	
SNP (= single nucleotide polymorphism)	اس.ان.پی.
snRNA (= small nucleolar RNA)	اس.ان.آر.ان.ای.
snRNP (= small nuclear ribonucleoprotein)	اس.ان.آر.ان.پی.
snurposome	اسنورپوزوم
social opportunity cost	هزینه فرصت اجتماعی
sociobiology	جامعه-زیست شناسی، زیست شناسی جامعه، سوسیوبیولوژی
SOD (= super-oxide dismutase)	اس.او.دی.
sodium	سدیم
sodium channel	مجرای سدیمی
sodium dodecyl sulfate	دودسیل سولفات سدیم
sodium lauryl sulfate	لوریل سولفات سدیم
sodium phosphate	فسفات سدیم
sodium pump	تلمبه سدیمی
sodium sulfite	سولفیت سدیم
soft agar culture	کشت آگار نرم
soft laser desorption	واجذب لیزری نرم
soft rot	پوسیدگی نرم
solanaceae	بادنجانیان، تیره سیب زمینی
solanine	سولانین
solenoid	سولنوئید
solid	استوار، بریسته، تام، توپر، جامد، جسم، جسم

English	Persian
	صلب، جماد، سه بعدی، صلب، فضایی، متفق، محکم، ناب، یکپارچه
solid support	پشتیبان جامد
solid waste	پسماند جامد
solid-phase synthesis	سنتز فاز جامد
soluble	حل شدنی، قابل حل
soluble fiber	فیبر قابل حل
solvalysis	حلال کافت
solvated	حلال پوشیده
solvation	حلال پوشی
soma cell	سلول سوما
somaclonal variation	تغییرات همسانه بدنی/سوماکلونال، تنوع تن تاگی
Somalia	سومالی
soman	سومان
somatacrin	سوماتاکرین
somatic	بدنی، پیکری، تنی، رویشی، غیر جنسی
somatic cell	سلول/یاخته بدنی/تنی/جسمی
somatic cell gene therapy	ژن درمانی سلول جسمی
somatic crossing over	کراسینگ اور سوماتیکی/بدنی
somatic mutation	جهش سوماتیکی/بدنی
somatic variant	تغییر تنی، تنوع بدنی، سویه سوماتیکی، واریانت یاخته غیر جنسی
somatoliberin	سوماتولیبرین
somatomammotropin	سوماتوماموتروپین
somatomedin	سوماتو مدین
somatoplasm	سوماتوپلاسم
somatostatin	سوماتواستاتین
somatotrophin (= somatotropin)	سوماتوتروفین
somatotropin	سوماتوتروپین، هورمون رشد
sonic hedgehog protein	پروتئین جوجه تیغی صوتی
sonography	سونوگرافی
sorbic acid	اسید سوربیک
sorbitol	سوربیتول

sorbose	سوربوز	spectrin	اسپکترین
sorcin	سورسین	spectrometer	طیف سنج
SOS protein	پروتئین اس.او.اس.	spectrophotometer	طیف نور سنج
SOS repair system	سیستم ترمیم اس.او.اس.	spectrum	بیناب، طیف
SOS response	واکنش اس.او.اس.	sperm	اسپرم
SOTE (= standard oxygen transfer efficiency)	بازاک استاندارد	sperm mother cell	سلول مادر اسپرم
		spermatid	اسپرماتید، نر زامچه
southern blot analysis	تحلیل ساترن بلات	spermatium	اسپرماتیوم، نر زامه بی تاژک
southern blot/transfer	ساترن بلات/ترانسفر	spermatocyte	اسپرماتوسیت
southern blotting	ساترن بلاتینگ	spermatogenesis factor 3	فاکتور اسپرم زائی 3
southern hybridization	هم تیرگی/دورگه گیری به روش ساترن	spermatogonium cell	اسپرماتوگونی، یاخته مادر اسپرم
southwestern blot	سات وسترن بلات	spermiogenesis	اسپرم زائی، اسپرمیوژنز، تشکیل اسپرم
soya bean	سویا		
SP₁ (= transcription factor)	اس.پی.1.	spermophilus	اسپرموفیلوس
SP₃ (= spermatogenesis factor 3)	اس.پی.3.	s-phase	فاز اس.
		spheroplast	اسفروپلاست، گویزه دش
SPaA (= surface protein antigen A)		sphingolipid	اسفنگولیپید
		sphingomyelin	اسفنگومیلین
spacer arm	شاخه حد فاصل	spin	اسپین
spacer DNA	دی.ان.ای. حد فاصل	spin trapping	اسپین تراپینگ
spacer gel	ژل حد فاصل	spina bifida	اسپینا بیفیدا
SP-B (= surfactant-associated protein B)	اس.پی.-بی.	spindle	دوکی شکل
		spinner culture	کشت اسپینر
		spinning	تاب دادن، تارتنی، تار ریسی، چرخاندن، ریسندگی، نخ تابی
specialization	تخصص		
specialized transduction	ترارسانی ویژگی یافته، ترانسداکشن تخصیص یافته	spinning cup protein sequencer	توالی گر پروتئین پیمانه گردان
speciation	تشکیل گونه، گونه زایی	Spinosad ™	اسپینوزاد
species	جور، گونه، نوع	spinosyns	اسپینوسین ها
species diversity	تنوع گونه	spiramycin	اسپیرامایسین
species richness	فراوانی گونه	spirochete	اسپیروکت، تار پیچان
species selection	انتخاب گونه	spirometer	دم سنج
species specific	مختص به گونه	spirometery	دم سنجی
specific activity	فعالیت اختصاصی	splice junctions	اتصال بهم تابیده
specific growth rate	نرخ رشد اختصاصی	splice variants	سویه های بهم تابیده
specific production rate	نرخ تولید اختصاصی	spliceosome	اسپلایسوزوم
specific volume	حجم مشخص	splicing	بهم تابیدن
specificity	اختصاصی بودن، ویژگی	splicing junction	اتصال بهم تابیده
spectinomycin	اسپکتینومایسین	SPM (= suspended	ذرات معلق

276

particulate matter)	
spontaneous assembly	تجمع فوری/خودبخودی، گردایش فوری/خودبخودی
spontaneous generation	پیدایش خودبخودی، تولید فوری/خودبخودی، نازیست زائی
sporangiospore	اسپورانژیوسپور
sporangium	اسپورانژیوم، هاگدان بر
spore	اسپور، تخم قارچ، تخم میکروب، هاگ، هاگه
sporozoite	اسپوروزوئیت، هاگ زیا
sport	ورزش
sporulation	اسپور زایی، هاگ آوری/زایی/گذاری/سازی
Spot-Formi cell	سلول اسپات-فرمی
spray	اسپری، افشک، افشاندن، افشانه پاشیدن
sprayer	اسپری، -افشان، افشانه، -پاش، دستگاه پاشیدن، ذره پاش
spread plate	سطح پخش
spreader	انتشار دهنده، اسپردر، پخش کن/کننده
squalamine	اسکوالامین
squalene	اسکوالن
SRBC (= sheep erythrocytes)	اس.آر.بی.سی.
SRF (= serum response factor)	اس.آر.اف.
SRP (= signal recognition particle)	اس.آر.پی.
SS (= suspended solids)	مجم
SSB (= single-strand binding protein)	اس.اس.بی.
SSCP (= single-strand conformational polymorphism)	اس.اس.سی.پی.
ssDNA	اس.اس.دی.ان.ای.
SSLP (= simple sequence length polymorphism)	اس.اس.ال.پی.
stab culture	کشت استبی/سوزنی

stability	پایداری، ثبات، دوام
stabilizing selection	گزینش استوار کننده/پایدار
stable RNA	آر.ان.ای پایدار
stacchyose (stachyose)	استاکیوز
stachyose	استاکیوز
stacked gene	توده ژنیزن تجمع یافته
stacking	انباشتن، توده کردن
staggered conformation	انطباق/سازش پله ای
staggered cut	برش پله ای
standard	استاندارد، الگو، پرچم، حد مطلوب، درفش، سنجه، متعارف
standard deviation	انحراف معیار
standard error	خطای معیار
standard oxygen transfer efficiency (SOTE)	بازده انتقال اکسیژن استاندارد
standing crop	محصول سرپا/قد کشیده، موجودی گیاهی
Stanley Bostitch oil-free air compressor	کمپرسور هوای غیر روغنی استانلی بوستیچ
stanol ester	استانول استر
stanol fatty acid ester	استر اسید چرب استانل
staphylococcal	استافیلو کوکی
staphylococcal enterotoxin	روده زهرابه استافیلو کوکی
staphylococcal toxin	سم استافیلو کوکی
staphylococcus	استافیلو، کوکخوشه گوییزه
staphylococcus aureus	استافیلو کوک اورئوس
staphylococin	استافیلوکوسین
staphylokinase	استافیلوکیناز
starch	آهار، نشاسته
start codon	کدون آغازین
starter	شروع کننده
startpoint	نقطه شروع
starvation culture	کشت گرسنگی
statin	استاتین
stationary culture	کشت ثابت
stationary phase	مرحله ثابت
statistics	آمار
staurosporin(e)	استارواسپورین
STDs (= sexually transmitted disease)	اس.تی.دی.

English	Persian
stearate	استئارات
stearic acid	اسید استئاریک
stearidonate	استئاریدونات
stearidonic acid	اسید استئاریدونیک
stearoyl-acp desaturase	استئاروئیل ای.سی.پی. دیساتیوراز
stem cell	سلول/یاخته اصلی/بنیاد/پایه/دودمانی
stem cell growth factor	عامل رشد سلول پایه، عامل رشد یاخته دودمانی
stem cell one	یاخته بنیاد یک، یاخته دودمانی یک
step aeration method	روش هوادهی پله ای
stereoisomer	ایزومر/همپار فضایی، استریو ایزومر
steric hindrance	مانع فضایی
sterile	استریل، سترون، ضد عفونی شده، عقیم، نازا
sterile water	آب استریل
sterility	عقیمی
sterilization	سترون/عقیم سازی، ضد عفونی
steroid	استروئید
steroidogenesis	استروئیدوژنز
sterol	استرول
stevioside	استوویوزید
STIC (= serum-trypsin-inhibitory capacity)	اس.تی.آی.سی.
sticky end (= cohesive end)	انتهای چسبنده/ناصاف
stigmasterol	استیگماسترول
stillbirth	مرده زائی
stochastic	آماری-تصادفی، اتفاقی، تصادفی
stock culture	کشت استاک/پایه
stock solution	محلول استاک/پایه
stomatal pore	منفذ روزنه ای
stop codon	رمز پایانی
stopped-flow technique	تکنیک جریان متوقف
storage protein	پروتئین انبار/اندوخته/ذخیره
stormy fermentation	تخمیر طوفانی
strain	تنش، جوره، دگروشی، دودمان، رقم، سویه، فشار، کرنش، کشش، نژاد، نسل
strand	تار، رشته، مو، نخ
streaking	استریکینگ
streptavidin	استرپتا ویدین
streptococal antigen	آنتی ژن استرپتوکوکی
streptococcal	استرپتوکوکی
streptococcal enterotoxin	روده زهرابه استرپتوکوکی
streptococcus	استرپتوکوک، پیچ گوبیزه
streptococcus mutans	استرپتوکوک میوتان (ها)
streptogenin	استرپتوژنین
streptokinase	استرپتوکیناز
Streptolydigins	استرپتولایدیژین ها
streptolysin	استرپتولایزین
streptomyces	استرپتومایس ها
streptomycin dependence	وابستگی به استرپتو مایسین
streptomycin resistance	مقاومت به استرپتو مایسین
streptonigrin	استرپتونیگرین
streptothricin	استرپتوتریسین
stress activated protein	پروتئین فعال شونده با تنش
stress proteins	پروتئینهای تنشی
stringency	شدت، حدت، قاطعیت
stringent control	مهار قاطع
stringent factor	عامل قطعیت
stringent plasmid	پلاسمید/دشتیزه سخت
stroma	استروما، بستره
stromelysin	استروملیسین
strong promoter	پیش محرک قوی
strong sustainable development principle	اصل توسعه نیرومند پایدار
STRs (= short tandem repeats)	اس.تی.آر.اس. ها
structural biology	بیولوژی/زیست شناسی ساختاری
structural chromosome	کروموزوم ساختاری
structural gene	زادژن، ژن ساختاری
structural genomics	ژن شناسی ساختاری
structural proteomics	پروتئین شناسی ساختاری

structure	ساخت، ساختار، ساختمان، ساختن، سازمان دادن، نظام، نظم	sulfatide lipidosis	سولفاتید لیپیدوز
		sulforaphane	سولفورافین
		sulfosate	سولفوسات
structure-activity model	مدل فعالیت ساختاری	sulphobromo phtahalein	سولفوبروموفتالئین
structure-functionalism	کاربرد گرایی ساختاری	sulphonated	سولفودار شده
strychnine	استرکنین، استریکنین	sulphonation	سولفودار شدن/کردن
STS (= sequence tagged site)	اس.تی.اس.	superactivated	فوق فعال
		superantigen	ابر پادگن
stuffer fragment	تکه/قطعه استافر	supercoil	ابر مارپیچ
subarachnoid space	فضای ساب آراکنوئیدی، محفظه نخاعی	supercoiled plasmid	پلاسمید فوق مارپیچ
		supercoiling	ابر مارپیچ شدن
subcellular	زیر سلولی	supercritical carbon dioxide	دی اکسید کربن فوق حیاتی
subclinical	نیمه تشخیصی/کلینیکی	supercritical fluid	مایع فوق بحرانی
sub-clone	نیمه کلون	supergene	ابر زاد، ابر ژن، برون زا
subcloning	نیمه همسانه سازی	supernatant	جسم شناور، رو شناور، شناور، مایع رویی
subculture	زیر-کشت، نیمه کشت		
sublethal gene	ژن نیمه-کشنده	supernumerary chromosome	کروموزوم مازاد
sub-mandibular glands	غدد زیر فکی/آرواره ای	superovulation	ابرتخمکگذاری
submerged culture	کشت غرقی	superoxide dismutase	سوپر اکسید دیس موتاز
submetacentric	زیرپس میانپار، سابمتاسنتریک	super-oxide dismutase	دیسموتاز ابر-اکسیدی
submunition	جنگ افزار/کلاهک/مهمات خوشه ای/فرعی	superparamagnetic nanoparticle	ریز ذره ابر فرا مغناطیسی
		support	پشتیبانی
subspecies	زیرگونه ای	supportive care	مراقبت پشتیبانی کننده
substantial equivalence	هم ارزی اساسی، هم ظرفیتی بنیادی	suppressor	بازدارنده، سرکوبگر
		suppressor gene	ژن بازدارنده
substantially equivalent	اساسا برابر	suppressor mutation	جهش بازدارنده
substitution	تعویض، جانشین سازی، جانشینی	suppressor t cell	یاخته تی بازدارنده
substrate	بستره، بنیاد، زیربستر، زیر ساخت، گوهر مایه، ماده اولیه، ماده زمینه	suppuration	ترشح چرک/ریم، تولید چرک
		supramolecular assembly	ساختار مولکولی بسیار بزرگ
		surface antigen	آنتی ژن سطحی
		surface culture	کشت سطحی
subtilin	سوبتیلین	surface plasmon	دشته مونه/پلاسمون سطحی
subtilisin	سوبتیلیزین		
subunit	زیر واحد	surface plasmon resonance	تشدید پلاسمون سطحی
succus entericus	ترشح روده، ایسوکوس انتریکوس	surface protein antigen A	آنتیژن ای. پروتئین سطحی
sucrase	سوکراز	surfactant	سورفکتانت، فعال کننده سطح، کاهنده کشش سطحی، ماده فعال سطحی
sucrose	سوکروز، قند نیشکر		
sugar molecule	ملکول قند		
sulfate reducing bacterium	باکتری کاهنده سولفات	surfactant-associated protein	پروتئین بی. وابسته به

English	Persian
B	سورفکتانت
surplus embryo	جنین اضافی/مازاد
surrogate market	بازار جایگزین
suspended particulate matter (SPM)	ذرات معلق
suspended solids (SS)	مواد جامد معلق (مجم)
suspension culture	کشت سوسپانسیونی
sustainable	پایدار، به اندازه، در حد معقول، قابل دوام
sustainable development	توسعه قابل دوام/پایدار
sustainable intensification of animal production systems	سیستمهای افزایش پایدار تولید حیوانات
sustainable use	کاربرد قابل دوام/پایدار
SV (= specific volume)	اس.وی.
SV40 (= simian virus 40)	اس.وی.40
SVI (= sludge volume index)	اس.وی.آی.
swarming	ازدحام/تجمع کردن، سوارمینگ
sweepstake route	مسیر بخت آزمایی/سوئیپ استیک
Swingfog ™	سوئینگ فاگ
swinging-bucket rotor	پره سوئینگینگ-باکت، پره سطل آونگین
switch gene	ژن سوئیچ/کلید
switch proteins	پروتئین های سوئیچ
switching	جابجائی، تعویض، سوئیچ کردن
syk protein	پروتئین سایک
Symba process	پروسه/روش/فرآیند سیمبا
symbiant	سیمبیانت
symbiosis	همزیستی
symbiotic	دارای همزیستی، هم زیست، هم زیستانه
sympatric	همبوم، همجا، همنیاک
sympatric speciation	گونه زایی همبوم/همجا
symport	سیمپورت
symptom	علامت، نشانه، نشانه بیماری
symptomatic	علامتی، نشانه ای، نشانه بیماری

English	Persian
synapse	ارتباط دو نرون، پیوندگاه نرونی،، سیناپس، همور
synapsis	پیوند همور
synaptic cleft	شکاف هموری/سیناپسی
synaptic gap	فاصله سیناپسی/هموری
synaptinemal complex	کمپلکس سیناپتینمال، همتافته نوار همموردی
synaptobrevin	سناپتوبروین
synaptotagmin	سیناپتوتگمین
synchronous culture	کشت همگاه/همزمان
syncytium	پیوسته یاخته، سینسیتیوم
syndactyly	پیوسته انگشتی، سینداکتیلی
syndecan	سیندکان
syndesine	سیندزین
syndesis	سیندز، همور
syndrome	سندروم، عارضه، نشانگان
synergid	سینرژید، هسته قرینه/کمکی
synergistic effect	اثر گذاری دوگانه/مضاعف، اثر هم نیرو زادی/هم افزا
synergy	سینرژی، نیرو زایی
synexin	سینکسین
syngamy	سینگامی، هم زامی
syngraft	سینگرافت، هم پیوند
synkaryon	سیناکاریون، هم نقشه
synovial	بندی، مفصلی
syntenic genes	ژن های هم مکان/سینتنیک
synthase	سنتاز
synthesis	آمایش، ساخت، سنتز
synthesizer	ترکیب/سنتز کننده، سنتسایزر
synthesizing	آمیختن، ترکیب سازی
synthetase	سنتتاز، لیگاز
syntomycin	سینتومایسین
syntrophism	سینتروفیسم، هم پروری
syntrophsysteminy	سینتروفیستمینی
syrup of ipecac	شربت اپیکا، شربت الیون کوکی
systematic activated	مقاومت فعال شده روش مند

English	Persian
resistance	
systematics	روش مند، سیستماتیک، منظم، نظام مند
systemic acquired resistance	مقاومت اکتسابی سازگانی/سیستمیک
systemic inflammatory response syndrome	نشانگان واکنش التهابی سازگانی
systemic lupus erythematosus	لوپوس اریتماتوز سیستمی
systeomics	بنیادی، درون تنی، ریشه ای، سیستمیک، عمومی

T

English	Persian
T (= thymine)	تی.
T cell receptor alpha	گیرنده سلول تی. آلفا
t lymphocytes	لنفوسیت های تی.، لنف یاخته های تی.
T4 DNA ligase	لیگاز دی.ان.ای. تی.4
T4 DNA polymerase	پلیمراز دی.ان.ای. تی.4
T4 RNA ligase	لیگاز آر.ان.ای. تی.4
tabun	تابون، عامل جی.ای.
TACF (= telomer-associated chromosome fracionation)	تی.ای.اف.سی.
tachykinin	تاکی کینین
tachypnea	تند نفس (ی)،تنفس سریع، نفس تندی
tag	اتیکت، برچسب (زدن)، نشانه (گذاری)، وابستگی
tag polymerase	پلیمراز برچسبی
tagging	برچسب زدن
tailing	دنبال کردن/رفتن
tandem affinity purification tagging (TAP tagging)	برچسب زدن پاکسازی کشش متوالی
tandem repeats	تکرار توام/متوالی
tannin	تانن، جوهر دباغی، جوهر مازو، مازو
TAP (= transporter associated with antigen presentation)	تی.ای.پی.
TAP tagging (= tandem	تی.ای.پی. تگینگ

English	Persian
affinity purification tagging)	
TAPVR (= total anomalous pulmonary venous return)	تی.ای.پی.وی.آر.
taq DNA polymerase (= Thermus Aquaticus Polymerase)	پلیمراز تی.ای.کیو. دی.ان.ای......
taq polymerase(= Thermus Aquaticus Polymerase)	پلیمراز .تی.ای.کیو....
target	نشانگاه، نشانه، هدف
target DNA	دی.ان.ای. هدف
target gene	ژن هدف
target organism	ارگانیزم هدف
target site (= recognition site)	جایگاه هدف
target validation	تائید هدف
targeting sequence	توالی هدفگیر
target-ligand interaction screening	غربال برهم کنش هدف- لیگاند
TATA box (= Cis-acting sequence in RNA polymerase II promoters)	جعبه تی.ای.تی.ای.
TATA homology	همسانی تی.ای.تی.ای.
TATA-binding protein (TBP)	پروتئین اتصال تی.ای.تی.ای.
taurine	تورین
taurocholate	توروکولات
taxol	تاکسول
taxon	آرایه، واحد آرایه شناختی، واحد سیستماتیک، تاکسون
taxonomy	آرایه شناسی، تاکسونومی، رده بندی شناسی، طبقه بندی، علم رده بندی
Tay-Sachs	تای-ساکز
Tay-Sachs disease	بیماری تای-ساکز
TBG (= thyroxine-binding-globuline)	تی.بی.جی.
TBP (= TATA-binding protein)	تی.بی.پی.
TCA (= tri-carboxyllic acid	تی.سی.ای.

English	Persian
cycle)	
T-cell	سلول تی.
T-cell dependent mechanism	سازوکار وابسته سلول تی.
T-cell growth factor	عامل رشد سلول تی.
T-cell independent mechanism	سازوکار مستقل سلول تی.
T-cell modulating peptide	پپتید تعدیل کننده سلول تی.
T-cell receptor	گیرنده سلول تی.
TCR (= T-cell receptor)	تی.سی.آر.
TCRA (= T-cell receptor)	تی.سی.آر.ای.
TDA (= thymus-dependent area)	تی.دی.ای.
TDF (= testis determining factor)	تی.دی.اف.
T-DNA	تی.دی.ان.ای.
TDS (total dissolved solids)	کم جم
TE buffer	بافر/میانگیر تی.ئی
technical testing	آزمایش فنی
technology	افزار سازی، تکنولوژی، دانش فنی، روش فنی، علم صنعت، فن، فناوری، فن شناسی
technology protection system	نظام حمایت تکنولوژی
tektin	تتکین
teliospore	تلیواسپور
telocentric	تلوسنتریک
telomerase	تلومراز
telomere	اولیگومر، پایانه پار، تلومر چند پار
telomere repeat factor	عامل تکرار تلومر
telomere-associated chromosome fractionation	خردشدگی/انکسار کروموزومی وابسته به تلومر
telomeric DNA	دی.ان.ای.تلومریک
telophase	پایانه چهر، تلوفاز
temperate phage	باکتری خوار/فاژ ملایم
temperature	تب، درجه حرارت/گرما، حرارت بدن، دما
temperature-sensitive	حساس به دما
temperature-sensitive	جهش حساس به دما

English	Persian
mutation	
template	الگو، شابلون، قالب، قواره، کالبد، مدل
template strand	رشته الگو
teosinte	تئوزینت
teratogen (= teratology)	تراتوژن
teratogenesis	تراتوژنز، ناقص الخلقه سازی، ناهنجار سازی
teratogenic compound	ترکیب ناهنجاری زایی، ترکیب هیولا سازی
teratology	تراتولوژی، ناهنجار شناسی
teratoma	تراتوما
terminal repeat	تکرار پایانه
terminal transferase	ترانسفراز پایانه
terminalization	پایانه سازی
termination codon	رمز اختتام، رمز انقطاع، کدون پایانی
terminator	پایان دهنده، ختم کننده
terminator cassette	کاست پایان دهنده
terminator region	منطقه پایان دهنده
terminator sequence	توالی پایان دهنده، رمز اختتام
terpene	ترپن
terpenoid	ترپن نما، شبه ترپن
terramycin	ترامایسین
tertiary base pair	جفت باز ترشیاری/سه گانه
tertiary period	دوره سوم/سه گانه
tertiary structure	ساختار/ساختمان سومین
tertiary structure of protein	ساختار سومین پروتئین
test cross	تقاطع آزمونی
testicular feminization syndrome (FTS)	سندروم فمینیزه شدن بیضه ای
testis determining factor	عامل تعیین بیضه
testosterone	تستوسترون
test-range	دامنه آزمایش
tetanus toxin	سم کزاز
tetanus toxoid	توکسوئید کزاز
tetracycline	تتراسایکلین
tetrad	تتراد
tetrahydrofolic acid	تترا هیدروفولیک اسید
tetralogy of Fallot	تترالوژی فالوت

tetraploid	تتراپلوئید، چارلا، چهار لا
tetrasome	تترازوم
tetrasomic (= tetrasome)	تترازومی
tetro-hydro-folate	تترو-هیدرو-فولات
TF (= transcription factor)	تی.اف.
TF II (= transcription factor for control of RNA polymerase II)	تی.اف.2
TFN (= transferrin)	تی.اف.ان.
TGF-β (= transforming growth factor-beta)	تی.جی.اف.بتا
thalamus	نهنج
thalassemia	تالاسمی
thale cress	تیل کرس
theoretical plate	سطح فرضیه ای
theory of local existence	نظریه وجود محلی
therapy of antisense mRNA	درمان ام.آر.ان.ای. بی معنی
thermal	حرارتی، گرم، گرمایی
thermal denaturation	تغییر ماهیت حرارتی
thermal hysteresis protein	پروتئین پسماند حرارتی، پروتئین هیسترزیس ترمال
thermoduric	مقاوم به گرما
thermoinducible lysogen	لیزوژن القا شونده با حرارت
thermophile	گرما دوست/خواه
thermophilic bacteria	باکتری گرما دوست
thermus aquaticus	ترموس آکواتیکوس
Thermus Aquaticus Polymerase	پلیمراز ترموس آکواتیکوس
theta replication (θ)	همانند سازی تتا
THF (= tetro-hydro-folate)	تی.اچ.اف.
thiamine	تیامین
thiamine pyrophosphate	پیروفسفات تیامین
thioesterase	تیواستراز
thiol group	گروه تیول
thioredoxin	تیوردوکسین
Thr (= threonine)	
threatened species	گونه تهدیدشده
threonine	ترئونین
threshold concentration	تراکم آستانه ای
thrombin	ترومبین
thrombolytic agent	عامل حل کننده/ترومبولیتیک

	لخته خون
thrombomodulin	ترومبومودولین
thrombosis	ترومبوز، تشکیل لخته
thrombus	ترومبوز، لخته
thyamine pyrophosphate	پیروفسفات تیامین
thylakoid	تیلاکوئید
thymidine kinase	تیمیدین کیناز
thymidine phosphorylase	فسفوریلاز تیمیدین
thymidylate synthase	تیمیدیلات سینتاز
thymine	تیمین
thymine-thymine dimer	دیمر تیمین-تیمین
thymoleptics	تیمولپتیک
thymus-dependent area	منطقه متکی به تیموس
thyroid releasing factor	عامل رها سازی تیروئید
thyroid stimulating hormone	هورمون محرک تیروئید
thyroxine-binding-globuline	گلوبولین اتصال دهنده تیروکسین
Ti plasmid	پلاسمید تی.آی.
TIL (= tumor-infiltrating lymphocyte)	تی.آی.ال.
tissue	بافت
tissue array	آرایه بافت
tissue culture	کشت بافت
tissue engineering	طراحی بافت
tissue plasminogen activator	فعال کننده پلاسمینوژن بافتی
tissue-necrosis	مرگ نسجی، نکروز بافتی
tissue-type plasminogen activator	فعال کننده بافت مانند پلاسمینوژن
titer	تیتر، عیار
TK (= thymidine kinase)	تی.کی.
Tk (= thymidine kinase)	تی.کی.
t-lymphocytes	لنفوسیت های تی.
T_m (= melting temperature)	تی. اندیس ام.
Tm (= melting temperature)	تی.ام.
tmRNA (= transfer-messenger RNA)	تی.ام.آر.ان.ای.
TMV (= tobacco mosaic virus)	تی.ام.وی.
TNF (= tumor necrosis factor)	تی.ان.اف.

English	Persian
tobacco mosaic virus	ویروس موزائیک توتون
tocopherols	توکوفرول
tocotrienols	توکوترینول ها
TOF (= tetralogy of Fallot)	تی.او.اف.
togaviridae	توگاویروس ها
togavirus	توگاویروس
tolerance	بردباری، تحمل، تولرانس، خطای مجاز، طاقت، مقاومت به دارو، رواداری، روایی
toll-like receptor	گیرنده های شبه باج گیر
TOPAS aerosol	افشانه تی.او.پی.ای.اس.
topoisomerase	توپوایزومراز
topotaxis	جای آرایی
total anomalous pulmonary venous return	برگشت کاملا غیر عادی سیاهرگی ریوی، برگشت کامل وریدی ریوی غیر عادی
total cell DNA	تمام دی.ان.ای. سلولی، دی.ان.ای.تمام سلولی
total dissolved solids (TDS)	کل مواد جامد محلول (کم جم)
total economic value	ارزش کامل اقتصادی
total environmental value	ارزش کامل زیست محیطی
total internal reflection fluorescence	بازتابش شب نمایی درونی کامل
total solids (TS)	کل مواد جامد (کمج)
totipotency	بس توانی، پر توانی توانمندی
totipotent stem cell	سلول پایه پر توان، یاخته دودمانی پر توان
toxemia	توکسمی، مسمومیت خونی
toxic molecule	ملکول سمی
toxicity	زهر آگینی، سمیت، مسموم کنندگی
toxicity characteristic leaching procedure	ویژگی سمیت فرآیند آبشویی
toxicogenomics	ژن سم شناسی
toxin	ترکیب زهردار، توکسین، داروی سمی، زهر، زهرابه، سم

English	Persian
toxin agent	عامل توکسین/زهرابه
toxin weapon	سلاح توکسینی/زهرابه ای
toxoid	زهر مانند، سم غیرفعال، سم گونه، شبه سم
toxoplasma	توگزوپلاسما
TP (= thymidine phosphorylase)	تی.پی.
TPA (= tissue plasminogen activator)	تی.پی.ای.
tPA (= tissue-type plasminogen activator)	تی.پی.ای.
TPP (= thiamine pyrophosphate)	تی.پی.پی.
TR (= terminal repeat)	تی.آر.
TRA (= T cell receptor alpha)	تی.آر.ای.
tra gene	ژن تی.آر.ای.
tracer	پی یاب، ردگیر، ردیاب، نشانه گذار
traditional breeding method	روش اصلاح نژاد سنتی
traditional breeding technique	فن اصلاح نژاد سنتی
trailer segment/sequence	توالی/جزء دنباله رو
trait	خصلت، خصوصیت، صفت، نشانویژه، ویژگی
trans fatty acid	اسید چرب تبدیلی
trans-acting protein	پروتئین ترانس اکتینگ
transactivating protein	پروتئین فعال ساز ترانس
transactivation	فعال سازی ترانس
transaldolase	ترانس آلدولاز
transaminase	ترانس آمیناز
transamination	انتقال آمین، انتقال بنیان آمین، انتقال گروه آمین، تبدیل آمین
transboundary harm	آسیب فرا مرزی/سرحدی
transboundary movement	جابجائی فرا مرزی/سرحدی
transboundary release	انتشار/رهائی فرا مرزی/سرحدی
transboundary transfer	انتقال فرا مرزی/سرحدی
transcapsidation	ترانس کپسیداسیون
transcript	رونوشت، نسخه

284

transcript analysis	تحلیل نسخه		سازی، دگرگونی
transcriptase	ترانسکریپتاز	transformation efficiency	قابلیت تبدیل/تغییر شکل
transcription	الگو برداری، رونویسی،	transformation frequency	بسامد تغییرشکل
	نسخه برداری	transforming growth	عامل تغییر رشد بتا
transcription activator	فعال کننده نسخه برداری	factor-beta	
transcription factor (TF)	عامل رونویسی	transforming oncogene	ژن توموری تغییر یابنده
transcription factor binding	جایگاه پیوند عامل رونویسی،	transgalacto-oligosaccharide	ترانس گالاکتو-اولیگوساکارید
site	محل اتصال عامل	s	
	رونویسی	transgene	تراژن، ترانس ژن
transcription factor for	عامل نسخه برداری برای	transgenesis	ترانس ژنز
control of RNA	مهار پلیمراز 2 آر.ان.ای.	transgenic	دارای ژنهای پیوندی، تراژنی،
polymerase II			واریخته
transcription unit	واحد نسخه برداری	transgenic animal	جانور تراژنی/واریخته
transcriptional activator	فعال کننده رونویسی	transgenic plant	گیاه واریخته/تراژنی
transcriptional profiling	نمای رونویسی	transgenosis	ترانسژنوز
transcriptional repressor	بازدارنده رونویسی	transgressive segregation	جداسازی ترانسگرسیو
transcriptome	ترانسکریپتوم	transgressive variation	سویه ترانسگرسیو
transcutaneous	ترانسکوتانوس	transit peptide	پپتید انتقالی/گذرا
(percutaneous)		transition	انتقال، گذار، تحول
transdermal	تراپوستی	transition state	حالت
transducing phage	باکتری خوار تبدیل کننده،		انتقال/گذار/عبور/واسطه/
	باکتری خوار ترارسان، فاژ		میانی
	ترانسدیوسر	transition-state intermediate	واسطه وضعیت انتقال
transduction mapping	بازنمایی ترا رسانی	transketolase	ترانس کتولاز
transfection	ترا آلودگی	translation	برگردان، ترجمه
transfer	انتقال، ترافرست، حامل،	translocation	انتقال، تراجایی، جابه جایی
	ناقل، نقل	transmembrane protein	پروتئین تراشامه ای
transfer DNA	دی.ان.ای. حامل	transmembrane regulator	پروتئین تنظیم کننده تراشامه
transfer factor	عامل انتقال	protein	ای
transfer RNA (tRNA)	آر.ان.ای انتقالی/حامل،	transmissibility	قابلیت انتقال
	اسید ریبونوکلئیک ناقل	transmission	انتشار، انتقال، پخش،
transferase	ترانسفراز		تراگسیل، سرایت، مخابره
transferin	ترانسفرین	transmission of infection	انتشار عفونت، تراگسیل
transfer-messenger RNA	آر.ان.ای.پیامبر انتقالی		آلودگی
transferred DNA	دی.ان.ای حمل شده	transplantation	پیوند، پیوند عضو، تراکاشت،
transferrin	ترانسفرین		قلمه زنی، نشاکاری
transferrin receptor	گیرنده ترانسفرین	transport mechanism	ترانسپورت مکانیسم،
transformant	تبدیل کننده، تغییر شکل		مکانیسم نقل وانتقال
	دهنده	transport proteins	پروتئین های انتقالی
transformation	تبدیل ترا ریختی، تراریخت	transporter associated with	ناقل وابسته به تظاهر آنتی
	سازی، تغییر شکل، دگر	antigen presentation	ژن

transposable element	عنصر جا به جا شدنی/ترانهادی
transposable genetic element	عنصر ژنتیکی جا به جا شدنی
transposase	ترانسپوزاز
transposition	تبدیل شدن، ترانهش، ترجمه شدن، جا به جا شدگی
transposon	ترانسپوزون، عنصر جا به جا شدنی
transversion	برگشتگی، تراگشت، جایگزینی ناهمجنس
treatment investigational new drug	داروی جدید درمان تحقیقی
treatment system	سیستم درمان
trehalase	تر هالاز، ترهالوز
tremorgenic indole alkaloid	آلکالوئید ترمورژنیک ایندولی
TRF (= telomere repeat factor)	تی.آر.اف.
TRF (= thyroid releasing factor)	تی.آر.اف.
triacyglycerides	تریاسیل گلیسرید ها
triacylglycerol	تری اسیل گلیسرول
tricarboxylic acid cycle	چرخه اسید تری کاربوکسیل
tri-carboxyllic acid cycle	چرخه اسید تری-کربوکسیلیک
trichoderma harzianum	تریکو درما هارزیانوم
tricho-rhino-phalangeal	تریکو-رینو-فالانژیال
trichosanthin	تریکو سانتین
trichothecene	تریکوتسن
trichothecene mycotoxin	تریکوتسن مایکوتوکسین
trichothene mycotoxin	تریکوتن مایکوتوکسین
trickle filter	صافی/فیلتر چکه ای
tricothene mycotoxin	تریکوتن مایکوتوکسین
triglyceride	تری گلیسرید
trihybrid	تری هایبرید
trinucleotide repeats	تکرار تری نوکلئوتیدی
triplet	سه تائی، سه حرفی، سه قلو، رمز سه حرفی
triploid	تریپلوئید، سه بخشی، سه جزئی، سه قسمتی، سه گان، سه گانه، سه لاد

tripple cropping	کشت سه گانه
tris-acetate buffer (TAB)	میانگیر تریس-استات
trisomy	تریزومی
trisomy-18 syndrome (Edward syndrome)	سندروم تریزومی 18
triticale	چاودم
tRNA (= transfer RNA)	تی. آر.ان.ای.
tRNA deacylase	دآسیلاز تی.آر.ان.ای.
trombose	ترومبوز
trophic	پروره ای، تغذیه ای، خوراکی، غذایی
trophic level	تراز پروره ای، سطح غذایی
tropism	تروپیسم، کشش، گرایش
tropomyosin	تروپومیوزین
troponin	تروپونین
TRP (= tricho-rhino-phalangeal)	تی.آر.پی.
Trp (= tryptophan)	
trucho-thio-dystrophy	دیستروفی تروکو-تیو
truck	حمل کردن، کامیون، واگن باری
true hermaphrodism	هرمافرودیسم واقعی
trypsin	تریپسین
trypsin inhibitor	بازدارنده تریپسین
trypsinogen	تریپسینوژن
tryptophan	تریپتوفان
Ts (= temperature-sensitive)	
TS (= thymidylate synthase)	تی.اس.
TS (= total solids)	کمج
TSD (= Tay-Sachs disease)	تی.اس.دی.
TSH (= thyroid stimulating hormone)	تی.اس.اچ.
T-shell	پوسته تی.
TT (= tetanus toxoid)	تی.تی.
T-T (= thymine-thymine dimer)	تی.-تی.
TTD (= trucho-thio-dystrophy)	تی.دی.دی.
tuberculin	توبرکولین
tuberculosis	باسیل مرض سل، بیماری سل، تب لازم، سل

tubulin	توبولین
tularemia	تولارمی
tumor	آماس، برآمدگی، تورم، تومور، دشبل، غده
tumor DNA	دی.ان.ای. توموری
tumor marker	نشانگر تومور
tumor necrosis factor	عامل نکروز دهنده تومور
tumor virus	ویروس تومور
tumor-associated antigen	پادگن مربوط به تومور
tumor-inducing plasmid	پلاسمید مسبب تومور
tumor-infiltrating lymphocyte	لنفوسیت نفوذ گر در تومور
tumor-suppressor gene	ژن سرکوبگر تومور
tumor-suppressor protein	پروتئین سرکوبگر تومور
turbidimetry	تربیدیمتری، تیرگی سنجی
turbidity	تاری، تیرگی، کدری، درجه کدر بودن، کدورت، گل آلودگی، مه آلودگی
turgor pressure	فشار ترگور
turner syndrome	سندروم ترنر
turnover number	عدد بازیابی/تبدیل/روگشت/نو سازی
twins	دوقلو
two-dimensional	دوبعدی
two-hybrid system	سیستم دورگه
tyndallization	تیندالیزاسیون
type specimen	درست مونه، نمونه استاندارد/شاخص
typhimurium (= salmonella tymphiumurium)	تیمفیموریوم
typhoidal tularemia	تولارمی شبه تیفوئید
Tyr (= tyrosine)	
tyrosine	پاراهیدروکسی فنیل آلانین، تیروزین
tyrosine kinase inhibitor	بازدارنده تیروزین کیناز

U

U (= uracil)	یو.
UAG (= termination codons)	یو.ای.جی.
ubiquinol (=reduced coenzyme Q)	اوبیکوئینول
ubiquinone	اوبیکینون
ubiquitin	اوبیکیتین
ubiquitinated	اوبیکیتین شده
ubiquitin-proteasome pathway	مسیر اوبیکیتین -پروتئازوم
UDP (= uridine di-phosphate)	یو.دی.پی.
UDPG (= UDP-glucose)	یو.دی.پی.جی.
UDP-glucose	گلوکز- یو.دی.پی.
ulceration	تشکیل زخم
ulceroglandular tularemia	تولارمی زخمی-غده ای
ultracentrifuge	فرامیان گریز، فوق سانتریفوژ
ultrafilter	فرا صافی/فیلتر
ultrafiltration	فرا پالایش/صافی کردن
ultrasonic	زبر/فرا صوتی، مافوق/ماوراء صوت
ultrasonication	فرا صوتی کردن
ultraviolet	فرا/ماوراء بنفش
ultra-violet (=ultraviolet)	ماوراء بنفش
ultraviolet microscope	میکروسکوپ فرابنفش
ultraviolet rays	پرتو/اشعه های فرا بنفش
ultraviolet spectroscopy	اسپکترواسکوپی/طیف نمایی فرا بنفش
umbrella species	گونه چتری
UMP (= uridine mono-phosphate)	یو.ام.پی.
uncertainty factor	عامل عدم قطعیت
uncoating enzyme	آنزیم ضد روکش
uncompetetive inhibition	سرکوب غیر رقابتی
unconfined release	آزادسازی نامحدود/مهار نشده
uncontaminated water	آب غیر آلوده
uncoupling	آزاد/باز/جدا/رها کردن
undefined medium	محیط کشت/واسطه تعریف نشده
unicellular	تک سلولی، یونی سلولار
unidirectional externality	بیگانگی یک جهتی
unimmunized individual	فرد مصونیت نیافته
unintended release	رها سازی نا خواسته

unintended transboundary movement	جابجائی ناخواسته بین مرزی/سرحدی		کاربردن، کاربرد، فایده، مصرف کردن
uniparental disomy (UPD)	دیزومی تک والدی	use value	ارزش کاربرد
unipartite structure (chromosome)	ساختار تک قسمتی/منفرد	user	استفاده/مصرف کننده، کاربر
		uteroglobin	یوترو گلوبین
unit	ایوه، دستگاه، شمار، غشای واحد، میزان، واحد، یکان	utility function	عملکرد مطلوب، کارکرد مطلوبیت
unit factor	عامل واحد	UTP (= uridine tri-phosphate)	یو.تی.پی.
unit membrane	غشای واحد		
univalent	تک ارزشی، یونی والان	uv (= ultraviolet)	یو.وی.
unsaturated fatty acid	اسیدچرب اشباع نشده	UV (= ultraviolet)	یو.وی.
unscheduled DNA synthesis	سنتز دی.ان.ای. خارج از برنامه	uv rays	اشعه یو.وی.
		uv1	ماورای بنفش نوع اول، یو.وی. وان/1
untranslated region (UTR)	منطقه ترجمه نشده		
unwinding enzyme	آنزیم بازکننده پیچ وخم	uv-absorbance spectroscopy	اسپکترواسکوپی جذب یو.وی.
unwinding protein	پروتئین باز کننده مارپیچ		
upflow sludge blanket	اسلاج بلانکت آپفلو		
upstream	بالادست، خلاف جریان آب، در بالای رودخانه، واقع در قسمت علیای رودخانه	**v**	
		v antigen	پادگن/آنتی ژن ویروسی
upstream activator sequence (UAS)	توالی فعال گر بالادست	vaccine	مایه، مایه آبله، واکسن
		vaccinia	آبله گاوی
upstream promoter elements	عناصر پیشبرنده بالا دست	vacuoles	حباب ها، حفره ها، کریچه ها، واکوئل ها
uptake	برداشت، برگیری، جذب، فراجذب، میزان جذب	vagile	جنبنده، دارای تحرک، متحرک
		vagility	تحرک، جنبش، قابلیت انتشار
uracil	اوراسیل		
uracil fragments	تکه های اوراسیل	vaginosis	التهاب مهبل، واژنوز
urate oxidase	اکسیداز اورات	Val (= valine)	
urea cycle	چرخه اوره	validation	اعتبار بخشی، تایید کردن، معتبر سازی
urease	اوره آز		
uricotelic	دافع اوره، یوریکوتلیک	valine	والین
uridine	اوریدین	valuation	ارزش، ارزش گذاری، ارزیابی، بهاء، قدر
uridine di-phosphate	دی-فسفات یوریدین		
uridine mono-phosphate	مونوفسفات یوریدین	van	کامیون کوچک، واگن باری، وانت
uridine tri-phosphate	تری فسفات یوریدین		
U-RNA	یو.-آر.ان.ای.	vancomycin	وانکومایسین
urobilin	یوروبیلین	variable number of tandem repeats	تعدادمتغیر تکرار های توام
urobilinogen	یوروبیلینوژن		
urocanic acid	اسید یوروکانیک	variable region	منطقه متغیر
urokinase	یوروکیناز	variable surface	گلیکو پروتئین دارای سطح
use	استعمال، استفاده، به		

288

glycoprotein	تغییر پذیر
variable surface glycoprotein	گلیکوژن سطح متغیر
variance	پراش، پراکنش، تغییر پذیری، تنوع، نا پایداری، ناجوری، واریانس، ورداپی
variation	اختلاف، انحراف، تغییر، تنوع، جوراجوری، گوناگونی، ورداپی
varicella zoster virus	ویروس واریسلا زوستر
variegation	گوناگونی
variety	تنوع
vascular endothelium	تویوش/درون پوشه آوندی
vascular hemophilia	هموفیلی عروقی
vasoactive intestinal polypeptide	پلی پپتید روده ای وازواکتیو
vasoconstrictor	تنگ کننده عروق
vasodilator	رگ گشا، گشاد کننده عروق
vasopressin	وازوپرسین
vasotocin	وازوتوسین
VDD (= vitamin d-dependency rickets)	وی.دی.دی.
vector	بردار،حامل، ناقل، منتقل کننده
vector borne	ناقل برد، وکتور برن
vectorial processing	پروسسینگ ناقلی
vectorial translation	ترجمه ناقلی
vegetative cell	سلول گیاهی
vegetative propagation	نشر گیاهی
vegetative reproduction	تولید مثل گیاهی
vehicle	حامل، حمل کننده، ناقل
vent	روزنه، سوراخ، شکاف، مقعد، منفذ
vent DNA polymerase	پلیمرآز دی.ان.ای. منفذ
vernalization	باردهی، بهارگی، بهاره کردن
vertical gene transfer	انتقال ژن عمودی
vertical transmission	انتقال عمودی
very late antigen	پادگن/آنتی ژن خیلی دیر
very low-density lipoprotein (very-low- density lipoprotein)	لیپوپروتئین با غلضت خیلی کم

vesicle	آبدانک، تاول، تاولچه، حبابچه، ریز کیسه، وزیکول
vesicular transport	انتقال ریز کیسه ای
vesicule	وزیکول
v-gene	وی.-ژن
viability	زیستایی، زیست پذیری، قابلیت زیست
viable count	شمارش زیستی، شمار قابل رویش/زیست
vibrio cholerae	باکتری مژکدار وبا، ویبریو کولرا
vicariant pattern	الگوی جانشین
vicilin	ویسیلین
vinblastine	وینبلاستین
vincristine	وینکریستین
viral	ویروسی
viral agent	عامل ویروسی
viral encephalitis	التهاب مغز ویروسی
viral hemorrhagic fever	تب ناشی از خونریزی ویروسی
viral retroelement	پسا عنصر ویروسی
viral transactivating protein	پروتئین فعال ساز ترانس ویروسی
virion	ذره ویروس، ویروس رسیده، ویریون، ویشه
viroid	شبه ویروس
virulence	بیماری زایی، زهرآگینی، شدت، قدرت عفونت زایی
virulent	بیماری زا، تند، حدت دار، زهرآگین، سمی
virulent phage	فاژ بیماری زا
virus	ویروس، ویش
virus disease	بیماری ویروسی، ویش زدگی
virus replication	تکثیر/همانند سازی ویروسی
viscosity	چسبندگی، گران روی، ضریب گران روی/چسبندگی، غلظت، ویسکوزیته
visfatin	ویزفاتین
visible fluorescent protein	پروتئین تابناک مرئی

vitafood	ماده غذائی طبی
vitamer	ویتامر
vitamin	ویتامین
vitamin d-dependency rickets	ریکتزی متکی به ویتامین دی.
vivaparous	ویواپاروز
VLA (= very late antigen)	وی.ال.ای.
VLDL (= very low-density lipoprotein)	وی.ال.دی.ال.
VNTR (= variable number of tandem repeats)	وی.ان.تی.آر.
volatile suspended solids (VSS)	مواد جامد معلق فرار (مج مف)
volicitin	وولیسیتین
voltage-gated ion channel	منفذ ترا شامه ای یون
volume	بلندی صدا، حجم، ظرفیت، گنجایش، مقدار
volume rendering	ایجاد ظرفیت
volumetric flask	بالن حجم سنجی
volutin granule	گرانول ولوتین
vomitoxin	زهرابه، قارچ زهر، ومیتوکسین
von Willebrand factor	عامل فون ویلبراند
von Willebrand factor disease	بیماری عامل فون ویل براند
von Willebrand's toxin	سم فون ویلبراند
v-oncogene	انکوژن وی.، وی.-انکوژن
voucher specimen	نمونه واچر، واچر اسپسیمن
VSS (= volatile suspended solids)	مج مف
vulgaris	پست، تیره گسترده
	لوبیاهای خوراکی، عوامانه
VZV (= varicella zoster virus)	وی.زد.وی.

W

wagon	گاری چهار چرخ، واگن باری
wand	چوب جادوئی
WAP (= whey acidic protein)	دبلیو.ای.پی.

Wardenberg syndrome	سندروم واردنبرگ
warfare	جنگ
Warfarin	وارفارین
washer	پولک، ماشین شوینده، واشر
Wasserman reaction	واکنش واسرمن
waste	باطله، پسماند، دورریز، نخاله
waste management	پسمانداری، مدیریت مواد پسماند
wasteland	زمین مرده
wastewater	پساب، فاضلاب
wastewater treatment	تصفیه پساب/فاضلاب
water activity	اکتیویته آب
water soluble fiber	فیبر محلول در آب
Watson-Crick DNA	دی.ان.ای. واتسون-کریک
waxy corn	ذرت روغنی
weak interaction	کنش متقابل ضعیف
weapon	اسلحه، سلاح، جنگ افزار
weapon of mass destruction	سلاح تخریب گروهی/کشتار جمعی
weaponization	تبدیل سازی به سلاح
weaponize	تبدیل به سلاح کردن
western blot	وسترن بلات
western blotting	وسترن بلاتینگ
WF (= von Willebrand factor)	دبلیو.اف.
whey	کشک
whey acidic protein	پروتئین اسیدی کشک
whirlpool separator	جداکننده گردابی
whiskers	سبیل (در گربه و حیوانات)، جاروی کوچک
Whiskers™	ویسکرز
white blood cell	گلبول سفید خون
white corpuscle	گلبول/گویچه سفید
white mold disease	بیماری کفک/کپک سفید
whole-genome shotgun sequencing	توالی یابی شات گانی همه ژنوم
wide cross	آمیزش/تلاقی دور
wide spectrum	طیف گسترده
wild type	تیپ/مونه/نوع وحشی
willingness to accept	میل به پذیرفتن
willingness to pay	میل به پرداختن

wilms tumor	تومور ویلمز
wilson's disease	بیماری ویلسون
wind	باد، بو، نفخ، نفس
Wiskott Aldrich syndrome	سندروم ویسکوت آلدریچ
wnt proteins	پروتئین های دبلیو.ان.تی.
wobble	وابل، لرزش، لقوه، لنگ زدن
wobble hypothesis	فرضیه وابل
Wolff-Parkinson-White	ولف-پارکینسون-وایت
Wolf-Hirosch horn syndrome	سندروم شاخ ولف-هیروش
Wolman's disease	بیماری وولمن
worried well	سالم نگران
WPW (= Wolff-Parkinson-White)	دبلیو.پی.دبلیو.
Wright dust feeder	داست فیدر رایت

X

x chromosome	فام/کروموزوم تن اکس.
X chromosome linked blood group	گروه خونی متصل به کروموزوم اکس.
X- linked recessive	مغلوب مرتبط به اکس.
x receptor	گیرنده اکس.
xanthine oxidase	آنزیم گزانتین اکسیداز
xanthophyll	زردینه، زردینه برگ، گزانتوفیل
xanthopterin	زانتوپترین
xanthosine	زانتوزین
xanthosine mono phosphate	مونوفسفات زانتوزین
xanthoxin	زانتوکسین
xanthylic acid	اسید زانتیلیک
x-chromosome	کروموزوم ایکس.
XD (= X-linked dominant)	اکس.دی.
xenoantibody	زنوآنتی بادی
xenoantigen	زنوآنتی ژن
xenoantiserum	زنوآنتی سروم
xenobiochemistry	زنوبیوشیمی
xenobiotic compound	ترکیب بیگانه/دگر زی
xenogeneic organ	اندام نا مشابه
xenogeneic transplantation	پیوند زنوژنئیکی
xenogenesis	بیگانه زایی، تولید مثل متناوب، جوراجوری، ناجوری

xenogenetic organ	اندام زنوژنتیک
xenogenic organ	اندام بیگانه سرشت، عضو نا همسرشت
xenograft	بیگانه پیوند، پیوند بیگانه، پیوند دیگر سرشت، زنوگرافت
xenopus laevis	زنوپوس لائه ویس
xenotransplant	پیوند بیگانه/زنو
xenotropic virus	ویروس بیگانه گرا
xeroderma pigmentation	پیگمنتاسیون زرودرما
xeroderma pigmentosum	زرودرما پیگمنتوسوم
Xg (= X chromosome linked blood group)	
x-gal	اکس.-گال
x-inactivation	عدم فعالیت اکس.
XLH (= X-linked hypophosphatemic rickets)	اکس.ال.اچ.
X-linked disease	بیماری مرتبط با اکس
X-linked dominant	غالب متصل به اکس.
X-linked hypophosphatemic rickets	ریکتزی هیپوفسفاتمیکی مربوط به کروموزوم اکس.
X-linked recessive mental retardation	عقب ماندگی ذهنی مغلوب مرتبط به کروموزوم اکس.
XMP (= xanthosine mono phosphate)	اکس.ام.پی.
XO (= 45XO)	اکس.او.
XP (= xeroderma pigmentation)	اکس.پی.
XR (= X- linked recessive)	اکس.آر.
x-ray	اشعه/پرتو ایکس
x-ray crystallography	بلورنگاری اشعه ایکس
x-ray diffraction pattern	الگوی شکست اشعه ایکس.
XR-MR (= X-linked recessive mental retardation)	اکس.آر.-ام.آر.
x-trisomy (= 47XXX)	تریزومی اکس.
XXY (= 47XXY)	اکس.اکس.وای.
xylan	زایلن
xylene	زایلین

xylitol	زایلیتول	zein	زئین
xylose	زایلوز	zigzag DNA	دی.ان.ای دارای ساختمان
xylulose	زایلولز		زیگزاگی
XYY (= 47XYY)	اکس.وای.وای.	zinc finger protein	پروتئین انگشتی روی
		zinc leucine protein	پروتئین لوسین روی

Y

		zoaster	زوآستر
y chromosome	کروموزوم وای.	zona pellucida	پروتئین اتصال اسپرم زونا
Y long arm	رشته طولانی وای.	sperm-binding protein	پلوسیدا
YAC (= yeast artificial	وای.ای.سی.	zonal centrifugation	سانتریفوژ ناحیه ای
chromosome)		zone electrophoresis	الکتروفوروز ناحیه ای
yeast	بوزک، خمیر مایه، مخمر	zoo blot	زو بلات
yeast artificial chromosome	کروموزوم مصنوعی مخمر	zoogloea	زوگلوئیا
yeast episomal plasmid	پلاسمید اپیزومی مخمر	zoonoses	زونوز
yeast integrative plasmid	پلاسمید اینتگرتیو مخمری	zoonotic	زونوزی
yeast replication plasmid	پلاسمید همانند سازی	zooplankton	زوئوپلانکتون
(YPR)	مخمری	zootoxin	زیا زهر، سم جانوری
yeast two-hybrid system	سیستم دو رگه مخمری	z-ring	حلقه زد
yin-yang hypothesis (of	فرضیه ین و یانگ	zygospore	زیگواسپور
biological control)		zygote	بارورشده، تخم، تخمک،
yohimbine	یوهیمبین		سلول/تخم گشنیده/بارور
Yq (= Y long arm)	وای اندیس کیو.		زیگوت، سلول تخم، نطفه
		zygotene stage	مرحله زیگوتین

Z

		zygotic induction	اینداکشن زیگوتی
		zygotic lethal	کشنده زیگوتی
		zyme system	سیستم زایم
Z-DNA (= zigzag DNA)	دی.ان.ای. زد	zymogen	آنزیم ساز، پیش آنزیم،
zearalenone	زیرالنون		پیشتاز آنزیم، زیموژن،
zeatin	زیاتین		مخمر، مولد آنزیم
zeaxanthin	زیازانتین، زیاکسانتین	zymurgy	زیمورژی

Numbers,
Latin and Greek Letters

2,3-DPG (= 2,3-diphospho-glycerate)	3،2-دی.پی.جی.	5'-DFUR (= 5'-deoxy-5-fluorouridine)	5-دی.اف.یو.آر.'
2,3-diphosphoglycerate	3،2-دیفسفوگلیسرات	5-phospho-ribosyl-1-pyro phosphate	5-فسفوریبوسیل 1 پیروفسفات
cyclic 3e, 5e-adenosine monophosphate	3ئی. 5ئی.-آدنوزین مونوفسفات چرخه ای	7-methylguanosine	7-متیل گوانوزین
3PG (= 3-phospho-glycerate)	3پی.جی.	α- fetoprotein	آلفا-فیتوپروتئین
3-phospho-glycerate	3-فسفوگلیسرات	α- trans - inducing factor	عامل فرا تحریک آلفا
3-hydroxy 3-methyl glutaryl coenzyme A reductase	3-هیدروکسی 3-متیل گلوتاریل کوآنزیم آ. رداکتاز	α-TIF (= α- trans - inducing factor)	آلفا-تی.آی.اف.
45XO	45اکس.او.		
47XXX	47اکس.اکس.اکس.	β-galactosidase	بتا گالاکتوزیداز
47XXY	47اکس.اکس.وای.	β-sheet	ورقه بتا
47XYY	47اکس.وای.وای.	B-Tal (= β-thalassemia)	
5' (= five prime)	5'	β-thalassemia	بتا تالاسمی
5'-deoxy-5-fluorouridine	5-داکسی-5-فلئورورایدین'		
5'-deoxy-5 fluorocytidine	5-داکسی-5 فلئورو سایتیدین'	λ (lambda) phage	فاژ لاندا
5'-DFCR (= 5'-deoxy-5 fluorocytidine)	5-دی.اف.سی.آر.'	μ (= micro)	مو
		μm (= micro-meter)	